Biology of Ourselves
second edition

To:
Heather and Scott
and
Dawn, Katherine, and Stephen

Gordon S. Berry / David C. Lynn

Biology of Ourselves

second edition

John Wiley & Sons

Toronto · New York · Chichester · Brisbane · Singapore

Copyright © 1990 by John Wiley & Sons Canada Limited

All rights reserved. No part of this work covered by the copyrights hereon may be reproduced or used in any form or by any means – graphic, electronic, or mechanical – without the prior written permission of the publisher.

Any request for photocopying, recording, taping, or information storage and retrieval systems of any part of this book shall be directed in writing to the Canadian Reprography Collective, 379 Adelaide Street West, Suite M1, Toronto, Ontario M5V 1S5.

Care has been taken to trace ownership of copyright material contained in this text. The publishers will gladly receive any information that will enable them to rectify any reference or credit line in subsequent editions.

Aquisitions Editor: Wilson Durward
Managing Editor: Lee Makos
Project Editor: Julie E. Czerneda
Copy Editor: Lisa Stacey/Debbie Davies
Production Coordinator: Deborah Starks
Design: Michael van Elsen Design Inc.
Illustration: James Loates *Illustrating*
Assembly: Zena Denchik
Typesetter: Q Composition Inc.
Colour separation: Bomac–A division of the Laird Group
Printer: The Bryant Press Limited

Canadian Cataloguing in Publication Data

Berry, Gordon S., 1930-
   Biology of ourselves

For use in secondary schools.
Includes index.
ISBN 0-471-79526-7
1. Physiology. 2. Anatomy. I. Lynn, David. II. Title.
QP36.B47 1990   612   C90-093088-8

Printed and bound in Canada

10 9 8 7 6 5 4 3 2 1

The cover photograph is a computer-generated image of four individuals involved in physical activity. Some questions may come to mind as you look at the photograph. What allows these people to be so active? What are they doing to keep themselves fit and healthy? A theme of this book is healthful living. The book offers you the opportunity to learn more about the structure of your body, and how your body works.

Before you begin, think about your body. Write brief notes in point form about your beliefs. Here are some questions to help you focus your thoughts. How does my body work? What will help me keep fit and healthy? Am I truly what I eat and do? What systems work together to keep my body healthy and balanced? How might I avoid problems that come from abusing my body's sensitive balance? Am I responsible for the environment that I live in? When you have completed this course, read your notes again and compare them to what you now know about your body and how it works.

Cover Credit: The Stock Market Inc.

TABLE OF CONTENTS

To the Teacher — x
To the Student — xi
Features of the Text — xii
Safety in the Laboratory — xiv

CHAPTER 1
Getting Under Your Skin — 2

1.1 From Molecule to Body System: An Overview — 4
1.2 The Skin: Structure and Characteristics — 11
1.3 Keeping in Touch with the Outside World — 21
1.4 Skin Health and Hygiene: More than Looking Good — 25
1.5 A Case of Overexposure — 30
Health Concerns: Human Disorders Research Assignment — 36
Biology at Work: Career Research Assignment — 37

CHAPTER 2
The Body in Motion — 38

2.1 The Body's Support System: The Skeleton — 40
2.2 A Close Look at Bone — 40
2.3 Bone Formation — 43
2.4 The Skeleton: Support — 45
2.5 Skeletal Adaptations — 56
2.6 The Joints: Connecting Bones Together — 57
2.7 When Bones Break — 61
2.8 Muscle: Getting Moving — 65
2.9 The Muscle-Bone Connection — 70
2.10 The Body in Action — 75

CHAPTER 3
The Nervous System: In Control — 82

3.1 The Nervous System and Behaviour — 84
3.2 The Brain and Its Parts — 87
3.3 Protecting the Brain — 95
3.4 The Spinal Cord — 96
3.5 The Autonomic Nervous System — 100
3.6 The Endocrine System — 102
3.7 Foreign Chemicals and the Nervous System — 104
3.8 Behaviour — 107
3.9 Learning — 111

CHAPTER 4
Sensing the External Environment — 118

4.1 Processing Information — 120
4.2 Vision: The External Structure of the Eye — 120
4.3 Vision: The Internal Structure of the Eye — 124
4.4 Vision: The Characteristics of Sight — 129
4.5 Vision: Measuring the Ability to See — 131
4.6 The Ear and Hearing — 137
4.7 Hearing Loss — 141
4.8 The Organs of Balance — 143
4.9 The Other Senses: Taste, Smell, and Touch — 145

CHAPTER 5
Blood and Circulation — 154

5.1 What is Blood? — 156
5.2 Blood Types — 162
5.3 Moving Blood Through the Body: The Circulatory System — 164
5.4 The System in Action: The Heart — 168
5.5 The Regulation of Heart Rate — 173
5.6 To Your Good Health — 176

CHAPTER 6
Respiration — 184

- 6.1 The Need for Oxygen — 186
- 6.2 The Air-Conducting Structures — 186
- 6.3 The Mechanism of Breathing — 192
- 6.4 The Exchange of Gases — 197
- 6.5 Air Pollution and Healthy Lungs — 202

CHAPTER 7
You Are What You Eat: Diet and Nutrition — 208

- 7.1 What Determines a Diet — 210
- 7.2 Carbohydrates for Energy — 210
- 7.3 Fats for Storage and Vitamin Metabolism — 216
- 7.4 Proteins for Building and Repair — 219
- 7.5 Vitamins for Health — 222
- 7.6 Other Dietary Requirements — 225
- 7.7 A Daily Food Guide — 229
- 7.8 Being an Informed Consumer — 234
- 7.9 The Assessment of Body Size — 238

CHAPTER 8
Digestion — 250

- 8.1 The Process of Digestion — 252
- 8.2 Physical Digestion: The Teeth — 252
- 8.3 Other Aspects of Digestion in the Mouth — 258
- 8.4 Digestion in the Stomach — 261
- 8.5 The Small Intestine and Associated Organs — 266
- 8.6 Absorption of Nutrients — 271
- 8.7 Elimination: The Large Intestine and Associated Structures — 274
- 8.8 Living with Your Digestive System — 276

CHAPTER 9
The Excretory System — 280

- 9.1 What is Excretion? — 282
- 9.2 The Organs of the Excretory System — 283
- 9.3 The Nephron: Site of Filtration — 285
- 9.4 Kidney Transplant: Sharing for Life — 287
- 9.5 Urine and the Antidiuretic Hormone (ADH) — 289
- 9.6 The Collection and Release of Urine — 291
- 9.7 Some Disorders Affecting the Excretory System — 292

CHAPTER 10
Genetics: Blueprint for Life — 296

- 10.1 Information for Development — 298
- 10.2 Chromosomes and the Production of Gametes — 302
- 10.3 Inheritance — 305
- 10.4 The Transmission of Genes — 308
- 10.5 An Example of Inheritance Due to Multiple Gene Forms: Blood Type — 313
- 10.6 Special Information Carried by the Chromosomes — 315
- 10.7 Errors in the Genetic Blueprint — 317

CHAPTER 11
Reproduction: Producing a New Individual — 324

- 11.1 The Male Reproductive System — 326
- 11.2 The Female Reproductive System — 330
- 11.3 The Menstrual Cycle — 335
- 11.4 Copulation and Fertilization — 337
- 11.5 Pregnancy and Early Development — 338
- 11.6 Birth and Lactation — 343
- 11.7 A Healthy Baby: The Role of the Environment — 346
- 11.8 Taking Charge of Your Own Sexuality — 347

CHAPTER 12
Mammal Dissection — 358

12.1 An Introduction to Dissection — 360
12.2 Examining the External Features of the Fetal Pig — 362
12.3 Dissecting the Digestive System — 365
12.4 Dissecting the Excretory System — 371
12.5 Dissecting the Respiratory System — 373
12.6 Dissecting the Circulatory System — 374
12.7 Dissecting the Reproductive System — 377

CHAPTER 13
A Question of Survival: Managing Human Waste — 382

13.1 What is Waste? — 384
13.2 Sources of Water Pollution — 385
13.3 Testing Water Quality — 390
13.4 Water Purification: Water Fit to Drink — 392
13.5 Wastewater Treatment: Return to the Environment — 394
13.6 Solid Waste — 396
13.7 Alternatives to Disposal — 405
13.8 Industrial Hazardous Waste — 407
13.9 Airborne Wastes — 409
13.10 Acid Precipitation — 414
13.11 We Are Learning — 420

CHAPTER 14
Our Environment: What Will We Make of It? — 424

14.1 The Human Population — 426
14.2 The Study of Populations and the Environment — 430
14.3 Factors Affecting Population Size — 433
14.4 Food Consumption — 435
14.5 Increasing the Food Supply — 438
14.6 Waste Not, Want Not — 441
14.7 Pollution of the Oceans — 442
14.8 How Much is a Tree Worth? — 443
14.9 The Atmosphere — 445
14.10 Local Environmental Issues and the NIMBY Syndrome — 448

SPECIAL FEATURES

Biotech: Burns and Artificial Skin — 23
Health Concerns: Human Disorders Research Assignment — 36
Biology at Work: Career Research Assignment — 37
Biotech: Bone Transplants — 47
Biotech: Torn Cartilage and Arthroscopy — 60
Biotech: Brain Mapping — 89
Life Signs: Seizure — 92
Life Signs: Over the Limit — 106
Biotech: School and the Visually Impaired Student — 133
Life Signs: Anemia — 159
Biotech: Heart Attack — 178
Biotech: Exercise Programs — 180
Life Signs: Drowning — 198
Life Signs: Diabetes — 212
Biotech: Food Energy — 215
Life Signs: Body Mass — 243
Life Signs: Ulcer — 264
Biotech: Hepatitis — 269
Biotech: Alcohol and the Human Body — 272
Biotech: The Artificial Kidney — 288
Biotech: Cracking the Genetic Code — 300
Biotech: Karyotypes — 318
Life Signs: Genetic Counselling — 321
Life Signs: Sexually Transmitted Disease — 351
Biotech: Composting to Reduce Waste — 406
Biotech: pH and Acid Rain — 415

ACKNOWLEDGEMENTS

The Authors would like to express their appreciation to the staff at John Wiley & Sons for the support and encouragement that was offered during the preparation of the second edition of this book. We are grateful to Wilson Durward for his enthusiasm in launching this project.

A special note of thanks must go to Julie Czerneda, who served formally as our project editor, and informally as coach, cheerleader, and taskmaster.

Our thanks also to Lee Makos, Lisa Stacey, Debbie Davies, Oliver Salzmann, Jeffrey Aberle, Deborah Starks, and Zane Kaneps for their thorough and capable efforts in aiding this manuscript through to its production. Thanks to James Loates for his excellent and attractive drawings, and to Michael van Elsen for his imaginative layout and design.

We would like to thank the following classroom teachers and specialists who have reviewed the manuscript and have provided us with many thoughtful suggestions and comments:

Dr. Howard D. Cappell, Director of Social and Biology Studies Division, Addiction Research Foundation. Professor, Departments of Psychology and Pharmacology, University of Toronto.

Bill Chaplin, Biology Teacher, Leamington D.S.S., Essex County Board of Education

Molly Hart, Head of Science, Francis Libermann S.S., Metro Roman Catholic Secondary School Board

Tom Hensley, Head of Science, D. & M. Thompson C.I., Scarborough Board of Education

Henry Pasma, Assistant Chairman of Science and Technology, Cawthra Park S.S., Peel Board of Education

Margaret Redway, Science Safety Consultant, Fraser Awareness Inc.

Dr. Hermina Richter, Dental Surgeon, Mississauga, Ontario

Dr. Lynda Small, Physician (Family Medicine), Mississauga, Ontario

F.M. Speed, Head of Science, University of Toronto Schools, Faculty of Education, University of Toronto

Margaret Williams, Coordinator Special Education, Peel Board of Education

It is appropriate that we renew our thanks to all those who helped make the first edition of this text so successful. Foremost of these is Harold Gopaul, to whom we are also indebted for his expertise and help in this second edition. Although we cannot mention each and every person here, we must acknowledge and thank Trudy Rising for her tremendous assistance.

Finally, a very warm thank you to our families and friends who have been so supportive and understanding throughout the project.

TO THE TEACHER

Biology of Ourselves, Second Edition, is a text designed to introduce students to the study of human biology. It is intended to provide a foundation for the understanding of both the structure and the function of the human body. This foundation will serve as the basis for healthy lifestyle decisions. It may also serve as a stimulus to further study, or to encourage students to seek careers in related fields.

A common thread woven throughout the text is the theme of healthful living. The text requires the reader to consider not only the internal environment of the body, but also the impact which each one has, individually and collectively, on the local and worldwide environment. The decisions each one makes, about what to eat and drink, about what to purchase, or about which activities to participate in, have far-reaching effects. Consequently, throughout the text a number of references require the reader to think critically about issues which range in perspective from a personal focus to a global awareness.

An attempt has been made to present each concept in a clear, concise manner, and yet not limit the possibilities for creative extension and challenge for the keenly interested student. Features of the text, such as the case studies and "Some Things to Find Out" questions, serve as a springboard for extension and student research.

Throughout this text are practical applications and interest points, which are designed to illustrate the impact of lifestyle and technology on the maintenance of a healthy body, including in many cases the logical extension to the environment of which we are a part.

Each body system is inextricably linked to all others and this connectedness is an important concept in teaching human biology. The theme of the "whole" person links every chapter and topic, regardless of the order of use. At several points, summarized information is presented in visual and/or table form to help students see the interrelationships between the different topics.

Building a foundation of factual knowledge is both necessary and desirable. The student activities which supplement each topic permit students to participate in the discovery and application of this knowledge. However, such gathering of information must be seen as only a tool to better understand and to make wise decisions.

This text and its activities provide an avenue to students for the continuing development of skills essential to informed decision-making: observation, the organization and analysis of information, communication, and critical thinking. We believe that with a foundation of knowledge about themselves, and with the use of these skills, students will be better equipped for healthful living.

The Authors

TO THE STUDENT

This book gives you the opportunity to study what you may consider to be the most interesting topic in the world – yourself! It gives you a chance to learn about how your body works, what will help you to keep it fit and healthy, and how to avoid the problems that come from abusing its sensitive balance and systems. You could think of this book as a kind of owner's maintenance manual, a body user's guide to healthy living.

Have you ever declined to go swimming, skating, or another similar activity with a friend just because you didn't feel like it? You weren't sick. You weren't annoyed with your friend. Your life just seemed to lack some of the zest and spirit it should have. Why? Frequently, the answer can be found in the health habits that we follow each day, such as the type of food we eat, a lack of fibre in our diet, limited exercise, or low iron in our food. With only a few exceptions, feeling great depends upon you, what you understand about your body, and how you apply good health practices.

Understanding your body can extend your life span as well as improve your quality of life. It has been said that as a teenager you build the habits that will eventually cause your death! Smoking is one example; drug use, rich diets, and constant lack of exercise are others.

One of the things that you will discover as you participate in the activities in this book is that we are all very different from each other. We look very much alike, we are similar in age, height, and so on, but in other ways we differ sharply. For example, everyone has a unique set of fingerprints, and while we all have the same basic facial features, each face is so unique that we can learn to recognize every individual in the entire school. One individual may be more, or less, sensitive to heat, cold, or allergens than another. One person sweats heavily while another may have dry skin. Sometimes the differences can be very great indeed. If you find that your results in a personal experiment are quite different from those around you, don't change your results to conform, or believe yourself to be "abnormal". You have simply proved that you are unique. Being unique is a very special characteristic of human beings, one in which you can take pride and satisfaction.

The field of human biology offers a very wide range of job opportunities: nursing, ambulance attendant, laboratory technician, skin care, dietician, and food services to mention only a few. If you enjoy working with people, perhaps something in this course will stimulate your interest in a specific branch of this interesting and rewarding field.

Every year scientists discover more about the human body, how intricate and wonderfully adaptable it is, and also how susceptible it is to hazards in the environment. Each year the world becomes more crowded and polluted as chemical and toxic wastes affect our water, food, and air. If we, and our planet, are to survive, we must try to understand these problems and how they affect us and all living things. Understanding the biology of ourselves and the effect we have on the environment makes it possible for each of us to contribute constructively, not only to individual and community needs, but to the future well-being of the whole earth.

We hope you will enjoy your study of the human body and find this text a useful guide.

Sincerely, Gordon Berry
David Lynn

FEATURES OF THE TEXT

Beginning each chapter is an introductory photograph and paragraph intended to capture the students' interest. These introductions also act to encourage the students to view each new topic as part of a whole – the study of human biology. A **Key Ideas** list is then presented. This serves as a preview of the major ideas and concepts to be discussed in the chapter, giving the students focus and direction.

At the end of most numbered sections are brief **Checkpoint** questions relating to the content just covered. Students can test themselves with these questions to see if they have understood the new material. They are also a useful tool for review.

Activities are an essential part of any science course. The activities in this book are well-tested and varied, helping to develop and expand important ideas in a practical way. They are presented immediately following the pertinent text discussion, so that students have the information they need readily available. Within each activity, students are reminded of information which should be recorded in their notebooks by the presence of *italicized* print. Questions to guide observation and analysis are provided.

Safety being a key concern, a comprehensive **Safety in the Laboratory** section precedes Chapter 1. Within each activity, potential hazards are clearly described and marked with the **Caution** logo.

Appearing frequently throughout the text are small notes called **Focus on You**. These notes make each topic a very personal matter by extending ideas or pointing out how the material in the text is directly related to the student. Many will encourage students to critically evaluate their own lifestyle management.

At pertinent points in the text students will encounter **Biotech** articles. These features report on new advances in biotechnology as well as on the application of technology to human biology.

Also included in the text are numerous **Life Signs** articles. These are case studies in short story form, based on real life situations in which human biology plays an important role in someone's daily life. Life Signs articles are placed within complementary text to enhance the material being discussed.

Newly defined terms are presented in **boldfaced** type, with a pronunciation key provided for those words most likely to be unfamiliar to the average student. These terms are repeated for review at the end of the chapter and also appear in the **Glossary** at the end of the book.

The appendix, **Commonly Abused Drugs and Their Effects**, has been provided as additional information and as a useful teaching tool.

Topic-related disorders are listed near the end of each chapter in the **Health Concerns** feature. Chapter 1 contains a box detailing how to research and prepare an assignment on these human disorders.

The **Biology at Work** feature presents a list of additional careers which accompanies each chapter. These careers relate directly to the chapter topic and vary in type and amount of training required. In Chapter 1 an additional box describes how to find information about a particular career using resources available to students.

The following features occur at the end of each chapter:

The **Chapter Focus** is a brief summary of the chapter objectives, including all the major points that the students should know and understand.

The matching questions in **Some Words to Know** help students review the new terminology presented in the chapter.

The answers to **Some Questions to Answer** can be found by reference directly to material in the chapter. A variety of application questions increases student interest.

The questions in the **Some Things to Find Out** section challenge students to research and find out answers from other sources beside the text, including the library, magazines, pamphlets, etc.

SAFETY IN THE LABORATORY

Laboratory work is an enjoyable part of the biology program. It can also be potentially dangerous if proper precautions are not taken. Safety should be a priority for everyone in a science lab. You have a responsibility to yourself and others to be aware of the safe use of all equipment and chemicals in the lab. Following is a list of safety rules you should observe:

- Look for the caution symbol for special safety information concerning an activity. Do *not* attempt any activity unless a teacher is present to supervise you.

- Familiarize yourself with the location of all fire exits, fire extinguishers, eye wash stations, etc.

- Do *not* bring food or drink into the laboratory.

- Before coming to class, familiarize yourself with the equipment needed, the procedure to be followed, and the special safety precautions for each activity.

- For activities that require you to design your own procedure, keep safety in mind. Be sure to have your procedure approved by the teacher before beginning.

- Listen carefully to your teacher's instructions. He or she will provide you with additional information about the equipment and chemicals you will be using. Follow these instructions carefully.

- Tie back long hair and loose clothing before beginning any activity.

- Wear eye protection when appropriate. This includes any activity involving an open flame or potentially harmful chemicals.

- Protect yourself and your clothing by wearing a lab coat or old shirt.

- Make sure your work area is clean and uncluttered. A flowchart of the procedure, prepared data tables, blank paper, and a pencil are all you need to begin.

- Minimize movement around the laboratory as much as possible. Try to collect all equipment and chemicals at the beginning so that you do not interfere with the work of others.

- Do *not* leave your experiment unattended.

- Do *not* deviate from the procedure unless authorized to do so by your teacher.

- Report all injuries, no matter how small, to your teacher.

- Notify your teacher when any piece of equipment is broken. Do not put broken glass in the regular waste basket – it belongs in a separate container.

- When heating a test tube, make sure it is made of heat-resistant glass. Always point the mouth of the test tube away from yourself and others. If you are using an open flame, hold the test tube on an angle and move the test tube back and forth through the flame. Do *not* hold the test tube constantly in one place.

- When required to smell something, use your hand to direct the air above the substance toward your nose. If you cannot detect an odour, move the container a little closer and try again. Never put your nose directly above the substance being smelled.

- Never taste anything while in the laboratory unless instructed to do so.

- When obtaining a chemical from a stock bottle, read the label twice to confirm that it is the chemical you require. Take the smallest quantity necessary. *Never* return chemicals to the stock bottle; dispose of all unused chemicals as instructed by your teacher.

- *Never* pipette anything by mouth. Use a pipetting bulb or syringe.

- When using a microscope, try to keep both eyes open, even though you may only need one eye to look through it. This will help to prevent headaches.

- If live specimens are being used in a particular activity, observe them in their natural setting if possible. Use common sense so that you do not put undue stress on the animal. Keep in mind at all times the respect for life.

- Special caution must be exercised when dissecting an organism.
  (a) Make sure that the area you are working in is well ventilated.
  (b) Wear plastic gloves when dissecting to prevent harmful preservatives from irritating your skin.
  (c) When practical, use your fingers rather than sharp instruments for exploring the specimen.
  (d) Familiarize yourself with the safe and proper use of all dissecting instruments. Large scissors are used to cut through skin, muscle, and small bones. Bone scissors are used to cut through large bones. Probes are used to move organs out of the way to improve viewing of underlying structures. Scalpels are used when making fine or very precise incisions; they can also be used for separating one structure from another. Scalpels are *not* used for prying open rigid structures. If the blade of the scalpel breaks, do *not* try and replace it yourself; your teacher will do this for you.

- Dispose of all specimens, materials, chemicals, etc., as instructed by your teacher.

- Do not remove anything from the laboratory without your teacher's permission.

- When an activity is completed, clean up the work area.

- Before leaving the laboratory, wash your hands thoroughly.

GETTING UNDER YOUR SKIN

Do you know which is the largest organ in the body? Here are some clues. It contains one-third of the body's blood volume. It is replaced every few weeks. It comes in a variety of colours and textures. It is so unique that it can be used to identify one specific person out of several million.

In 1 cm^2, it contains 1 m of blood vessels, more than 100 glands, nearly 4 m of nerves, sensors to detect many environmental changes, and at least 3 million cells.

As you probably guessed, the organ is the skin. This vital structure provides an outer coat that protects us from the sun and many infections, regulates our body heat, monitors our environment, prevents fluid loss, and produces hair and nails, among its many important functions. Remarkable as it is, the skin will be only the beginning of your examination of human biology.

KEY IDEAS

- *The cell is the basic unit of all living things.*
- *The body is organized into cells, tissues, organs, and systems.*
- *The skin is the body's first line of defence against infections.*
- *The skin is the body's contact with the external environment. Its senses act as a warning system.*
- *The skin provides insulation and temperature control among other functions.*
- *Skin care and hygiene is an important contributor to good body health.*

1.1 From Molecule to Body System: An Overview

Chemical Composition

All living things are made of cells. In turn, cells are composed of only a few basic chemical elements. For example, although there are more than 100 known elements, only six of them are needed to make up 99 percent of the human body. (See Table 1.1.) These elements are bonded together in complex arrangements to form a wide variety of chemical compounds. Hydrogen and oxygen combine to form water, which accounts for a very large percentage of our body mass. The remaining one percent includes a few elements that are present in very small or trace amounts, yet these elements play a significant, even vital, role in the health of the body.

Table 1.1 The Elements Found in the Human Body

Element	Symbol	Percentage
Oxygen	O	65.00
Carbon	C	18.00
Hydrogen	H	10.00
Nitrogen	N	3.00
Calcium	Ca	2.00
Phosphorus	P	1.00
Potassium	K	0.35
Sulphur	S	0.25
Chlorine	Cl	0.15
Sodium	Na	0.15
Magnesium	Mg	0.05
Iron	Fe	0.004
Other elements		0.046

The Cell

You will have studied the cell extensively in other grades, but the following figures will provide you with a summarized review of the cell organelles, their structure and function. (See Figures 1.1 and 1.2.)

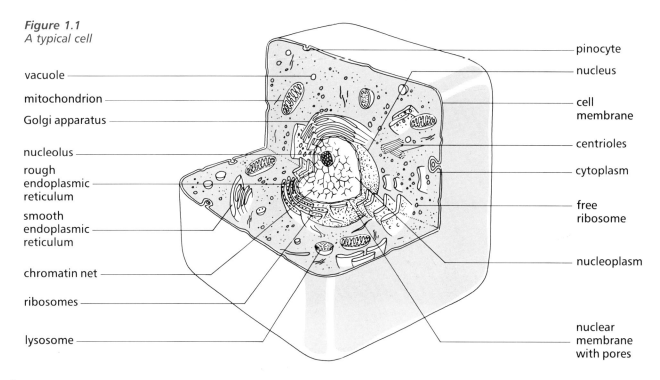

Figure 1.1
A typical cell

One of the characteristics of organisms composed of many cells is that the size and shape of each kind of cell varies greatly according to its function in the body. For example, red blood cells are round dimpled cells with large surface areas to pick up oxygen efficiently. Nerve cells have long extensions to carry impulses over large distances in the body. Skin cells are thin and flat to effectively cover the surfaces of the body and organs. As you read through this book, look for the unique design of each type of cell and how it matches its specialized function.

Figure 1.2
Review summary of cell organelles

Mitochondria
- *round or oval, two-layered membrane shell.*
- *inner membrane with shelflike folds and tubes.*
- *numbers present vary with energy needs of cell.*
- *power house.*
- *produces "packets" of energy for cell activities.*

mitochondrion

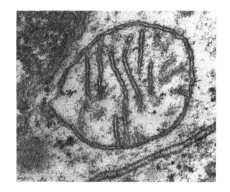

Endoplasmic Reticulum
- *a double membrane which connects the plasma and nuclear membranes in a series of folds.*
- *ribosomes present on rough E.R.*
- *no ribosomes on smooth E.R.*
- *transports fluids and chemicals throughout the cytoplasm.*
- *ribosomes contain granules of nucleic acids active in protein synthesis.*

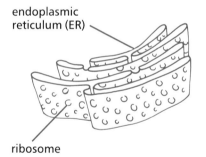

Golgi Apparatus
- *clusters of membrane-lined channels.*
- *produces or accepts lipids and enzymes, and packages these for export.*

Golgi apparatus

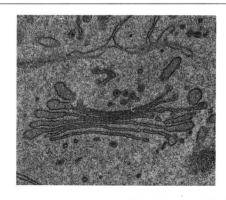

Lysosomes
- *sacs of enzymes surrounded by membranes.*
- *digest large molecules.*
- *can destroy cell if membrane is broken and contents escape.*

Vacuoles
- *bubblelike containers enclosed by a membrane.*
- *contain water and help maintain water balance.*
- *may also contain wastes or food.*

lysosome

vacuoles

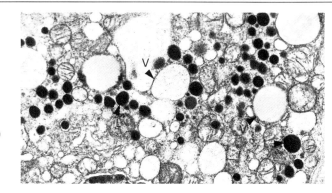

Nucleus
- *central sphere surrounded by a nuclear membrane and containing nucleic acids.*
- *directs cell activities.*
- *stores genetic material.*

Chromatin Net (Chromosomes)
- *show as dark strands during cell division and as dispersed granular patches in other phases.*
- *nuclear protein strands that hold the genetic code (DNA).*
- *the inherited storehouse of genetic information.*

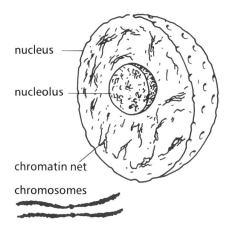

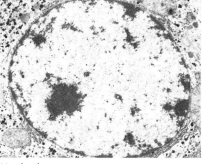

Nucleolus
- *a collection of loosely bound granules of RNA and protein.*
- *no membrane present.*
- *may be the site of RNA production.*

Cell membrane
- *an elastic envelope composed of three layers: protein, lipid, protein.*
- *it is semi-porous and has numerous globular proteins present.*
- *acts as a boundary for the cell.*
- *controls passage of materials in and out of the cell.*
- *globular proteins aid the passage of water-soluble molecules across the membrane.*

Cytoplasm
- *a complex mixture of materials, — water, gases, wastes, nutrients, raw materials, etc, used in cell processes.*
- *includes the cell organelles but not the nucleus.*
- *supplies cell needs, removes wastes.*

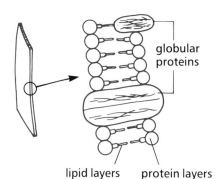

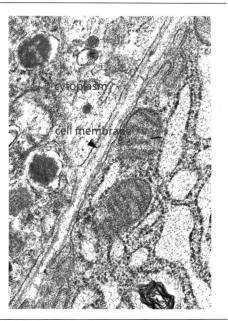

Centrioles
- *two present near nucleus.*
- *separate and migrate to poles during cell division.*
- *responsible for the organization of the spindles which attach to chromosomes during cell division.*

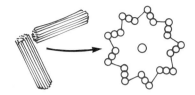
centrioles

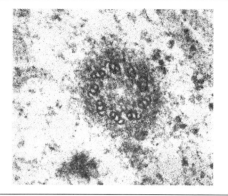

Microvilli
- *fingerlike projections of the cell membrane.*
- *increase cell surface area for more efficient absorption of materials into the cell.*

microvilli

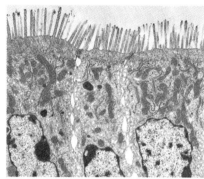

Cilia
- *fine hairlike structures which project from the surface of some cells.*
- *move rhythmically to move particles along a tube or over the surface of a cell.*

cilia

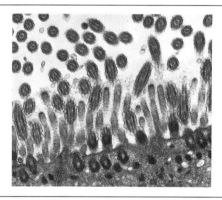

Pinocytes
- *tiny sacs extending from the cell membrane.*
- *part of a process by which larger particles are moved across the cell membrane.*

pinocyte

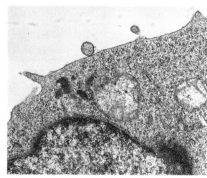

Cells Which Work Together: Tissues

Tissues are groups of similar cells that work together to carry out a particular job. Tissues in the human body can be grouped into four major divisions. (See Figure 1.3.)

Figure 1.3
Tissue summary

Epithelial tissue or covering tissue.
These tissues form a surface over organs, they form the skin and line the inside of tubes and chambers in the body.

Examples

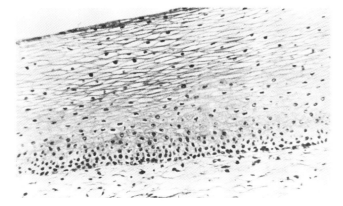

(a) Section through the skin on the hand showing thin, flat cells, with several layers to take care of wear and tear.

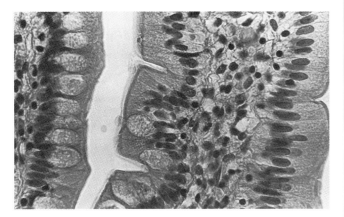

(b) Section showing cells lining the digestive tract. Some cells are for secretion and some for absorption of digested products.

Connective tissue.
These tissues bind and pack cells together, or act as supporting tissue.

Examples

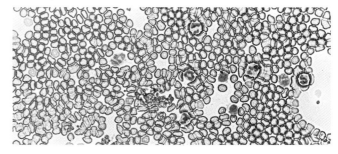

(a) Blood cells which carry oxygen and carbon dioxide and fight infections

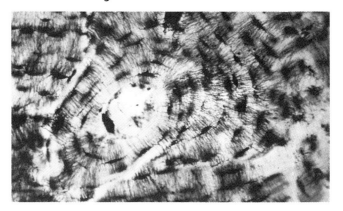

(b) Bone tissue, designed to support the body and provide attachment for muscles

Nerve tissue.
Nerve tissue is responsible for conducting impulses to and from the brain and for co-ordinating the various activities of the body.

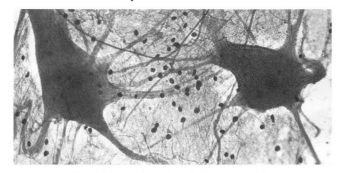

Nerve cells which receive stimulation, transmit impulses to be carried throughout the body

Muscle tissue.
Muscle tissue has the ability to contract, to shorten and thus bring the ends of the muscle closer together.

Examples

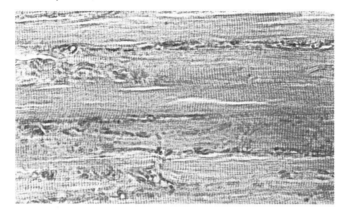

(a) Skeletal or voluntary muscle, which moves bones

(b) Smooth or involuntary muscle, which moves food along the digestive tract

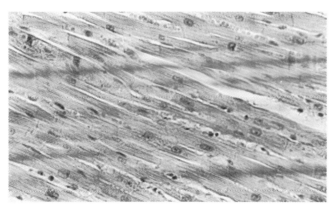

(c) Cardiac muscle of the heart, which pumps blood through the body

Organs and Systems

The several kinds of tissues already discussed may be found working together to perform some particular job. The heart, for instance, contains muscle tissue which contracts to provide the pumping action needed to circulate the blood through the body. It also has an inner and outer lining of epithelial cells. Nerve tissue stimulates the muscles and controls the rate at which the heart beats. Connective tissue gives the heart its structure and fills it with blood. All of the tissues forming the heart have a common purpose – to pump blood. When several tissues work together in this way, they form an **organ**. Organs are then grouped into specialized **systems**. (See Figure 1.4.) The circulation system, for example, has the job of providing a transport and delivery system. Each system has its own tasks to perform. (See Table 1.2.) In the next section, you will examine the body's largest organ – the skin, part of the integumentary system.

Table 1.2 Body Systems – Their Organs and Functions

Systems	Major Organs and Tissues	Functions of the System
Integumentary	skin, hair, nails, skin glands, receptors	protection, temperature control, environmental stimuli
Skeletal	bones, joints, cartilage	support, protection, stores minerals, produces blood cells
Muscular	skeletal muscles	body movement, heat production

Systems	Major Organs and Tissues	Functions of the System
Circulatory	blood, heart, blood vessels	transports nutrients, wastes, gases, circulates the blood
Lymphatic	lymph, lymph nodes, tonsils, spleen, thymus, lymph vessels	return tissue fluids to blood, aid immunity, form white blood cells
Respiratory	nose, throat, larynx, trachea, lungs	exchange of oxygen and carbon dioxide
Digestive	mouth, esophagus, stomach, liver, pancreas, intestines	breakdown and absorption of nutrients, excretion of wastes
Urinary or excretory	kidneys, ureters, bladder, urethra	excrete wastes and regulate blood composition
Nervous	brain, spinal cord, nerves, specialized sense organs	control of body activities, monitor internal and external changes
Endocrine	hormone secreting glands, pituitary, thyroid, adrenals, pancreas, ovaries, testes	controls metabolism, growth, co-ordination with nervous system
Reproductive	testes, ducts, and glands	produces sperm, semen, hormones
	ovaries, Fallopian tubes, uterus, vagina, breasts	produces ova, hormones, nourishes offspring

Figure 1.4
How the body is organized

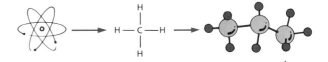

The smallest units of living things are molecules, chemical elements bonded together into compounds, e.g., carbon dioxide, glucose, amino acids, proteins.

Molecules are organized into microscopic functioning cell components called organelles (small organs), which carry out specialized processes in the cell, e.g., mitochondria, vacuoles, ribosomes, etc.

Organelles are contained within cells, e.g., muscle cells, nerve cells, etc.

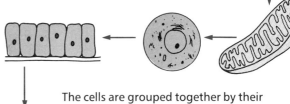

The cells are grouped together by their specialized structures and functions into tissues, e.g., epidermal tissues, nervous tissues, etc.

Tissues are assembled into organs, each with a special function to perform, e.g., stomach, heart, liver, kidney.

Organs are organized into systems. Each system has a major function, e.g., respiration, circulation, digestion.

The systems of the body are co-ordinated and integrated into the total living body. Each is important but dependent upon the others.

CHECKPOINT

1. What six elements make up most of the body's mass?
2. Explain the difference between tissues and organs and give three examples of each.
3. Name three systems in the body and give the general function of each.
4. Name ten organelles found within the cell and give the functions of each.

1.2 The Skin: Structure and Characteristics

The skin is elastic, stretching to accommodate each bulging muscle, then retracting as the muscle relaxes. It can expand to adjust to extra size during pregnancy or to cover a bulging bump on the head. The skin is semitranslucent, depending upon the amount of pigmentation present. The skin contains vast numbers of tiny blood vessels, the capillaries. The blood flowing through these vessels gives the skin a rosy colour. If these capillaries dilate (expand), the colour deepens. This is apparent when we blush or are very hot. Fear or cold causes the capillaries to contract. The amount of blood in the skin then decreases, giving the skin a pale or "blue" appearance.

The outer surface of the skin is constantly being worn away. After a shower or bath, if you rub your skin roughly with a towel, you can see the dead skin cells rub off. This is more obvious in summer when you have been exposed to the sun and may be sunburned. The speed with which cuts and wounds generally heal is evidence of the ability of skin cells to regenerate (replace themselves) after injury.

The texture and thickness of the skin varies according to where it is found on the body. Skin that is normally covered by clothing will be thinner than that exposed to air. Skin that receives considerable wear, such as on the hands and feet, is much thicker than that found on the arms or back. Age will also determine how soft the skin is. We are all aware of the tender texture of the skin of a healthy baby compared to that of adults.

The functions of skin are summarized in Table 1.3.

Table 1.3 The Functions of the Skin

- Protects soft inner cells from wear and tear
- Shields the body from the harmful rays of the sun
- Excretes small amounts of body wastes
- Acts as a defense against bacteria
- Indicates what is happening around us, e.g., temperature or pressure change, touch
- Heals wounds, replaces damaged cells around cuts
- Acts as an insulator when the weather is cold
- Releases heat when warm
- Produces hair and nails
- Indicates health, e.g., the body is paler when one is ill
- Produces vitamin D in sunlight
- Produces oils
- Prevents fluid loss
- Indicates our feelings, e.g., pale with fright, darker with rage
- Keeps out the rain, sheds water

The Structure of Skin

The skin is composed of two layers. The external sheath of cells, called the **epidermis** (ep-i-der-mis) is made up of layers of flattened epidermal cells. Immediately below the epidermis is a thicker layer of connective tissues called the **dermis**. This layer contains a mixture of fibres, fat cells, blood vessels, hair follicles, and specialized nerve endings. (See Figure 1.5.)

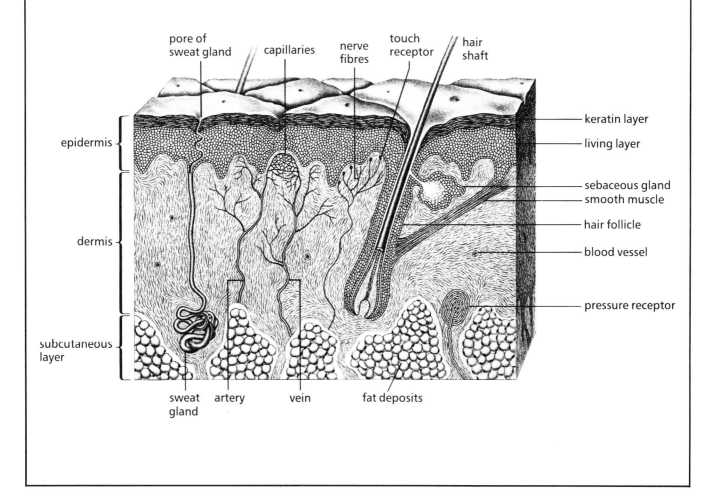

Figure 1.5
The structure of the skin. One square centimetre of skin may have 100 sweat glands, 12 hairs and sebaceous glands, 1.5 m of blood capillaries, and many hundreds of specialized nerve endings.

The two layers are quite firmly cemented together. Excessive rubbing of the skin, however, such as occurs when a shoe fits improperly and chafes the skin, may cause the layers to separate. When the upper layers of cells are forcibly separated in this way, the resulting space fills with fluid and a blister forms.

Activity 1A

The Human Skin

What features of skin can be observed by examining your hand?

Materials

dissection microscopes
centimetre ruler (transparent)

Procedure

1. Examine the skin surface of the palm of your hand with a binocular microscope. *Describe the magnified appearance of the ridges and furrows present.* Place a ruler across the ridges and count the number of ridges present in 1 cm. *Compare your result with those of other students. What useful purpose do the ridges serve?*

2. Examine the back of your hand. Can you discover any regular patterns there? *Describe what you see.*

3. *Is the skin on the back of your hand firmly attached to the underlying tissues, or can it be lifted away? How does this compare with the attachment of the skin on the palm of your hand? Try to explain the differences that you find.*

4. Look closely at the ridges on your fingertips and on the palm of your hand under the microscope. Rock your finger back and forth, or vary the position of the light, while you look for beads of perspiration. *Can you see the openings of the sweat glands? Describe their appearance and exact location.*

5. *Where does the skin appear to be thickest? Where is it thinnest? Why does the skin vary in thickness?*

6. *In your experience, do small cuts in the surface of the thick pads on the palms of the hands draw blood or cause pain? Explain your answer.*

7. Examine any cuts or scars that may be present on your skin. *Describe the appearance of these features.*

8. If your skin is dry, you may find scaly patches on the surface of the skin. *Examine this tissue and explain how the condition is caused. Is this a natural feature of the skin?*

9. On the back of your hand, you will find fine hairs. Examine the point at which the hair protrudes from the skin. *Describe the area immediately around the hair and the angle at which the hair leaves the skin.*

10. Some other features you may wish to examine are fingernails and the cuticle, a hair shaft (use a compound microscope – can you tell how blondes differ from brunettes?), a hair root, split ends, bruise, or a freckle.

The Epidermis

The outer layers of the epidermis, called the **stratum corneum**, which we touch, caress, and care for, are only dead cells! The layer of dead cells contains no blood vessels to bring needed nutrients. (See Figure 1.6.) The cells then die and undergo a chemical process that changes them from soft, easily damaged cells into hard, tough callus. A substance called **keratin** hardens the cells and makes them waterproof, thereby helping to prevent water loss from the body. These cells are, however, able to pass on sensations such as pressure to nerve endings in the layers below the epidermis.

The bottom layer of cells in the epidermis is composed of cells which divide to produce new cells. This deep layer of cells, the **stratum germinativum** (jer-min-a-tiv-um), constantly replaces the cells that are worn away from the surface of the body. It also produces **melanin**, the pigment that colours or tans the skin.

Melanin production is stimulated by exposure to ultraviolet light. Doses of sunlight thus cause the skin to tan. Prolonged exposure, however, may destroy several layers of epidermal cells and cause painful sunburn and blisters. Melanin also accounts for the dark colour of the tissue around the nipples and for freckles, which are small irregular patches of melanin.

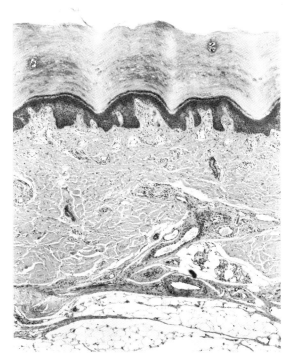

Figure 1.6
Section of the skin. Note the dry outer layers peeling away.

The Dermis

The **dermis** is composed of living tissues which perform a variety of functions. The structures in this layer are specialized to monitor the changes that occur in the environment immediately around the body. Blood vessels regulate body temperature in response to impulses from special heat and cold receptors. Touch, pressure, and pain receptors are located at different levels within the dermis. Glands, hairs, and fat cells are held in place by a matrix of collagenous and elastic fibres that give structure and elasticity to the skin.

Where the epidermis and the dermis meet, there is a wavelike layer formed of many tiny cones and ridges. These patterns show through to the surface of the skin on the hands and feet, some of them becoming fingerprints. The patterns emerge while the baby is developing in the uterus and never change, except with respect to size.

The Patterns of Skin

All humans share the same major features. The eyes, nose, and mouth are all located in the same approximate positions on the face, yet the diversity which exists in even these few features is striking. With the exception of identical twins, close observation will usually reveal sufficient variation in facial features to identify every individual. Also, we differ from others around us in more ways than facial appearance.

As everyone knows from watching police dramas on television or reading detective novels, fingerprints (or dermal prints) are different for every individual. Patterns on the skin of fingers can be classified into ten basic arrangements of loops and whorls. (See Figure 1.7.) Police officers use several additional, more advanced characteristics to identify a particular set of prints. The chance that someone will have eight of these special characteristics in common with you is more than 39 trillion to 1!

The patterns of whorls and loops may differ from finger to finger on each hand. Like the patterns on fingers and toes, the arrangement of ridges on the palm of each hand and on the sole of each foot differs among individuals. The print made by the lips and the pattern of hair follicles on the head are also unique.

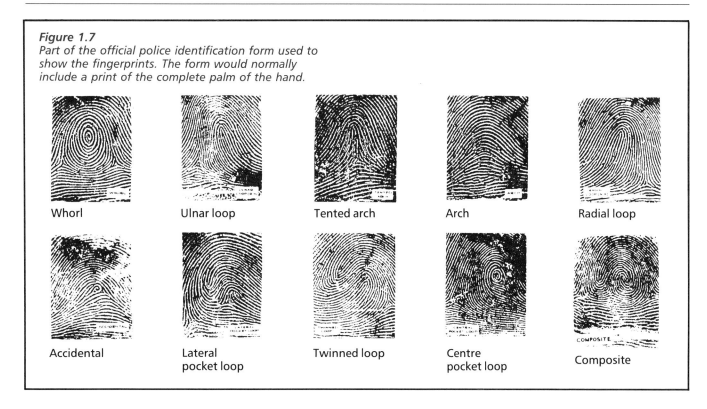

Figure 1.7
Part of the official police identification form used to show the fingerprints. The form would normally include a print of the complete palm of the hand.

Activity 1B

Fingerprints: A Unique Identification

Each human being is unique. One of the ways in which you differ from every other person is the pattern of your fingerprints.

Materials

inkpad
white unlined paper
soap
alcohol and paper towels
a photocopy of a set of fingerprints (first three fingers of a right hand) belonging to a person from the class whose identity will be kept secret

Procedure

1. *Copy the chart in Figure 1.8 in your notebook.*

2. On a scrap of paper, make a few trial prints. To do this, ink and print one finger at a time. Roll your fingertip on the inkpad. Roll it from side to side and up to the tip. Roll the finger on the paper, using light pressure. Roll it from one side to the other rather than simply press the pad of the finger down onto the paper.

3. When you have mastered the technique, make a print of each finger on the chart. Start with the left hand and keep the fingerprints in sequence in the appropriate space.

4. Clean your fingers with alcohol then soap and water.

5. Repeat step 3 with your right hand.

Chapter 1 / Getting Under Your Skin

FOR IDENTIFICATION PURPOSE ONLY – *AUX FINS DE L'IDENTIFICATION SEULEMENT*	BIN	
	BD	

SEX – *SEXE*	SURNAME (Include former names, maiden name, etc.) *NOM DE FAMILLE (Y compris noms utilisés précédemment, nom de jeune fille, etc.)*	FOR RCMP IDENTIFICATION SERVICES USE *À L'USAGE DU SERVICE DE L'IDENTITÉ JUDICIAIRE DE LA GRC*
☐ M		
☐ F	Given names *Prénoms usuels*	

Address – *Adresse*	Postal Code *Code Postal*	

Applicant for – *Demande aux fins de*

	THUMB – *POUCE*	INDEX	MIDDLE – *MÉDIUS*	RING – *ANNULAIRE*	LITTLE – *AURICULAIRE*	
R I G H T						D R O I T E
L E F T						G A U C H E

Figure 1.8
Draw a similar chart in your notebook.

---

6. Refer to the fingerprints in Figure 1.7 and identify each of your prints. *Record the type in the space below each print.* (When fingerprints are used by the police for identification purposes, there are many other refinements of these basic patterns that are used. Enlargements may be necessary to examine both the pattern of ridges and the patterns of tiny dots made by skin pores along these ridges.)

7. Compare your chart with your partner, or another student. Find a pattern type that you have in common, e.g., whorl, arch, etc., and study these prints carefully. Although these two prints belong in the same category, they are different. *List as many differences as you can between the two prints.* You may wish to use a dissecting microscope to enlarge the prints.

8. On the chalkboard, draw a master chart to record the generalized fingerprint data of each member of the class. *Enter your results in this chart.*

9. Using Figure 1.7, identify the print types on the photocopy of fingerprints you have been given.

10. Compare your findings with the master chart on the chalkboard and find out to whom the photocopied fingerprints belong.

Nails

Nails are hard structures which are slightly convex on their upper surface. They are produced by the epidermis, originating as elongated cells which later fuse into flat plates. The living cytoplasm of these cells is replaced with keratin which makes them hard and more durable. The nailbed constantly produces more cells which results in the elongation of the nail. Fingernails grow more rapidly than toenails. (See Figure 1.9.)

Figure 1.9
A longitudinal section through the finger and nail

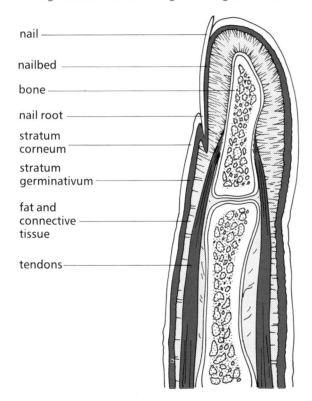

Figure 1.10
Features of the skin surface

Men and women have about the same amount of hair, but it is usually finer on women than on men. Hair is a stained (pigmented) shaft of keratinized cells. It is non-living. (See Figures 1.11 and 1.12.) Each hair follicle seems to have a growing period followed by a "rest period". Hairs with a bulbous base are in the resting phase.

Figure 1.11
A human hair showing the hair follicle, nerves, blood vessels, glands, and muscles associated with the hair

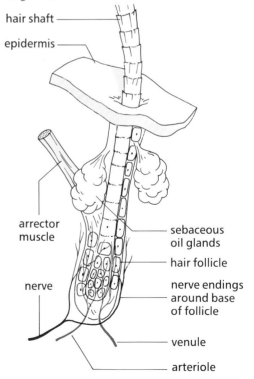

Hair

Humans do not appear to be particularly hairy, but, in fact, we have approximately the same number of hair follicles as chimpanzees. (See Figure 1.10.) Human skin, however, produces quite small, transparent hairs that are not easily seen.

Figure 1.12
A section through the skin showing several hair follicles

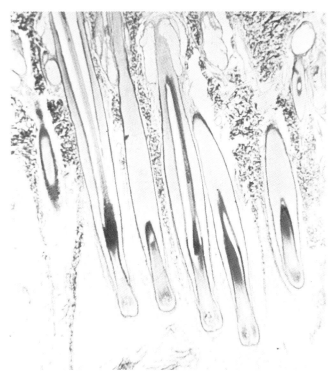

Hair is continually being shed and replaced. It is estimated that adults lose about 40 hairs each day. Permanent hair loss or baldness is a tendency inherited by men in particular. Loss of hair may be affected by the male hormone testosterone or be caused by illness, poor circulation to the hair follicles, or infections of the scalp. Baldness commonly progresses with age.

Hairs are sensory organs with nerve endings located around their roots. Hairs act as tiny triggers to indicate that some light object is in close contact with the body.

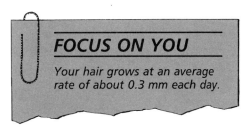

FOCUS ON YOU

Your hair grows at an average rate of about 0.3 mm each day.

Activity 1C

Tactile Discrimination

Does the same touch feel the same to different areas of the skin?

Procedure

Close your eyes. Draw a cotton thread across the palm of your hand, first with your eyes closed, then with your eyes open. *Record your impressions.* Now draw the thread across the back of your hand, or the back of your wrist. *Record the differences and explain why these differences occur.*

The Two Main Types of Skin Glands

Sebaceous (seb-ae-shus) glands occur over most of the body surface, with the exception of the palms and soles. The openings of these glands are found around hair follicles. (See Figure 1.11.) Sebaceous glands produce an oily secretion called **sebum** which helps to keep the hair from becoming dry and brittle. The oil also keeps the skin soft and helps to waterproof its surface. Underactive sebaceous glands cause the skin and hair to be dry, while overactive glands result in oily hair and skin. If the secretion of these glands accumulates inside a hair follicle, it oxidizes, forming a **blackhead**. If the glands become infected, they cause pimples or **acne**.

The **sweat** glands are found over the entire body surface, but are especially numerous under the arms, on the palms, soles, and forehead. These glands are located in the dermis and have tubes leading up to the surface through the epidermis. The sweat glands are stimulated by nerves that cause the glands to operate when the body temperature rises or when a person is experiencing nervous tension. (See Figure 1.13.)

Figure 1.13
During strenuous exercise, the body cools itself by producing large amounts of sweat.

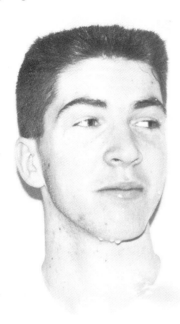

Sweat is a clear fluid composed mainly of water. It also contains salts and may contain organic compounds such as urea. When we sweat (perspire), the evaporation of the water in our sweat removes heat from the skin's surface. This cooling in turn removes heat from the blood in the vessels immediately beneath the skin, helping to reduce the body's overall temperature. A person who is badly dehydrated may not be able to sweat. Without assistance to cool the body, he or she may suffer a potentially fatal heat stroke as the body temperature soars.

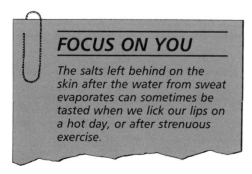

FOCUS ON YOU

The salts left behind on the skin after the water from sweat evaporates can sometimes be tasted when we lick our lips on a hot day, or after strenuous exercise.

Activity 1D

Sweat Glands

How are sweat glands distributed on the surface of the skin?

Materials

iodine solution (0.01 mol/L)
cornstarch
small paint brushes
dissection microscopes or hand lens
small pieces of bond or glazed paper
antiperspirant

Procedure

1. If your hands feel sticky or sweaty, wash them and dry them well.

2. Paint a small square with iodine on the palm of the hand and another on the wrist, neck, or forehead. The area should be about 1.5 cm². Allow the iodine to dry.

3. If your hands are cold, shake them vigorously or clap them together to warm them. Dust the painted areas with cornstarch and leave this on for several minutes. Blow off the loose starch.

4. Press the painted area of your hand firmly against a page taken from your notebook. Hold it there, without moving, for about 30 s. Use a small piece of paper for the neck or forehead prints. *Label the prints you have made to show from what part of the body they were taken.* Rows of small dots should appear on the paper. *Select a part of the print in which the dots show up clearly and draw a 1-cm square around them. Count the number of dots in each square and record your results. Which areas have the most sweat glands?*

5. Examine the prints and the painted areas of your skin carefully. *Are the dots found at regular intervals? Are they present on the ridges or in the valleys of the lines on the hand? Try a fingerprint if you are not sure. Describe how the patterns on the hand differ from those on the wrist or forehead.*

6. Treat the palm of one hand with an antiperspirant. Leave the other hand untreated. Perform some light exercise, such as running in place for 2 min, or running up a flight of stairs and along a hallway. After the exercise, repeat the test on each palm.

7. *Record your results. Compare these results with those produced in the first part of this activity. Account for any differences that you observe in your results.*
   Why should bond paper be used for the tests? If you are not sure try to do the test with a piece of paper towel.

8. Although sweat glands are not found in equal numbers on all parts of the body, a rough estimate of the total number of glands can be made as follows. You need two pieces of data: the number of sweat glands in 1 cm^2 (use your lowest score from the tests above); and the total surface area of the body. A person with a body mass of 55 kg and 155 cm in height has a surface area of about 1.5 m^2. *Find the approximate number of sweat glands in a person of this size.*

FOCUS ON YOU

Taking salt tablets during or after exercise is unnecessary and potentially dangerous. The average person already consumes more salt than the body needs, even during exercise in warm weather. Since your stomach must take water from the rest of the body to absorb salt, using salt tablets when your body needs water only makes matters worse.

FOCUS ON YOU

To maintain a constant internal body temperature, an average person exercising at a moderate to high level of intensity in warm weather may lose over 1 L of water per hour from his or her body in the form of sweat. This must be replaced, or dehydration, dizziness, nausea, and cramps can result. Drink plain, cool water before, during, and after exercise, even if you do not feel thirsty.

Other Skin Features

Moles are small spots of pigmented skin. They may form small bumps and may have a hair or hairs growing from them. Moles are best left alone; constant irritation of the mole may affect the tissue and lead to a cancerous condition. Colour changes, bleeding, or rapid growth of a mole should be reported to a physician. If necessary, the mole can be removed.

Birthmarks are made up of unusual patterns of capillaries, heavily pigmented skin, or raised, bumpy layers of skin. Some disappear during childhood. Other types can be removed by cosmetic surgery.

Freckles are small harmless spots or patches of pigmented skin. The colour is a result of melanin formed by the skin cells as a protection from the harmful effects of ultraviolet light present in the sun's rays. People with red or blonde hair are more prone to develop freckles than people with other hair colours. Freckles usually begin to appear at about seven or eight years of age.

Wrinkles are small folds in the skin, generally occurring as skin ages. Young skin can stretch and snap back into place. As the body grows older, special fibres in the skin gradually lose their elasticity. Over time, the elastic fibres break down, causing the skin to become loose and slack like an old elastic band.

Eventually, everyone gets wrinkles. Factors that hasten the arrival of wrinkles include long exposure to the drying effects of sun and wind. This can be seen in the faces of people who work outdoors such as farmers and fishermen. Your own body will already show a difference in the texture of the face and back of the hands compared to the covered areas of your body. Diets low in vitamins also affect the durability of the skin. (See Figure 1.14.)

Figure 1.14
Exposure to cigarette smoke speeds up wrinkling, even for non-smokers if they live or work in a smoky environment. Smoke components cause the blood vessels of the skin to constrict and the skin thus receives less oxygen and nutrients from its blood supply.

CHECKPOINT

1. *List five important functions of the skin.*
2. *List two kinds of glands found in the skin and state the function of each.*
3. *What is melanin and what does it do?*
4. *How does the stratum germinativum differ from the stratum corneum?*

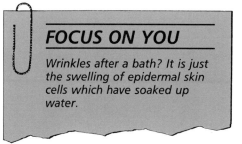

FOCUS ON YOU

Wrinkles after a bath? It is just the swelling of epidermal skin cells which have soaked up water.

1.3 Keeping in Touch with the Outside World

The skin provides a wide range of information about the immediate environment in which we live. Imagine what life would be like if, for example, you had to wear thick gloves all day. How would you be able to tell if you were holding something warm or cold, soft or hard, sharp or smooth? Fortunately, specialized nerve endings in the skin monitor conditions around the body and send impulses to the brain for interpretation.

Sense Receptors

Some nerve endings in the skin are sensitive to heat and cold, while others respond to touch and pressure. Pressure-sensitive nerve endings occur at a greater depth in the skin than do the touch sensors. These nerve cells are capable not only of indicating contact with some object, but also register the texture of the material. We can, for example, distinguish between the textures of wool and silk, glass and wood, or skin and fur. Sometimes temperature, texture, and pressure combine to identify a substance. (See Figure 1.15.) A heavy, cold, smooth piece of metal is easily recognized as being different from a piece of wood, which is lighter, rougher, and warmer to the touch.

Figure 1.15
The importance of touch

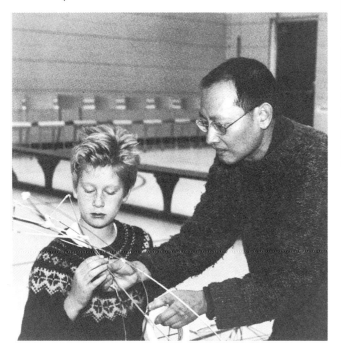

There are also many free nerve endings which are sensitive to pain. These are found just under the epidermis and around the hair follicles. Not every part of the skin has the same sensitivity. The forehead has about 200 pain receptors for every 1 cm^2 of skin. The breasts and lower arms are also well-supplied with nerve endings. The nose, however, has only about 50 receptors for every 1 cm^2, and the lobe of the ear contains even less.

Regulation of Body Temperature

The ability of the skin to detect the temperature surrounding the body is very important. Body temperature must remain within very narrow limits despite our environment. Variations of even 0.5°C can affect our feeling of well-being. By monitoring the "outside" as well as the internal temperature of blood and organs, the body is able to make adjustments to keep its temperature constant. This adjusting process is known as **homeostasis**.

There are more **cold** receptors than **heat** receptors in the skin, because cold is a greater threat to body temperature stability. Heat receptors respond best between 37°C and 40°C. Cold receptors have two peaks, working well at 15°C to 20°C and again at 46°C to 50°C. The latter range may indicate a drop in temperature. For instance, if you step out of a hot shower, you may suddenly feel cold, although the room temperature is quite normal. This is because the receptors are triggered to indicate a drop in temperature, not just "cold" as in a point on the thermometer.

The major heat-regulating centre of the body is located in the **hypothalamus** (hy-poe-thal-ah-mus) of the brain. One group of cells in this organ controls heat production; another group of cells is responsible for controlling heat loss. Nerve impulses transmitted from sensors throughout the body and the temperature of the blood reaching the brain are used by the hypothalamus to determine body temperature. Normally, about 87.5 percent of body heat is lost through the skin surface and another 10.7 percent through the lungs. The remaining heat is lost by excretion of waste products from the body. If too much heat is being lost, some of the small blood vessels in the skin close up (constrict) and some muscles may be activated to shiver, thus generating heat. Other stimuli may cause the skin to pucker into bumps of "goose flesh". These activities help to bring a lowered body temperature up again.

BIOTECH

Burns and Artificial Skin

Fires, hot water or steam, hot objects, or caustic chemicals can all cause burns. When skin is burned, it can no longer function as a protective suit of armour for the tissues underneath. The body's first line of defence against infection is breached, and serious complications and even death can result. In fact, burns are one of the leading causes of death for people between the ages of 1 and 40.

The severity of a burn is determined by the depth to which the tissue has been damaged. If the burn is mainly superficial and damages only the upper epithelial layer, it is called a *first-degree burn*. There are usually no lasting effects from burns of this type. If the burn damages all of the epidermis, and some of the underlying dermis layer, it is called a *second-degree burn*. Since the nerve endings in the dermis are affected, this type of burn is very painful. Blistering often accompanies second-degree burns, and sometimes scar tissue will form. If the burn destroys all of the epidermis and the dermis in an area, it is called a *third-degree burn*. Burns of this severity destroy all of the receptors and nerves in the skin. As a result, the victim can no longer feel anything in the affected area.

Major hospitals have developed burn treatment centres to treat serious burn victims. Of primary concern are the problems of infection and fluid loss. The first step in treatment is the removal of the dead skin, and treatment with antibiotics. The patients are bathed in special tubs of warm water to remove the layers of dead skin. A layer of a synthetic polymer, hydrogel, is then placed over the damaged tissue. This layer permits the exchange of gases, regulates fluid loss, and allows the absorption of antibiotics.

As soon as possible, skin grafting is attempted. These grafts may be taken from other undamaged areas on the patient's body, from family members, or even from cadavers. Pigskin may also be used. Since the grafts from sources other than the victim's own skin are made of proteins that are foreign to the body, they are normally rejected after a few weeks, by which time the victim's body should have begun to produce new skin.

Scientists have recently developed an artificial skin. Powdered shark cartilage mixed with a solution of collagen extracted from cowhide is poured into a pan and freeze-dried. It is then baked and bonded to a rubberlike plastic membrane, which is then also freeze-dried. The substance is stored until needed. This artificial skin is usually accepted by the body without the use of drugs to suppress the rejection reaction.

When this substance is put on the wound, the body cells migrate through it and begin to rebuild the skin. Nerves and blood vessels also grow into it. After a few months, the artificial layer breaks down and moves to the surface, where it will slough off. The rubberlike membrane is peeled off after about two weeks, and small patches of epidermis are then grafted onto the underlying layer.

FOCUS ON YOU

Deposits of fat beneath the skin help to maintain a normal body temperature by forming an insulating barrier to reduce heat loss. Women generally have a higher percentage of fat cells than do men. This fat serves to make female bodies more resistant to cold.

If the body gets too hot, blood vessels may expand (dilate), sweating may increase, and muscles will relax, thereby producing less heat. Unless we are ill and have a severe fever, the body is usually able to maintain a constant temperature with only minor variation.

Under normal conditions, young people rarely suffer from **hypothermia**, which involves the lowering of the body core temperatures; however, extreme cold can be encountered during accidents and is a potentially fatal threat to anyone. Heat energy can be lost or gained by three different methods – **conduction**, **convection**, and **radiation**. For heat to be transferred by conduction, there must be direct contact with another substance. If the body is in direct contact with the cold ground, without any insulating material present, for example, heat will be lost by conduction.

Convection requires the presence of a current of air or water. These substances absorb heat energy, become less dense, and as a result carry the heat away. Sweat evaporating from the skin surface is an example of heat loss by convection. Heat causes sweat (water) to turn to vapour. The heat required for this change is drawn from the skin and its loss rapidly reduces the body temperature.

The body absorbs heat by radiation when warm sunshine strikes the body. If you place your hand near the skin's surface when you have a sunburn, or have been exercising vigorously and are very hot, you can feel this heat.

If you are exposed to severe cold, there are certain basic principles which you should follow. These principles arise from the ideas discussed above. You may not always have the things you need to cope properly with an emergency situation; however, if you understand the problems involved, you may be able to improvise and avoid a potentially dangerous situation.

How to Conserve Body Heat in an Emergency

1. Find or make a shelter. Sit out of the wind. Cold, moving air quickly drains away the body heat.
2. Remove wet clothing. Body heat will be used up trying to dry wet clothing.
3. Wear *layers* of clothing. (See Figure 1.16.) The extra clothes trap insulating layers of air and help to prevent heat loss. If the body is exposed to extreme cold and the body core tempera-

Figure 1.16
Dress for the cold by wearing layers of clothing.

ture drops (hypothermia), it is necessary to raise the temperature of the skin before insulating it with more layers of clothing. This is similar in principle to placing frozen foods in an insulating bag. The layers, in this case, are keeping the cold in and preventing external heat from warming the frozen foods.

4. Keep moving. Heat is generated when muscles contract. Keep this movement within reason; you need to conserve energy, so sufficient movement to maintain good circulation is necessary without needless waste of energy reserves.
5. Stay awake. We generate less heat when we are asleep.

CHECKPOINT

1. Name four types of information about the environment detected by specialized nerve endings found in the skin.

2. What areas of the body surface have the greatest numbers of nerve endings? Why might this be so?

3. Temperature sensors in the skin cannot measure "temperature". What is it that they do measure or tell us?

4. What is meant by homeostasis? Explain, using an example.

5. Explain the meaning of the following terms:
   (a) convection
   (b) conduction
   (c) radiation.

6. List four ways you can conserve heat in a cold weather emergency.

1.4 Skin Health and Hygiene: More Than Looking Good

We are often concerned about our appearance, whether we are looking for approval from our friends or we want to make a good first impression at a job interview. This concern leads us to have two views of ourselves: the way we think we look; and the way we would like to look. The way we would like to look may determine some quite drastic decisions about ourselves, including those which may be costly or even painful. A new weight loss program, a skin graft or structural changes to the nose, hair grafts, or even a change of hairstyle are examples. The appearance of our skin and hair are major contributors to the personal image we present, both to ourselves and to others. A good image promotes self-confidence and radiates health. One way to achieve this is with good skin care.

Taking Care of Your Skin

- **Wash frequently with warm soapy water.** Sweat, sebum, dead cells, dirt, and dust accumulate on the skin surface. This moist, warm environment is an ideal place for bacteria to breed, and this mixture produces unpleasant body odours. Frequent washing removes the dirt, and the soap removes the oils which cause the dirt and dust to stick to the skin. The *gentle* rubbing action of the towel removes the dead cells from the surface of the body, stimulates the skin, and increases circulation. Moderate use of deodorants and antiperspirants are useful, but it is necessary to remove the source of body odours, not simply cover up their presence. Heavy use of some antiperspirants can block the pores of the skin.

- **Clean any wound or break in the skin and use an antiseptic lotion.** The skin forms a suit of armour over the body and provides a defence against the invasion of bacteria. Cuts, scratches, pimples, and wounds produce a gap in these defences through which bacteria can enter the body. Antiseptics inhibit the action of the bacteria and form a temporary barrier. Cover wounds and scrapes with a suitable bandage.

- **Treat burns seriously.** Hold a burn under cold running water for several minutes. This will remove the heat, help limit the damage, and reduce the pain. Do not use creams or oils on a burn. Small burns should be lightly covered with a sterile bandage. More serious burns should be treated at the hospital.

- **Don't fuss with blemishes.** There is a great temptation to squeeze spots and pimples.

When you do this, you can transfer the infections to other areas, damage the tissues around the affected area, cause the infection to be pushed into the tissues around the site, or open up the surface to further infection. There are a number of antiseptic and proprietary lotions on the market which can be used to help control skin problems. If these are not effective, get medical advice.

Acne and Other Common Skin Problems

Acne, which troubles many teenagers, is not necessarily caused by lack of cleanliness. Acne is caused by bacteria which infect blocked hair follicles. (See Figure 1.17.) The surrounding skin becomes swollen and tender. A severe infection may result in permanent scarring of the skin.

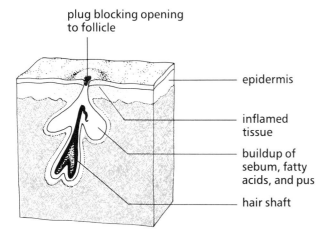

Figure 1.17
A blocked hair follicle traps the sebum below the skin surface, leading to an acne infection. What can be done to help prevent this condition?

Various types of food have been blamed for many of the acne problems suffered by adolescents. Recent research, however, has found no evidence to support the suggestions that chocolate or fatty foods, for example, are responsible for, or aggravate, the condition. A well-balanced diet ensures the skin will have the nutrients required to heal.

The main culprit appears to be **androgens** – sex hormones. These hormones are first secreted at the age of puberty (the time at which both sexes become functionally capable of reproduction) and not long after that, acne makes its appearance in most young adolescents. Many studies have shown that the quantity of androgens produced by the sex glands influences the amount of oily secretions from the skin. Gentle cleansing of the skin, coupled with a healthy diet and lifestyle, will help. Severe cases should be treated by a physician.

Boils are round, tender, reddened areas of skin containing a central core of pus and bacteria (staphylococci). Some boils erupt and release the central core before disappearing. Others simply regress and disappear. Boils should not be squeezed, as this may force some of the bacteria from the core into the surrounding tissue thus initiating other sites of infection. Boils should be covered to prevent the spread of the infection. A small boil will usually heal without treatment. Severe cases, involving several boils, should be treated by a doctor who may prescribe an antibiotic salve.

Athlete's foot is a fungus infection that frequently begins in the warm, moist areas between toes. It produces a painful itching or burning sensation. The fungus can be picked up in public showers, gyms, or swimming pools, or by using other people's infected footwear. The infection causes the skin to peel, leaving a red, shiny layer exposed. It may spread to cover the whole sole of the foot. Treatment requires that the foot be kept dry with pads of cotton and powder between the toes and frequent changes of fresh socks. Careful drying between the toes after a bath or shower is important. The application of non-prescription remedies is usually sufficient to cure mild cases. Serious cases of athlete's foot require medical attention.

Dandruff is caused by the shedding of the outer layers of the scalp, which produces rather obvious white flakes of skin. Poor circulation can increase the rate at which this shedding takes place. Infections, poor diet, lack of regular washing, strong shampoos, or insufficient rinsing of the hair after washing, may all promote this scalp condition.

> **FOCUS ON YOU**
>
> Warts are caused by viruses which make the skin's epidermal cells grow rapidly. Your doctor is your best source of advice on their removal.

Skin Cancer

Prolonged exposure to the sun increases your risk of skin cancer. Hundreds of thousands of cases are reported in North America each year. If caught in time, most skin cancers can be removed quite easily. Common indications that skin cancers are developing include skin sores that do not heal, changes in the size or colour of a wart or mole, and the discovery of unusual patches of coloured skin developing. The face, neck and hands are the most common sites for skin cancers, as these areas receive the most sun. Although the symptoms usually occur after age 30, the habits of earlier years may be the cause of the problem. Skin cancers are rarely dangerous when treated early. Unfortunately, some forms can spread to the rest of the body and be fatal if left untreated.

Cosmetics

Cosmetics are aids that people use to smell better, or to improve the appearance of their skin or hair. Cosmetics may also be used to hide a blemish or to give a more youthful appearance. (See Figure 1.18.) Often these products work well, but it is important to read and to follow the directions on the label, especially when using eye cosmetics.

Some people are very sensitive to certain substances used in the manufacture of cosmetics. After applying a product containing these chemicals to their skin, they may become itchy, their skin may become red and sore, their eyes may run, or their nose become congested. More serious allergic reactions such as hives can also result. Before using a new cosmetic or hair preparation, try a small amount on your arm, the back of your hand, or a small patch of your scalp. Leave it there for a few hours. If you notice any irritation, don't use that product, as the full treatment may cause an even more severe reaction. To prevent clogging of skin pores, make-up should always be removed every night. Cleansing creams may be used, but their residue should then be removed with warm, soapy water.

Figure 1.18
Common skin cosmetics. Test a small amount on your skin before using.

Taking Care of Your Hair

Well-groomed hair is a very attractive part of the image we present to our friends and the people we meet. Many students spend a great deal of time and money on keeping their hair looking nice.

Shampoos and Hair Conditioners
Many shampoo advertisements claim to make your hair shinier, healthier, sweeter-smelling, and more manageable. They also claim that the product will nourish your hair and make you more attractive. How valid are these claims? (See Figure 1.19.)

Figure 1.19
Who to believe?

More than 95 percent of your hair is dead tissue. Only the root is living and growing. Dead tissue cannot acquire nourishment, so the root hair cells are nourished by the bloodstream from within. The hair care product you use may contain proteins or amino acids, but your hair cannot use these effectively as nutrients.

One important component is the pH. Your hair will have a pH range between 5 and 6. If the shampoo is too basic, it can cause the outer layers of the hair shaft to swell and become softer. This will produce dull, lifeless hair with little curl. Acidic products will harden and shrink the cuticle, producing "bounce" and shine in the hair; however, in more acidic shampoos, these values are offset by the hair becoming more brittle and easily broken, especially at the ends.

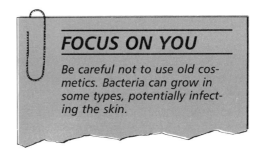

FOCUS ON YOU

Be careful not to use old cosmetics. Bacteria can grow in some types, potentially infecting the skin.

Activity 1E

Comparing Hair Care Products

What are the pH values of some common hair shampoos and conditioners?

Materials

samples of several shampoos and hair conditioners
Hydrion pH paper or pH meter
small test tubes

Procedure

1. *Prepare a chart listing all the products to be tested. Beside this list make a column where you can record the pH of each product.*
2. Test each product with the Hydrion paper or the pH meter and *record your results.*

Questions

1. Examine your data and make a new list of the products, putting the most acidic (lowest numbers) first, and the most basic last.
2. What range of pH did you discover? What was the most common value? How does this compare with the normal pH for hair?
3. Read the advertising on the products and see if any particular pH is recommended for your type of hair.
4. Were there any products that had high pH ranges and that you would rather not use on your hair? Why would you avoid these products?

Here are some suggestions for good hair care.

- **Wash your hair regularly.** It is not necessary to wash it every day, as this removes all the natural oils and tends to make the hair dry and brittle. Once every three days is usually enough, but if the hair is very dry, four to five days is preferable.
- **Choose hair care products carefully.** Use mild shampoos containing little detergent so that not all the oils are removed. Conditioners help if the hair is very dry, or if you are bleaching or colouring your hair. Rinse out all the shampoo from your hair thoroughly. Be sensible and sceptical about the many advertisements for shampoos, conditioners, cures for baldness, spilt ends, etc. They are enticements to buy, and most of the claims are exaggerated.
- **Massage the scalp.** Massaging the scalp gently with the fingertips during washing helps to stimulate the skin and hair roots. Avoid vigorous rubbing of the hair when it is wet. Pat the hair dry with a towel instead. Hair dryers and curling irons dry and damage hair, so use them sparingly.
- **Eat a balanced diet.** Diet plays an important role in hair conditioning. Hair, like other body tissues, requires a good supply of the right nutrients. Vegetables and fruits contain many vitamins that help produce good hair growth.
- **Brush your hair regularly.** Regular brushing helps distribute the hair's natural oil from the base to the ends, adding gloss and shine. Do not lend your brush or comb to friends or borrow theirs.
- **Limit exposure to the sun.** Avoid long exposure to the sun without a hat, as this dries the hair and also makes it brittle.

CHECKPOINT

1. *List four ways in which you can take care of your skin.*
2. *What are some of the problems associated with the use of cosmetics?*
3. *List two positive things you can do to take care of your hair, and two things you should avoid doing to your hair. Explain your choices.*

1.5 A Case of Overexposure

Canadians spend so much of the year wrapped up against the cold that, when summer comes, the image of a golden tan is almost irresistible. Tanning salons in Canada are estimated to generate more than $300 million a year. Add to this the money spent on tanning lotions, sunscreens, and the cost of holidays designed to promote a tan, and that "sunshine look" is obviously an important facet of Canadian lifestyles. (See Figure 1.20.)

Figure 1.20
How much is too much?

The benefits of a tan are, however, purely cosmetic and psychological. There is no evidence that a tan has any health benefit at all. On the other hand, tanning has definite risks which include premature aging of the skin, as well as increased risk of skin disorders, such as skin cancer, phototoxic reactions, burns, blisters, redness, itching, and hives. The culprit is ultraviolet radiation.

Ultraviolet Radiation

Ultraviolet radiation has a shorter wavelength and higher energy than visible light. (See Figure 1.21.) Just as visible light has different wavelengths producing different colours, so different wavelengths of ultraviolet light produce different effects on the skin. Ultraviolet light has three different wavelengths – UV-A, UV-B, and UV-C.

Figure 1.21
Ultraviolet radiation and your skin

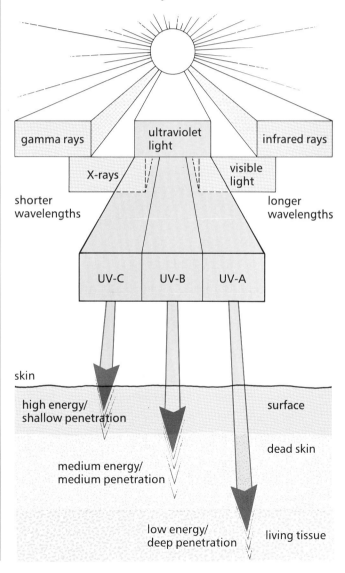

- UV-A tans the skin by oxidizing existing melanin, thus making the skin darker. It also stimulates the production of small amounts of melanin. It starts the tanning process before the skin burns and has the least cancer risk.
- UV-B has three effects: it oxidizes and darkens the melanin present; it increases the production of melanin; and it increases the number of pigment producing cells (melanocytes). This tan is darker and longer lasting, but tends to produce burns before it tans. It is a known cancer risk.
- UV-C has the most energy but the least penetration. Dead layers of skin stop UV-C radiation, but it is dangerous to living cells causing burns and destroying cells as well as having carcinogenic effects. It is therefore very important to protect the eyes with special goggles when exposed to this type of radiation.

All three types of UV radiation are present in sunlight. "Safe" tanning machines are supposed to produce only UV-A radiation. (Some UV-B may be present.) When using these machines, the time of exposure may be controlled by timers, and the person tanning can use protective goggles; however, in the last five years, some evidence has been produced to indicate that even UV-A may pose some risks.

The diminishing ozone layer around the earth has produced grave concerns around the world. As this layer thins, more UV light is allowed to pass through and reach the ground. This may be the reason for the steady increase in the number of cases of skin cancer in recent years. Common sense may indeed be the best protection your skin can receive when you go out in the sun.

People at Special Risk

Sun tanning, whether natural or artificial, holds special risks for people in the following categories.

- People who burn easily or do not tan at all (These people usually have light skins, are very fair and produce little melanin.)
- People with large moles (dysplastic nevi)
- People with sunburn or chronically damaged skin
- People with a family history of skin cancer
- People who have photosensitivity disorders (produce photoallergic responses)
- People who are taking medications that are photosensitive (These include many antibiotics, anticonvulsants, antidepressants, antidiabetic, diuretic prescriptions, and tranquilizers, as well as a number of cosmetic products.)

Quick-tanning Products

There are two types of quick-tanning products. Both have been on the market for several years. The first comes in tablet or capsule form. There is little restriction on these products, as they come under the heading of cosmetics. These products are basically dyes that colour the skin. They contain substances such as beta-carotene, found in carrots and tomatoes, and they give the skin an orange-yellow or orange-brown tinge. The second tanning product is used on the skin. This

FOCUS ON YOU

Doctors sometimes use UV light to treat certain skin disorders, but such treatment is carried out under strict medical supervision.

type contains pigments that react with the skin proteins to darken the skin. Neither product offers any protection against the dangerous effects of UV radiation.

Sunscreens

Sunscreens are applied to the skin and contain PABA (para-aminobenzoic acid) and zinc oxide. These substances, especially PABA, absorb ultraviolet light and therefore offer some skin protection. Sunscreens allow some tanning to take place, whereas **sunblockers** prevent or control the amount of UV light penetrating the skin.

The higher the Sun Protection Factor (SPF) listed on the product, the greater the amount of sun protection offered. For example, using a sunscreen with an SPF of 3 means it will take three times longer for your skin to burn than if it were not protected. Using a sunscreen with an SPF of 8 will extend the time it takes for you to burn by a factor of 8. People that burn easily should use a product with an SPF factor of 15. Products with an SPF greater than 15 have caused skin irritation in some people.

Advice to Sun Lovers

- **Respect the power of the sun.** Avoid the noonday sun, and start the sunbathing season with shorter periods early or late in the day. Watch out for hazy days – only complete cloud cover will provide protection from sunburn. Reflected sunlight from water, sand, snow, and ice can produce very painful burns.
- **Know your own skin sensitivity.** Be sensible about your own skin characteristics. Don't stay out in the sun just because others, who may be darker than you, are staying out.
- **Avoid excessive exposure to the sun.** Long exposure to the sun is inviting dry skin, early wrinkles, sunburn, and possible skin cancer. Enjoy the sun on your body, but limit the time to sensible periods of exposure.
- **Use a sunscreen.** Be sure that the lotion you use contains PABA and that it contains a suitable factor for your skin.
- **Use protection.** Use a hat or protective clothing. Wear sunglasses to protect your eyes from UV light. Use lip protection to prevent dry, chapped lips.

CHECKPOINT

1. What are the three types of ultraviolet radiation? Which of these is the most dangerous?
2. What three effects does UV radiation have on the skin?
3. List three types of people at special risk from overexposure to the sun.
4. What substance, found in sunscreens, is currently believed to be the best protection against UV radiation?
5. What SPF value is recommended for persons with fair skin?

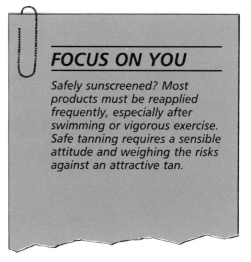

FOCUS ON YOU

Safely sunscreened? Most products must be reapplied frequently, especially after swimming or vigorous exercise. Safe tanning requires a sensible attitude and weighing the risks against an attractive tan.

Chapter 1 / Getting Under Your Skin

CHAPTER FOCUS

Now that you have completed this chapter, you should be able to do the following:

1. Name the parts of the cell and give the general structure and function of each organelle.
2. Describe the different levels of organization of the body into cells, tissues, organs, and systems, and give an example of each.
3. List the four major groups of tissues.
4. Name at least four systems of the body and list the organs each contains.
5. List the functions of the skin.
6. Label the parts of a skin section diagram and give the functions of each part.
7. Describe the two types of glands found in the skin and state their functions.
8. List the specialized sensory nerve endings found in the skin.
9. Describe the role of the skin in maintaining body temperature.
10. Describe how proper skin care can be maintained.
11. Describe how the effects of exposure to ultraviolet light can be hazardous to the skin and how skin problems can be avoided.

SOME WORDS TO KNOW

Match each of the descriptions given in the left-hand column with a word shown in the right-hand column. DO NOT WRITE IN THIS BOOK.

CELL REVIEW

1. "Bubbles" of water, etc., surrounded by a membrane
2. Small sacs which help to pass large particles across the cell membrane
3. Structures in which energy is converted for cell activities
4. Membrane-lined channels which produce lipids and enzymes and package them for transport
5. Double-layered membranes often found with ribosomes present
6. Thin membranes made up of protein and lipid layers containing small pores
7. Found in the nucleus, contains genetic material
8. Mixture of fluids, structures, and materials found outside the nucleus, but inside the cell membrane
9. Thin, fingerlike extensions found on some cells which help to increase the surface area of the cell for more efficient absorption
10. Two cylinderlike structures, found just outside the nucleus, which are active during cell division

A cell membrane
B nucleus
C mitochondrion
D endoplasmic reticulum
E Golgi bodies
F cilia
G vacuoles
H nucleolus
I ribosomes
J centrioles
K microvilli
L pinocyte
M chromosomes
N cytoplasm
O lysosomes

33

THE SKIN

1. Thick, underlying layer of the skin containing nerve endings, glands, fat cells, etc.
2. Colour pigment found in the skin
3. Outer layer of dead epidermal cells
4. Tubular structures that secrete sebum
5. Layer of cells that constantly replaces the outer layers of dead epithelium
6. Protein that hardens the cells of the skin, nails, and hair
7. Tissues that line the surfaces of the body and organs
8. Glands that secrete oil

A epithelium
B keratin
C dermis
D epidermis
E stratum germinativum
F stratum corneum
G melanin
H hypothermia
I sebaceous glands
J sweat gland
K acne

Select any two of the unmatched words and, for each, give a proper definition or explanation in your own words.

SOME QUESTIONS TO ANSWER

1. Draw a simple cell diagram from memory and label the parts.
2. Explain why the skin is such an important organ.
3. What are the dangers of overexposing the skin to the sun's rays?
4. How can the dangers of sun tanning be minimized?
5. (a) How are wrinkles formed?
   (b) What effect does smoking have on the premature aging of skin?
   (c) How can you help reduce wrinkling?
6. A child has just been pulled from freezing water and is suffering from hypothermia. What can you do to help?
7. Explain how you might advise a friend on what sunscreen to purchase.
8. What advice can you give to persons with allergies, when buying a new cosmetic or hair care product?
9. Why can water at 20°C feel warm at one moment and cool a moment later?
10. List seven functions of the skin.
11. Explain how and why skin cells are replaced.
12. In what ways is your skin unique and different from that of other people?
13. What causes body odour?
14. What changes take place in the skin
    (a) when you are angry?
    (b) when you are very hot?
    (c) when you are very frightened?
    (d) when you have been out in the hot sun for several hours?
15. Why do the numbers of sweat glands and the numbers of sensory nerve endings vary from one place in the body to another?
16. Why is the elbow rather than the hand a better part of the body to use when testing the temperature of a baby's bath water?
17. Element-E, Tissue-T, Organ-O, Compound-C, System-S, Cells-CE. Select one of the letters from the preceding list which you think identifies each of the following terms.
    (a) columnar epithelium
    (b) neuron
    (c) cardiac muscle
    (d) fat
    (e) liver
    (f) circulation
    (g) oxygen
    (h) digestion
    (i) protein.

SOME THINGS TO FIND OUT

1. What effect do frequent "perms" have on the hair?

2. What new techniques are being used to remove unwanted birth marks?

3. Research the role of the skin in controlling body temperature. Select a person who smokes and another (a control) who does not. Then determine how smoking affects skin temperature. Graph your results. Take the temperature of the hand at regular intervals for about half an hour after the cigarette has been smoked by the person who smokes. Take similar readings with the non-smoker control.

4. Determine what causes "goose flesh". You might also investigate how it is caused and used in other animals.

5. The most effective sunscreen is reported to be PABA (para-aminobenzoic acid). Do a survey of the creams available and of the data available on their effectiveness. Can you think of a means of testing their effectiveness by using several creams on your own arm?

HEALTH CONCERNS: HUMAN DISORDERS RESEARCH ASSIGNMENT

Introduction

The human body is a very efficient machine. Most of the time it works well, adjusting to changes in the environment, resisting infections, and adapting to meet a host of potential dangers; however, the body cannot always cope with a particular infection, or its own systems may fail and sickness result. This assignment will give you an understanding of one disorder in detail. It should make you aware of sources of information about human disorders and of the many support programs that are available for some diseases. It should also give you a greater understanding of the problems that face many disabled or afflicted persons.

Procedure

1. Select one of the disorders listed and prepare a research assignment on the topic. The assignment should be about three pages in length and either typed or neatly written.

2. The following points should be considered in your report:
   (a) a brief description of the disorder by way of introduction
   (b) the cause or causes of the disorder
   (c) the symptoms
   (d) treatment
   (e) any side effects or associated problems
   (f) an explanation of the problem, if known
   (g) prognosis and possible future treatments
   (h) other relevant factors.

3. When choosing your topic, you may wish to select a disorder with which you are familiar, either directly or indirectly. Perhaps you, or a member of your family, have suffered from a particular disorder and it would be useful to learn more about the disease. Or you might consider visiting a person suffering from arthritis (or other ailment) and conduct an interview which would show how that person copes with difficulties or how the disorder has affected his or her lifestyle. If you approach such a person, be considerate and sensitive to his or her feelings. Such people are often quite willing to talk about their disorder and you may gain some very special insights into their problems.

Here are some disorders for you to consider. You will find more listed at the end of each chapter.

HEALTH CONCERNS

acne	food poisoning
scoliosis	crib death
herpes	emphysema
Alzheimer's disease	premenstrual
cystic fibrosis	syndrome (PMS)
psoriasis	rabies
hay fever	acquired immune
cataracts	deficiency
anorexia nervosa	syndrome (AIDS)
allergies	
asthma	
arteriosclerosis	
anemia	
tetanus	
alcoholism	

BIOLOGY AT WORK: CAREER RESEARCH ASSIGNMENT

Every student must eventually choose a career. This assignment will help you discover where career information can be found, including what courses are required to enter a specific profession, what colleges and programs are available, and what working conditions, salary ranges, and employment opportunities exist. Even if your own career ambitions do not lie in the field of human biology, working on this assignment will show you how to find information about any career choice.

Procedure

1. Prepare a report about three pages in length.

2. Research and organize the information as follows:
   (a) what the person does
   (b) training required
   (c) high school credits needed and post-secondary training required
   (d) possible salary ranges
   (e) hours on the job
   (f) advantages and disadvantages of the job
   (g) current job opportunities and employment possibilities.

3. You may collect and include pamphlets, pictures, etc., to illustrate your report.

Where to look for information:

- School or public library. Most libraries keep files of useful information on a variety of topics.
- Student services office. Many schools have a job search computer program as well as useful files.
- Pamphlets are available in drug stores, dental and medical offices, health clinics, city health units, etc.
- You can find associations for many professions listed in the yellow pages of city telephone directories, and you can write for information.
- Visit or telephone people in the field. Be thoughtful and polite with your request. State what it is you want and ask if you can call back at a time that is convenient to them.

Here are some possible jobs you might investigate. More are listed at the end of each chapter.

BIOLOGY AT WORK

ALLERGIST
AMBULANCE ATTENDANT
PHARMACIST
HEALTH OFFICER
PHYSICIAN
OCCUPATIONAL THERAPIST
LAB TECHNICIAN
SPEECH THERAPIST
KINESIOLOGIST
PHYSIOTHERAPIST
CHIROPRACTOR
DENTAL HYGIENIST
X-RAY TECHNICIAN
OPTOMETRIST
DENTURE THERAPIST
FITNESS INSTRUCTOR
MASSEUR/MASSEUSE
SURGICAL NURSE
FOOD INSPECTOR
GENETIC COUNSELLOR
MEDICAL SECRETARY
MEDICAL ILLUSTRATOR
NUTRITIONIST
COSMETICIAN

THE BODY IN MOTION

Think for a moment about your favourite way to exercise. Whether it is walking, running, swimming, or playing a sport, parts of your body will be moving as you exercise. Some of the movements, such as walking, seem so easy to do that we don't even have to think about them. Others, such as hitting a ball with a racquet, require some practice. All of these movements require two basic things. One is a framework which supports the body and its parts. This is the skeletal system. The other provides the way in which this framework, and the body parts attached to it, are moved. This is the muscular system.

Consider the nimble fingers of a keyboard musician, or the speed and power of a hockey player's shot, or even the ability to open a door while balancing an armload of books.

Working together, the skeletal and muscular systems provide the human body with an astonishing range of motion.

KEY IDEAS

- *In order to move, the body requires support and force.*
- *Bones provide the body with support and a system of levers; muscles supply the force to move the bones.*
- *Bones are living tissues, constantly changing, growing, and being reconstructed.*
- *The skeletal system does more than provide the framework for the body; it also produces blood cells and stores minerals.*
- *Bones vary in size, shape, and function.*
- *There are three main kinds of muscle tissue, each performing quite different roles.*

2.1 The Body's Support System: The Skeleton

The skeletal system is a remarkable and versatile bony structure which performs a number of important functions. It provides a framework for the attachment of muscles which support the body and make movement possible. Bone protects some delicate organs, such as the brain. It produces blood cells and also acts as a reservoir for the storage of minerals, such as calcium and phosphorus. The entire skeleton is a built-in shock absorber, taking the brunt of impact when a person jumps or falls.

Although their internal structure is the same, the bones of the skeletal system vary greatly in size and shape. This means they can be grouped into four different categories – flat, long, short, and irregular. (See Figure 2.1.) *Flat bones*, such as those in the skull, provide protection for soft tissues and points for muscle attachment. The *long bones*, which are found in the arms and legs, provide strength and support. They also serve as levers. *Short bones* are found in the wrist and ankle. They are not much longer than they are wide. These bones provide flexibility. *Irregular bones*, such as those found in the backbone, or the middle ear, have unique shapes related to a specialized function. For example, a vertebra, one of the bones of the backbone, has small extensions for muscle attachment and an opening to allow the spinal cord to pass through it.

CHECKPOINT

1. List the four categories of bone shapes.
2. (a) Which group of bones provides protection?
   (b) Where are some of these bones located in the body?
3. List four functions of the skeletal system.

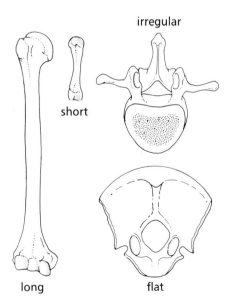

Figure 2.1
The four main types of bones. How is the shape of each bone related to its function? From which part of the body does each bone shown here come?

2.2 A Close Look at Bone

A casual examination of bone gives the impression that it is dead material. Actually, bone is far from dead and inactive. In spite of its solid and rigid appearance, it is constantly changing, growing, and being reconstructed. It is supplied with blood vessels and nerves just like any other body tissue.

There are two kinds of bone tissue. **Compact bone** is hard and very strong. **Spongy bone** is lighter and full of tiny spaces; in fact, it looks like a sponge. Spongy bone is also very strong but contains many more blood vessels than compact bone. Both kinds of bone tissue may be found within one individual bone. (See Figure 2.2.)

Bones are not completely solid. If they were, they would be so heavy that movement would become difficult. In the centre of most bones is a hollow canal containing bone **marrow**. The bone marrow is made up of many blood vessels, fat cells, and some blood-forming tissues. Near the end of a bone is an area called the **epiphyseal cartilage**, or growth plate. New bone tissue is formed here, increasing the length of the bone.

Covering the bone is the **periosteum**, a thin, double membrane that contains blood vessels, nerves, and cells which form bone. It is this membrane that controls the development of bone.

Covering the ends of most bones is a smooth, but very tough layer called the **articular cartilage**. This layer provides a shock-absorbing cushion, as well as protection for one bone as it moves against another.

Cartilage

If you touch the tip of your nose, or gently bend your ear, you can feel a framework under the skin. This framework is made of **cartilage**, a strong, but flexible type of connective tissue. A microscopic examination of cartilage would show that it is made of long fibrous strands surrounded by a jellylike substance. The fibres provide strength, and the jellylike filler acts as a shock absorber. What advantages are there to having ears and the tips of our noses made from cartilage instead of bone?

Activity 2A

Bone Structure

What does the inside of a bone look like?

Materials

a fresh leg bone (femur) which has been cut lengthwise
a dried leg bone cut lengthwise
a whole leg bone
magnifying glass or lens
probe
knife or scalpel
(Beef bones are best and should be obtained from a butcher.)

Procedure

1. Compare the diagram of the parts of a bone (Figure 2.2) with the freshly cut bone. *Locate and describe the parts which are labelled on the diagram.*

2. Feel the texture of the different parts of the bone (use both the freshly cut and dried bones). *Describe any differences in texture between (a) compact bone and spongy bone, and (b) compact bone and articular cartilage.*

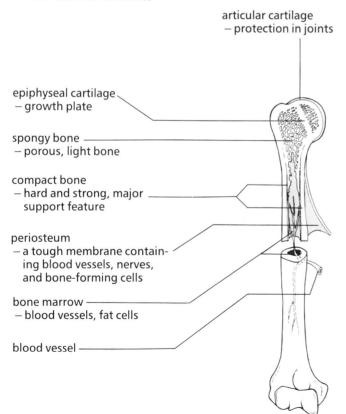

Figure 2.2
A section through a long bone, showing its internal and external structures

- articular cartilage — protection in joints
- epiphyseal cartilage — growth plate
- spongy bone — porous, light bone
- compact bone — hard and strong, major support feature
- periosteum — a tough membrane containing blood vessels, nerves, and bone-forming cells
- bone marrow — blood vessels, fat cells
- blood vessel

3. Use the blade of the knife to scrape a small section in different parts of the bone. *Which part of the bone seems to be the hardest?* Try to remove a small piece of articular cartilage and periosteum. *How thick are they?*

4. Remove some of the bone marrow from the freshly cut bone section. *Describe it.*

5. Carefully examine the inside and outside of the dried bone, looking especially for any small openings. *Why would there be a small opening on the bone?*

6. Compare the thickness of the compact bone along the length of the bone. *How does the middle of the bone compare with the ends?*

Activity 2B

The Properties of Bone

What allows a bone to be strong and flexible?

Materials

two rib bones from a butcher (the longest possible) or two chicken leg bones
small pieces of uncooked bone (must fit into crucible)
eye/face protection
hydrochloric acid (15 percent)
Bunsen burner
crucible
tongs
retort stand
ring support

Procedure

CAUTION!
You must wear eye/face protection for this activity.

1. Clean as much of the soft tissues from one of the ribs as possible. Immerse the bone in a jar of 15 percent HCl solution. Soak for at least two days. When the bone becomes pliable, use the tongs to remove it from the acid, and wash it thoroughly in running water.

CAUTION!
If you accidentally spill hydrochloric acid on yourself or the lab area, flood the affected area with water and notify your teacher immediately.

2. Compare the treated rib with the untreated rib. *What differences do you notice?* Locate the thin covering on the outside of the untreated bone. *What is this called?* Try to remove this covering. *Why is it so difficult to remove?* Try to tie the treated bone in a knot. *How has the structure of the bone changed?*

CAUTION!
Do not touch the crucible until you are told to do so by your teacher.

3. Take a small piece of another bone and place it in a crucible. Use a hot Bunsen burner flame to heat the bone. Continue heating until no more smoke comes off. *Record your observations.* (Note: The treatment in step 1 will remove the inorganic materials from the bone, leaving the protein. The treatment in step 3 will remove the protein and leave the inorganic material.)

Questions

1. (a) What effect does the acid have on the bone?
   (b) Does it alter its hardness or flexibility?

2. (a) What properties do inorganic substances give to bone?
   (b) What are the inorganic substances?

3. Does heating alter the shape, hardness, or flexibility of the bone?

4. What is the function of the protein in bones?

CHECKPOINT

1. How is compact bone different from spongy bone?
2. Why is epiphyseal cartilage called the growth plate?
3. (a) Describe the structure of the periosteum.
   (b) What is its function?

2.3 Bone Formation

In the embryo, a model of the future skeleton develops from cartilage and fibrous tissues. This model acts as a pattern for each bone as it is produced. Even at two months of age, the embryo may start to develop bone cells called **osteocytes** (os-tee-oe-site) in this mold. (See Figure 2.3.) Once started, the process of **ossification** (bone formation) continues throughout life, although the major part of it occurs during our first 20 years or so.

As they develop, bone cells become buried in a mixture of minerals, mainly calcium and phosphorus. The cells are separated within a network of mineral deposits, but are connected to each other by tiny canals. These little canals deliver blood supplies to the bone cells and are connected to large tubelike vessels, the **Haversian canals**. In compact bone, tissue is laid down in a circular fashion resembling the cut section of a tree with its annual rings. (See Figure 2.4.)

The **Haversian system** connects with large blood vessels that bring new supplies of oxygen, as well as calcium, phosphorus, and other essential nutrients, to the bone tissue. Blood vessels also carry away carbon dioxide and cells produced in the bone marrow. The minerals, too, may be taken away if they are urgently needed elsewhere. For example, if pregnant women eat an inadequate diet (low in quantities of milk, for instance), the bone tissue may break down to supply needed minerals for the baby. A mother with a young baby may find that her teeth need special attention after the baby is born. The demands of the growing fetus for calcium may have had priority over the needs of her own body.

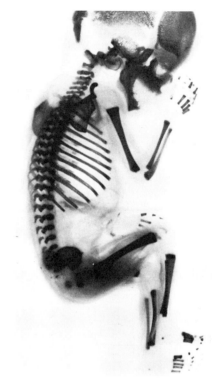

Figure 2.3
Bone formation in a fetus of about 16 weeks. Note that the bones do not meet, but are extending from the centre of each bone as they grow.

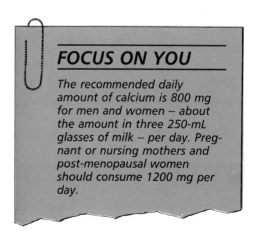

FOCUS ON YOU

The recommended daily amount of calcium is 800 mg for men and women – about the amount in three 250-mL glasses of milk – per day. Pregnant or nursing mothers and post-menopausal women should consume 1200 mg per day.

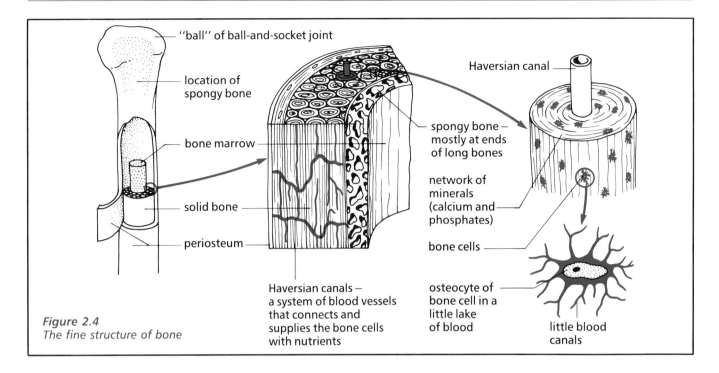

Figure 2.4
The fine structure of bone

Osteoporosis

Osteoporosis is a common bone disorder in which the spaces in the bone tissue become enlarged. (See Figure 2.5.) The bone takes on a porous appearance, and becomes weaker as a result. Even a minor injury can cause the bone to break. In some cases, the bones are weakened to the point where they may break simply from the weight of standing!

The disease is directly related to a lack of calcium in the diet, and seems to be aggravated by cigarette smoking and a lack of exercise. Women over age 45 are more likely to suffer from osteoporosis. The hormonal changes which take place at menopause make the bones lose calcium at a faster rate.

Figure 2.5
On the left is normal bone, with its dense outer layer and inner network. On the right is a bone showing the effects of osteoporosis. Note the thin outer layer and the larger inner spaces. It is easy to see why this bone is more fragile and is likely to break.

CHECKPOINT

1. What substance forms the model of the future skeleton?
2. What happens to bone cells as they develop?
3. How does an adequate supply of oxygen get to the bone cells?
4. What may cause a loss of minerals from bone?

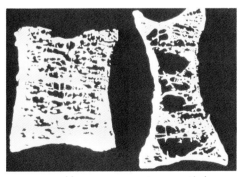

2.4 The Skeleton: Support

The skeleton is actually made up of two parts. The **axial skeleton** contains the following bones: the skull, including tiny bones of the ear (29 bones); the vertebral column, sacrum, and coccyx (26 bones); and the rib cage (12 pairs of vertebrae plus the sternum). The **appendicular skeleton** includes: the arm, hand, and pectoral girdle, including the scapula (shoulder blade) and clavicle (collar bone) (32 × 2 bones); and the leg, foot, and pelvis (31 × 2 bones).

For most of us, it is more important to understand the function of bones than to memorize their names; however, the names are useful for identifying bones. Knowing these names reduces the amount of description needed to identify each bone. (See Figure 2.6.)

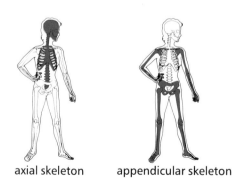

axial skeleton appendicular skeleton

Figure 2.6
The major bones of the skeleton ▶

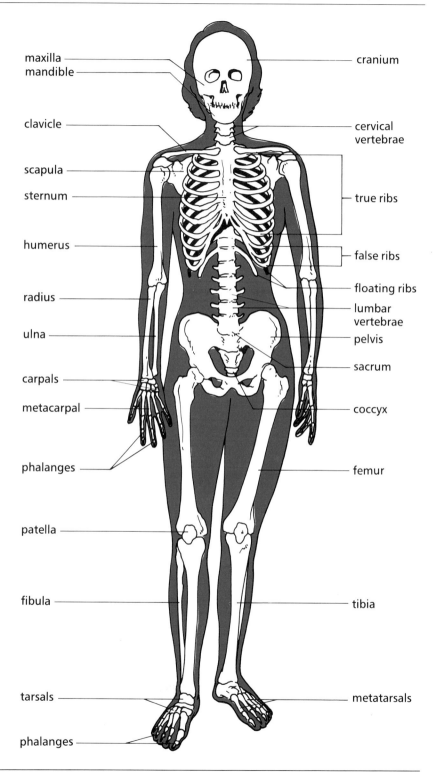

The Bones of the Skull

At first, the skull appears to be made up of just two bones, a large, irregular, ball-shaped part and a loose fitting movable jaw. In fact, the skull is made up of many bony plates joined permanently together by **sutures**. If you look at the diagram of the skull, you can see some of the wavy, zigzag lines. (See Figure 2.7.) These are suture lines, which mark the points where the bony plates have fused together. In some places, where three plates come together, there are small triangular depressions. In a newborn baby, these indicate areas in which the separate bony plates have not yet completely joined. It is important that the skull bones do not become fused too early in life. Passage through the birth canal is made easier because the separate plates

Figure 2.7
The major bones of the skull

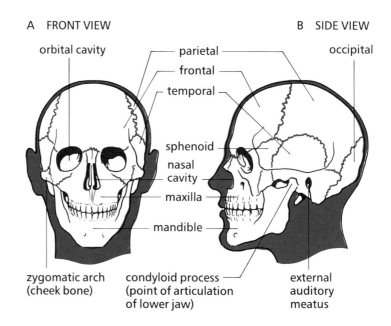

Front and side views of the skull showing the major bones and suture lines.

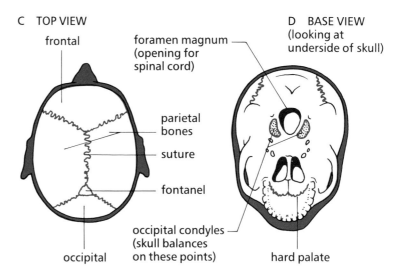

A top view of the skull showing the principal sutures joining the bones of the cranium. The right-hand diagram shows the two occipital condyles which rest on the top bones of the spinal column. It also shows the large opening through which the spinal cord passes in to join the brain.

can glide over each other slightly without injuring the brain. Also, the skull must increase its size after birth to make room for the growing brain. The bones fuse when brain growth is complete. (See Figure 2.8.) If you feel along the centre of your skull, moving your fingers back toward the crown, you can probably still find a small depression called the **fontanel**.

The bones of the skull provide a good example of how their structure is specialized to serve a particular function. (See Figure 2.9.) As you continue in this chapter, look for other examples of shape being suited to function.

Figure 2.8
X-rays of the skull. Note the fine suture lines that indicate the fusion of the cranial bones.

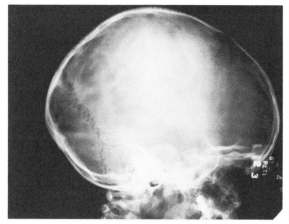
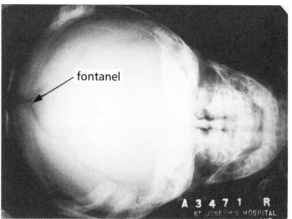

BIOTECH

Bone Transplants

Most people have heard of heart transplants, or even cornea transplants, but few people are aware that bone transplants are becoming routine operations. These operations are used to replace cancerous bone, or to repair arthritic joints.

The body does not reject transplanted bone tissue. In fact, the bone, which is supplied from donors who have willed their bodies to medicine, seems to release a substance or a signal to the nearby living bone to make new bone cells. These new cells help to fuse the transplanted bone to the existing bone. This improves the chances for a successful operation, and allows a wide range of people to be considered for such an operation.

Many research hospitals across North America have bone banks. The bones are deposited in various forms. Some are whole, while others have been ground up, or are in tiny pieces. The bones are kept frozen at temperatures of $-80°C$. Some have been used successfully after five years of frozen storage.

The transplanted bones may be attached to the existing bone with a stainless steel splint and screws during an operation which may last from 5 to 18 h. In other operations, the finely powdered bone is packed into areas, such as the jaw, which have been damaged by disease.

Two problems are associated with bone transplants. The first is the loss of a large volume of blood during the operation. Transfusions of 4 to 6 L of blood may be necessary. The most common cause of an unsuccessful transplant is the second problem, bone infection. Patients are given massive doses of antibiotics for months after the operation to prevent infections. Even with these problems, success rates of up to 85 percent have been reported.

Figure 2.9
The functions of the main bones of the skull

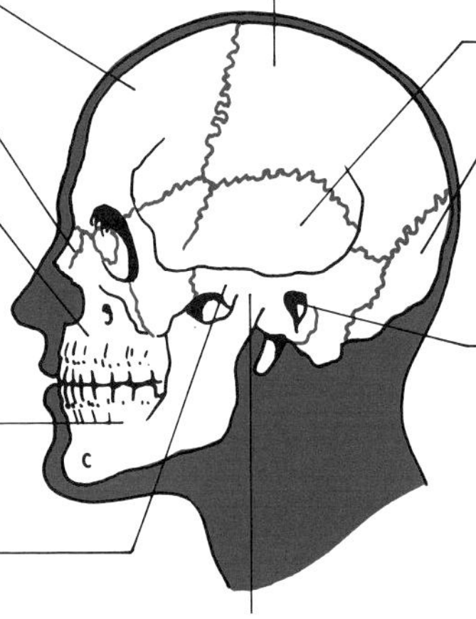

FACIAL BONES

Frontal bone forms the arches over the eye socket and determines the shape of the forehead.

Nasal bones form the bridge of the nose and vary in shape in different people.

Maxilla is a fixed bone, and does not move for speech or for chewing.

Mandible is the only major, freely moving bone in the skull. (There are small moving bones inside the inner ear.)

Zygomatic bones give shape to the cheeks and form lower part of the eye socket.

BONES OF THE CRANIUM

Parietal bones, one on each side, meet at the top of the skull, and form most of the sides and top of the head.

Temporal bones form the lower sides of the skull.

Occipital bone forms the back of the skull and curves under where it has a hole through which the spinal cord passes to connect with the brain.

Auditory meatus is the opening for sound vibrations to reach the inner ear.

Condyloid process, one on each side, is the point at which the lower jaw pivots with the temporal bone behind the zygomatic arch.

The Vertebral Column

The **vertebral column**, also called the spine, is made up of 26 small, irregularly shaped bones. Each bone, or **vertebra** (ver-te-brah), has a round, drum-shaped body with three winglike projections, two lateral and one projecting at the back. The vertebrae are tied together by bands of ligaments. (See Figure 2.10.)

The vertebrae provide the main support column for the body. The round, drum-shaped portion of each vertebra carries the weight. The projections at the side and back provide a place for the attachment of muscles and ribs.

In the middle of each vertebra is a hole, through which the spinal nerve cord passes (rather like a string connecting a bony necklace). The vertebrae protect this vital nerve cord from knocks, but, because each vertebra can move a little and slide sideways, a severe blow can cause a shearing action and damage the cord. If the cord is severed, paralysis of the body below this point would result. It is therefore most important not to move an accident victim if any back injury is suspected. Damaged vertebrae could sever the cord.

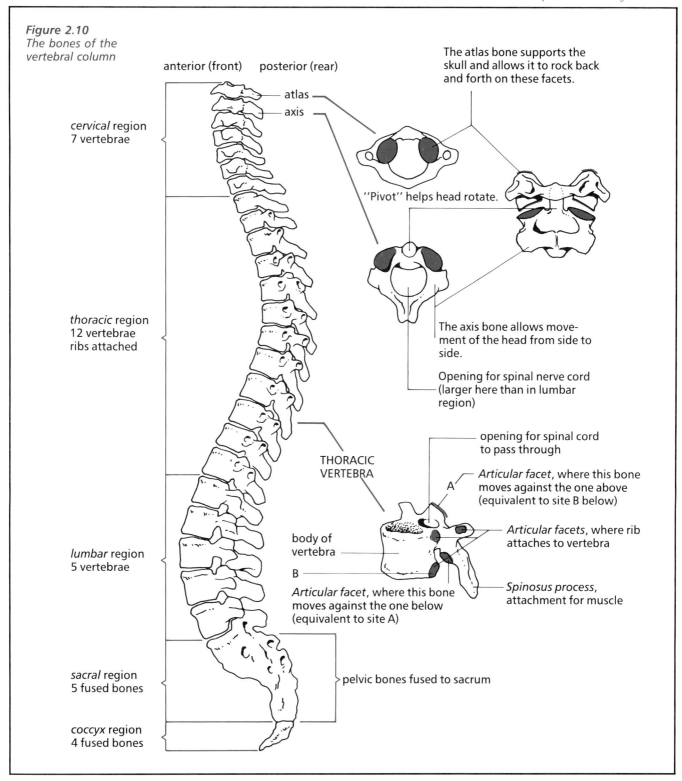

Figure 2.10 The bones of the vertebral column

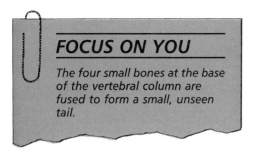

FOCUS ON YOU

The four small bones at the base of the vertebral column are fused to form a small, unseen tail.

Intervertebral Discs

Between the vertebrae are cushions of compressible fibrous cartilage called the **intervertebral discs**. These discs act as shock absorbers and separate the bones of the vertebral column. At times, they are subjected to considerable stress. Sometimes this pressure can cause the disc to be displaced. When this occurs, the displaced disc may slip backward causing some nerves in the spinal cord to be pinched between the vertebrae. This condition can be very painful. (See Figure 2.11.)

Figure 2.11
Compressed spinal nerve due to an injury or poor posture habits. The pain experienced will depend upon the amount of displacement of the disc.

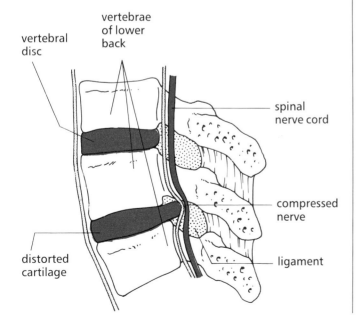

The bones of the vertebral column have limited sliding action, only enough to permit the whole spine to bend in a shallow curve to the left or right. The column has excellent forward bending ability but very limited capacity for twisting and bending backward. (See Figure 2.12.)

Figure 2.12
This diagram illustrates why we can bend further forward than backward.

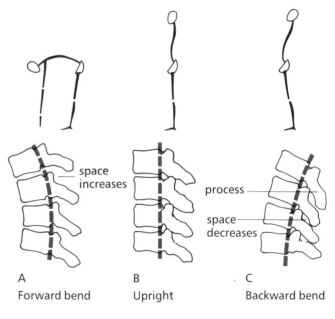

A — Forward bend
B — Upright
C — Backward bend

The curves of the spine are very important. It might seem that a perfectly straight spine would be more efficient; however, it is the curves which provide much of the body's springlike resilience and its resistance to shock when running or jumping. If the human body were supported by four legs, rather than being upright on two legs, the curve of the vertebral column would arch upward like a bridge. (See Figure 2.13.) The weight of the abdomen would then be slung naturally beneath the spine where it could be borne easily. The added weight carried during pregnancy could also be handled efficiently. Our upright posture forces the vertebral column to arch backward in order to bear such weight. The

Figure 2.13
The direction of force experienced when body mass is supported by a four-legged or two-legged animal in an upright position. Note the curve of the spine.

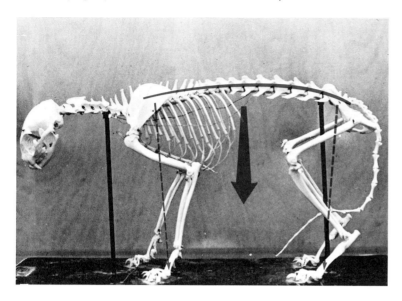

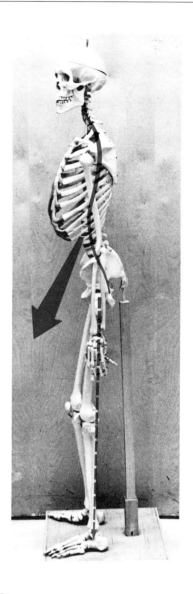

pelvis is also tipped in this direction. The weight of the abdomen thus falls on the lower back. During pregnancy, women frequently suffer strain and feel discomfort in the small of the back. Figure 2.14 shows some of the disorders that distort the normal curvature of the spine.

Figure 2.14
Some disorders of the spine

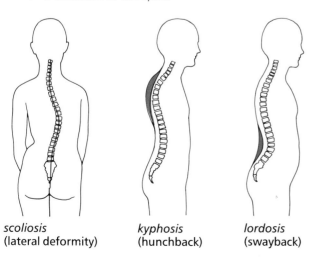

scoliosis
(lateral deformity)

kyphosis
(hunchback)

lordosis
(swayback)

FOCUS ON YOU

Lower back pain is, in fact, a very common ailment in both men and women. Our sedentary way of life has contributed greatly to the weakness of the support muscles and the ligaments of the spinal vertebrae.

The Sternum and Ribs

The ribs are flat, curved bones. The first seven pairs are attached by cartilage to the **sternum** at the front of the body. (See Figures 2.15 and 2.16.) The next three pairs lie below the sternum; each pair is attached by cartilage to the pair of ribs above. The last two pairs are called floating ribs, because they do not complete the circle and are not attached in front. For this reason, they are more easily broken; even an exceptionally hard hug can crack one of these springy, floating ribs.

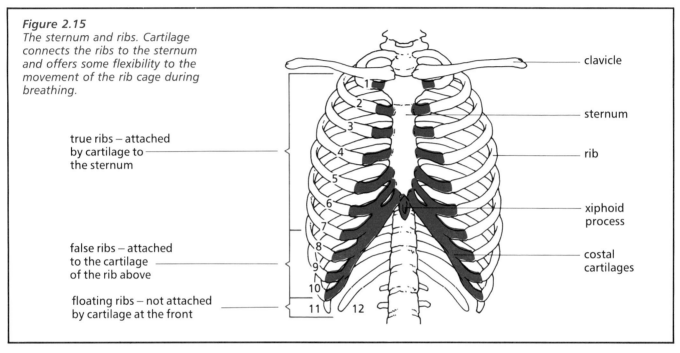

Figure 2.15
The sternum and ribs. Cartilage connects the ribs to the sternum and offers some flexibility to the movement of the rib cage during breathing.

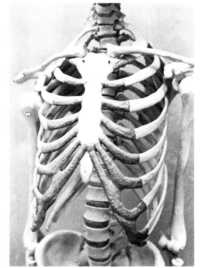
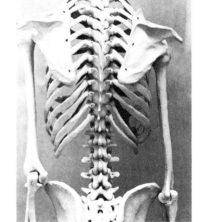

Front   Back

Figure 2.16
Bones of the chest. Front view: Note the attachment of the ribs to the sternum by cartilage. Rear view: Note how the scapula (shoulder blade) is loosely attached over the ribs, which allows more arm flexibility.

The Pelvis

The **pelvis** or **pelvic girdle** resembles a bowl-shaped dish without a bottom. The pelvis provides attachment for the bones of the lower limbs. It is extremely strong and bears most of the weight of the body, as well as the thrust of all leg movements (running, jumping, walking, etc.). (See Figures 2.17 and 2.18.)

Figure 2.17
The human female pelvis is broader and lighter than the male's. The opening within the pelvis is appreciably larger than in the male. Why do you think this is so?

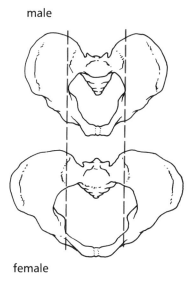

Figure 2.18
The pelvic girdle

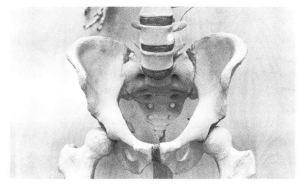

The Lower Limbs

The **femur**, or thigh bone, is the longest and strongest bone in the body. Its round, smooth head fits into a socket formed in the pelvis. The lower end of the femur connects with the larger of the two bones of the lower leg, the **tibia**. A flat, disc-shaped bone, the **patella** or kneecap, is found just in front of this joint. It is loosely attached (you can move it a little with your fingers) and offers a protective pad for the joint. The **fibula** is the other bone found in the lower leg. It improves stability.

The ankle is made up of seven **tarsal** bones. These bones provide a sliding joint which enables the foot to be extended and flexed with every step we take. The foot consists of **metatarsals**, which are the larger bones of the foot, and the **phalanges** (fa-lan-jeez), the small bones at the ends of the toes. (See Figure 2.19.)

The bones of the foot should curve in two directions forming natural arches. One arch spans the ball of the foot and the heel; the other is at right angles to this, across the width of the foot. (See Figure 2.20.) These arches provide the "spring" in our step.

If these arches break down and lose their muscle tone, they are sometimes called fallen arches. The result is a painful ache in the foot. This disorder may be caused by a number of things, such as poor prenatal nutrition, poor posture, extra body mass, improperly fitted shoes, or other factors. Well-fitted shoes are very important, especially for young children. Extreme shoe styles can also result in permanent distortion of the foot and thus the spine.

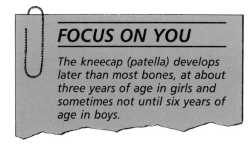

FOCUS ON YOU

The kneecap (patella) develops later than most bones, at about three years of age in girls and sometimes not until six years of age in boys.

Figure 2.19
Bones of the lower limb

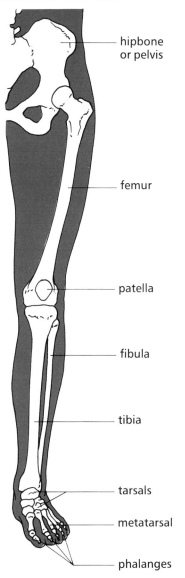

Figure 2.20
The human foot

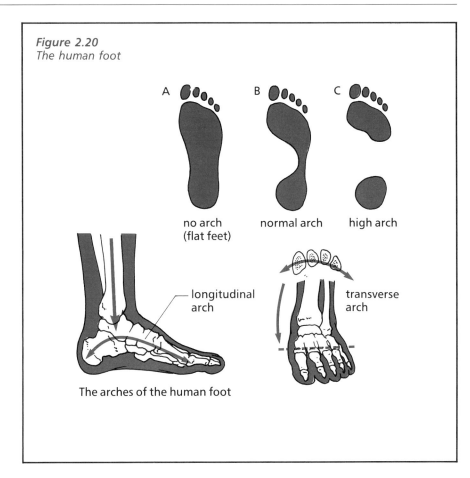

The arches of the human foot

FOCUS ON YOU

There is no such thing as "breaking in" a pair of shoes. What gets altered, and sometimes painfully, is your foot! New shoes will feel comfortable right away if they really fit.

The Upper Limbs

The bones of the arms are attached to the **pectoral girdle**. The **clavicle** (clav-i-kul), or collar bone, connects at one end with the top of the sternum and at the other end with the **scapula** (skap-yoo-lah), or shoulder blade. The clavicle is just below the skin surface and you can feel its outline quite clearly. The scapula, which is also close to the surface, lies on top of the ribs. This bone is only very loosely attached. If you ask a partner to move an arm in a circular motion, you can feel the movement of the shoulder blade on the upper back. Because the scapula is so loosely attached, it is held in place by ligaments and muscles; thus, the arm has great freedom of movement in almost any direction.

The ends of the clavicle and the scapula together form part of a socket for the bone of the upper arm. The rest of the socket is composed of a capsule of cartilage. The bone of the upper arm is called the **humerus**. It connects with the scapula at the shoulder and with the two bones of the lower arm at the elbow. (See Figure 2.21.)

The **ulna** is the main supporting bone of the forearm. It is attached to the humerus at one end and the bones of the wrist at the other. Beside it is a slightly shorter bone called the **radius**, which rotates around the ulna so that the hand can be turned. If you stretch out your forearm with the palm of your hand facing up, then turn your hand over, you can get an idea of how the radius moves by watching the skin of your arm.

The wrist is made up of eight small bones or **carpals**, which are joined to the five **metacarpals** that form the hand. The bones forming the fingers and thumb are the phalanges. There are three phalanges in each finger and two in the thumb. (See Figure 2.22.)

CHECKPOINT

1. *What are the two major divisions of the skeleton?*
2. *How can you tell that the skull is made up of more than one bony plate?*
3. *Why aren't the bones of the skull already fused together at birth? Give two reasons.*
4. *What may happen if one of the vertebra in the spinal column is pushed out of place?*
5. *What is the function of an intervertebral disc?*
6. *How are the bones of the lower limbs attached to the body?*
7. *What is the function of the smaller of the two bones found in the lower leg?*
8. *The pectoral girdle attaches the arm to the rest of the body. Which structure makes up the socket that holds the humerus?*

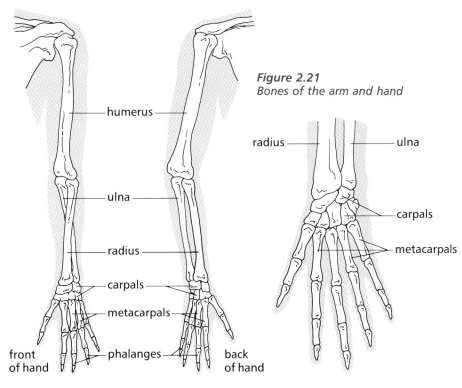

Figure 2.21
Bones of the arm and hand

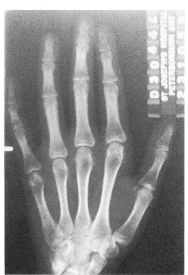

Figure 2.22
An X-ray of the bones of the hand

2.5 Skeletal Adaptations

The skeletons of most vertebrates are remarkably similar; however, each vertebrate's skeleton has its own characteristic variations related to the different functions that must be performed. (See Figure 2.23.) For example, the bones which support the wings of a bird are hollow, lighter, smaller, and shorter than the bones found in the forelimbs or "arms" of other vertebrates. Also, birds do not have fingers or toes on their wings. Instead, the phalanges (the bones corresponding to those in our fingers) are joined together to make a place for feather attachment.

Animals which can run very fast have leg bones which are relatively long in relation to the rest of the body. This increases the length of each stride and the speed of the animal. The bones of the foot are much longer than in other vertebrates. To make the leg even longer, some animals, such as a dog or a deer, are really walking on their toes.

Horses have fewer bones in their feet than dogs or deer. The bones have lengthened, and some have fused together to form a hoof. This reduces weight without reducing strength. Horses also have a springing action in their feet. A special ligament straightens the foot at each step, and in effect pushes the leg upward like a spring that has been compressed.

Figure 2.23
The differences between these skeletons reflect the different functions required by each species.

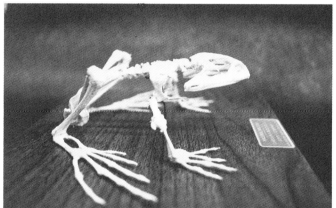
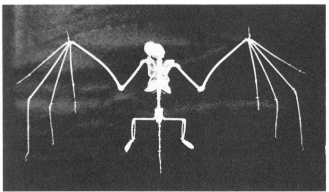
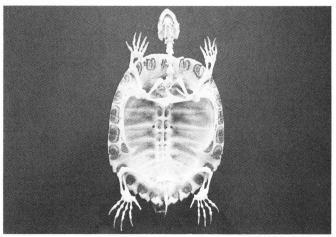

Activity 2C

Vertebrate Skeletons

How do the skeletons of other animals with backbones compare to that of a human?

Procedure

Your teacher will provide you with the skeletons of other vertebrates. Examine the structure of each one, and compare it to the human skeleton. Use the diagram of the human skeleton to assist you. (See Figure 2.6.) As you examine each skeleton, try to answer these questions:

1. *How is the skeleton of this animal similar to mine?*
2. *How is it different?*
3. *Why is it different?* (Hint: What does this animal do that you don't?)

2.6 The Joints: Connecting Bones Together

Where two bones come together, they form an **articulation**, or **joint**. There are three kinds of joints:

- **Immovable Joints**. These joints are fused so that no movement occurs between the bones thus joined. Immovable joints are found, for example, between the fused bones of the skull.
- **Slightly Movable Joints**. In these joints, a small amount of movement is possible, and the articulating surfaces are protected by a pad of cartilage. Vertebrae, for example, are connected by this type of joint.
- **Fully Movable Joints**. These joints provide great freedom of movement such as is found in the joints of the shoulder, hip, knee, and elbow. The joint is contained within a special capsule, with a **capsular ligament**, helping to hold the bones in place. The chief function of these movable joints is to permit the body the mobility needed for running, jumping, reaching, bending, etc.

Fully Movable Joints

These joints are held in place by straplike **ligaments**. These connective tissue structures are flexible and have a small amount of elasticity. They bind the bones together and help to prevent dislocations while still allowing the bones to move. (See Figure 2.24.)

Figure 2.24
Ligaments are strong bands of tissue which hold bone to bone. They are flexible and allow small amounts of movement between the bones.

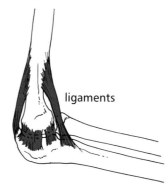

If two bones move freely, the ends might be expected to rub together causing friction and wear. This is prevented in fully movable joints by covering the ends of the bones (the articulating surfaces) with **articular cartilage**. Inside the capsule is a smooth **synovial** (sy-noe-vee-al) **membrane** that lines the capsule and secretes **synovial fluid**. This fluid lubricates the joint, bathing the tissues, and helping to reduce the friction produced by the moving bones. The capsule forms a tight seal around the joint to prevent the fluid from escaping. (See Figure 2.25.)

Figure 2.25
The hinge joint of the elbow and the ball-and-socket joint of the hip

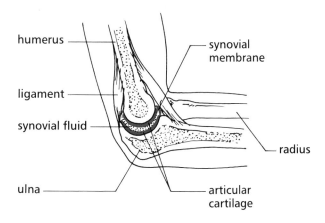

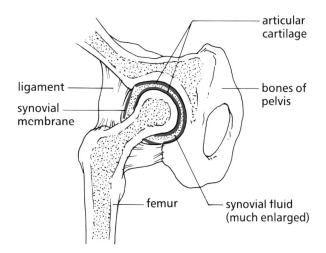

In the synovial capsule, the fluid lubricates the joint. The membrane keeps the fluid from escaping and the articular cartilage makes a smooth bearing surface and prevents the bones from rubbing together.

Joints are also often grouped as follows according to some special characteristic that they possess. (See Figure 2.26.)

- **Ball-and-Socket Joints**. This type of joint allows movement in almost any direction. The rounded, ball-shaped head of one bone fits into the hollow depression found in another. Sometimes more than one bone is involved in forming the socket. Ball-and-socket joints are found in the shoulder and the hip.

- **Hinge Joints**. These joints allow movement in one plane only. The knee and elbow joints are examples of hinge joints.

- **Gliding Joints**. These joints are more limited, allowing only small sliding movements of one bone over another. Gliding joints are found in the wrist and ankle.

- **Pivot Joints**. Pivot joints permit radial or circular motion. An example of a pivot joint is in the forearm where the radius articulates with the humerus.

Rheumatoid Arthritis

The term **arthritis** means "joint inflammation". It is used to describe many conditions with the symptom of pain in the joints. Those who suffer from **rheumatoid arthritis**, the most severe form of arthritis, have swelling and pain in the joints. For some reason, the body's immune system attacks the cells lining the synovial cavity, and eventually the cartilage and bone in a joint. New bone tissue is laid down in an attempt to repair the damage, but this may cause the joint to be unable to move.

Aspirin is the most common treatment for arthritis, since it reduces pain and inflammation in the joint. Arthroscopic surgery may also be used to remove damaged tissues. (See Special Feature Box on page 60.)

Figure 2.26
Types of joints found in the human body

Immovable, fused joints **between bones of skull**

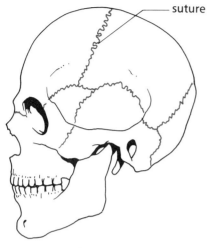
suture

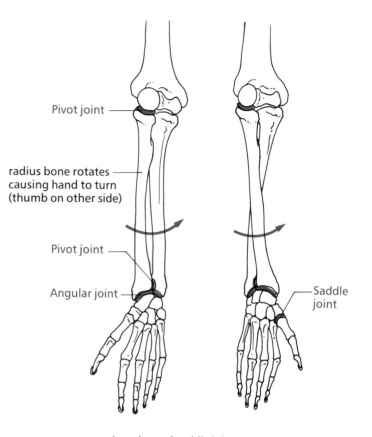

Pivot joint

radius bone rotates causing hand to turn (thumb on other side)

Pivot joint

Angular joint

Saddle joint

Angular and saddle joints are similar, having a concave surface moving on a convex surface. Movements are mostly in two directions only.

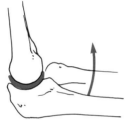

Hinge joint of the elbow – allows only two-way movement

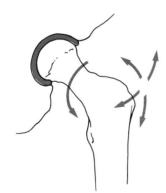

Ball-and-socket joint of hip – very free movement in almost any direction

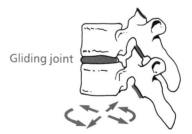

Gliding joint

One part slides over another. The amount of movement possible at each vertebra is quite small.

BIOTECH

Torn Cartilage and Arthroscopy

The cartilage covering the end of a bone in a joint acts as a shock absorber and stabilizer. A joint such as the knee is often subjected to intense twisting movements which, if combined with a great deal of pressure, may tear the cartilage. Since it does not have its own blood supply, cartilage has difficulty repairing itself. tip is a light, and a rod lens capable of producing a magnified image of 10 times to 20 times life-size. Around 20 000 fibre optic strands connect it to a video camera, and the image is then projected on a video monitor.

Instead of an opening of the entire synovial cavity, only three small holes are needed for arthroscopy. One is required for an *inflow cannula*, a small tube to carry

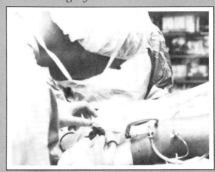

A surgeon using the arthroscope in knee surgery

The torn tissue remains in the joint, and may cause discomfort and further damage, much like a stone that gets into your shoe. If the tear is a serious one, the damaged cartilage must be removed. The only option used to be to open the knee and remove the entire pad of cartilage, the *meniscus*. This major surgery required a hospital stay, five weeks on crutches, and intensive therapy. In addition, complications, often even arthritis, would follow.

Today, *arthroscopic surgery* is used to remove only the damaged portion of the cartilage. An arthroscope is an instrument resembling a steel straw, with a diameter less than 5 mm. At its

Arthroscopic surgery on the knee requires three small openings, one each for the arthroscope, the shaver, and the tube that supplies fluid.

fluid into the joint. This fluid inflates the joint, providing more room to work, and also provides a fresh supply of fluid to the joint. A second hole is required for a small cutting tool that revolves 30 000 times per second. The third hole allows the arthroscope to illuminate the area, and permits the surgeon to see the damaged tissue.

The surgeon can now move the cutting tool to the damaged area, and remove just the torn area of the cartilage. The less cartilage that is removed, the better the chances for recovery without the complications of arthritis. The bits of severed tissue are removed from the joint by suction lines attached to the cutting tool and the arthroscope.

In addition to providing a much clearer view of the details in the joint, arthroscopy shortens the recovery period dramatically. Usually the patient requires crutches for only two to five days and may be able to resume normal activities in two weeks.

Activity 2D

Fully Movable Joints

How is a fully movable joint constructed?

Materials

fresh knee joint from a chicken
scalpel
probe
hand cream

Procedure

1. Whenever two objects move against one another, there is friction. If you place the palms of your hands together, and rub them back and forth for a short time, you will notice the result of friction. *What happens?* Try rubbing your hands together again, but this time place some hand cream on the palms before you begin. *What happens this time?*
   A fully movable joint must have some way of reducing the friction between the bones as they move past one another.

2. Cut away some of the material wrapping the chicken joint. *What is this material called? Describe it. How is it attached to the bone?*

3. Place your finger into the joint and touch the end of a bone. *Describe what you feel. Where does this substance come from? What purpose does it serve? What kind of a joint is this? How does it differ from a hip joint?*

Questions

1. Locate each of the following, and briefly describe its appearance:
   (a) cartilage   (d) ligament
   (b) bone   (e) tendon.
   (c) muscle

2. (a) How is the muscle arranged?
   (b) What keeps it in place?

CHECKPOINT

1. *Give three examples of how the structure of an animal's skeleton may be related to a function the animal performs.*

2. *Name and give one example of each of the three main types of joints or articulations.*

3. *What is the function of*
   (a) a ligament?   (b) cartilage?

4. *How is friction reduced in a joint?*

5. *How is the movement of a ball-and-socket joint different from that of a hinge joint?*

2.7 When Bones Break

Fractures are broken or cracked bones. A descriptive name is given to each particular kind of fracture. **Green-stick fractures** usually occur in young children. In this case, the bone does not separate completely, but breaks like a green, sap-filled stick. **Simple fractures** refer to broken bones which do not pierce the skin. **Compound fractures** are breaks in which the ends of the bones push out through the skin. Such breaks are very serious, because of the possibility of infection. (See Figure 2.27.)

Dislocations occur when the bones of a joint are pulled out of alignment. The ligaments that hold these bones together are stretched, distorted, or torn. Dislocations are thus extremely painful.

A **sprain** results from a temporary separation of the bones, after which the bones return immediately to their normal alignment. The injured area usually swells rapidly and is painful because of the stretching or tearing of ligaments.

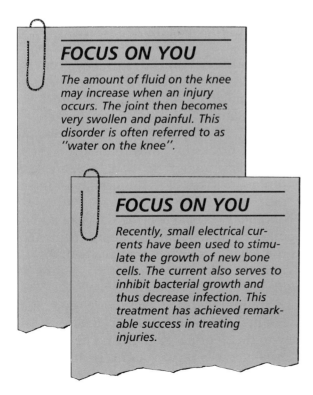

Figure 2.27
Common types of fractures

simple fracture (does not break skin)

compound fracture (breaks skin)

green-stick fracture (in young children)

Football players, basketball players, and dancers are frequently affected by knee injuries. Sometimes the cartilage on the end of the femur or tibia tears. The joint then becomes very painful and difficult to move. Repair of torn cartilage may require surgical treatment.

FOCUS ON YOU

The amount of fluid on the knee may increase when an injury occurs. The joint then becomes very swollen and painful. This disorder is often referred to as "water on the knee".

FOCUS ON YOU

Recently, small electrical currents have been used to stimulate the growth of new bone cells. The current also serves to inhibit bacterial growth and thus decrease infection. This treatment has achieved remarkable success in treating injuries.

The Healing of Broken Bones

The time that bones take to heal may vary considerably, not only with the size of the bone and the type of fracture, but also depending on the efficiency with which minerals are made available to repair the tissue. The bones of young people heal much more rapidly than those of adults. The bones of elderly people are very slow to mend.

Sometimes bones do not recover as rapidly as expected. In some exceptional cases, the pieces of bone may fail to rejoin, even after a year.

Steps in the Repair of Broken Bones

- STEP 1
  Immediately after a break occurs, blood clots form at the break in the bone.

- STEP 2
  After the doctor has "set" the bone and the parts are properly aligned again, these blood clots are absorbed by the body and replaced with new connective tissue called **procallus** ("pro" means *before* or *first*). This substance fills up the spaces in the tissue and absorbs the dead bone cells. It may take from one to eight weeks for this phase to be completed.

- STEP 3

  The procallus is gradually transformed into **callus**, a harder substance, which seals the broken ends of the bone, providing some support as new bone is formed; however, the callus is not nearly as strong as bone. To make up for this, it forms a band of thickened material, like a cuff, around the break site.

- STEP 4

  Finally, new bone tissue replaces the callus. It grows out from the periosteum (a special membrane that nourishes the bone) until the bone finally achieves its original shape and size. To avoid disturbing the action of the procallus and callus during healing, the bone must be immobilized. The doctor, after checking that the parts of the broken bone are correctly aligned, will enclose them in a rigid cast. This prevents the muscles and tendons from pulling the bone out of shape. For most breaks this also involves immobilizing the joints at either end of the broken bone. During the period that the bone is in a cast, the ligaments tend to lose their elasticity and become stiff, while the muscles lose their strength and become soft.

After the cast is removed, the limb must be exercised to bring it back to full use.

Better Safe Than Sorry

Bones are remarkably strong and flexible. The femur, for example, can withstand a force of over 100 times your body weight before it would break; however, on some occasions, we participate in activities which subject the bones, and other tissues, to forces which may cause serious injury. Even riding in a car is potentially dangerous, but there are ways to reduce the risks.

The use of a seat belt restrains the body in the case of an accident, and prevents the bones (and other tissues) from receiving a sharp blow in a localized area. A helmet worn while cycling or skating, or a hard hat worn at work, can prevent serious injuries to the head. The helmet cushions a blow to the head, and distributes the force of the blow over a wider area. (See Figure 2.28.) The same principle applies to the use of shin pads, shoulder pads, chest protectors, and even safety shoes. What other examples of personal safety equipment can you think of?

Figure 2.28
The proper use of seat belts can prevent injuries.

First Aid for Bones

It has been estimated that about one in every 100 people will suffer a fracture during the year. If we knew in advance who these people were, perhaps we could prevent some of these injuries. But the best we can do is be prepared to give proper emergency care. This can ease the pain and possibly help to avoid further injury to the surrounding tissues. For this reason, it is a good idea to receive some formal first-aid training. When you enrol in one of these courses, you will learn much more than how to treat skeletal injuries. The discussion here will provide you with some general guidelines for emergency care of fractures. These procedures should only be performed by those who have formal first-aid training.

A fracture may be accompanied by swelling and extreme sensitivity at the place where the bone is broken. In some cases, the injured part of the body will look unusual. A person with a suspected fracture should not be moved, or asked to move. The part of the body with the suspected fracture must first be immobilized with a splint. (See Figure 2.29.) After the splint has been applied, medical attention should be obtained for the injured person as quickly as possible.

If the person has a suspected fracture of the spine or neck, medical help should be called immediately, and the injured person should not be moved. It is important to remain calm, and reassure the injured person.

On other occasions, the bone may not be fractured, but there may be some other tissue damage. In these cases, use the **R.I.C.E.** treatment. R.I.C.E. stands for Rest, Ice, Compression, and Elevation. Ice should be applied to the injury for about 10 to 20 min, and be repeated every 2 to 4 h. The injury should be rewrapped securely with an elastic bandage after each treatment with ice. It is important to maintain circulation, so the bandage should not be too tight. The injured part should also be elevated above the level of the heart, where possible. Obviously, it is easiest to do this if the patient is lying down. As always, get medical attention and advice for serious or persistent problems.

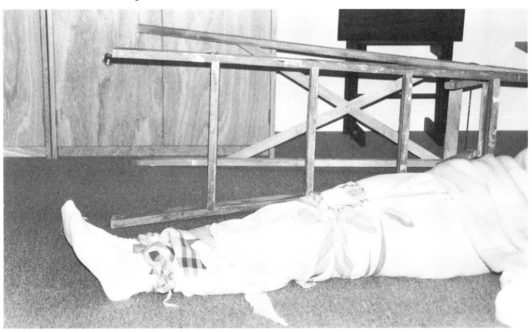

Figure 2.29
Why must a broken bone be immobilized?

Activity 2E

X-Rays Add Something to the Picture

X-rays give us a look at parts of the living body that we cannot normally see. This is especially important if a physician suspects that a part of the skeleton has been injured.

Examine the X-rays below. Make your own diagnosis of the problem based on the information provided in the preceding section.

Figure 2.30
X-rays

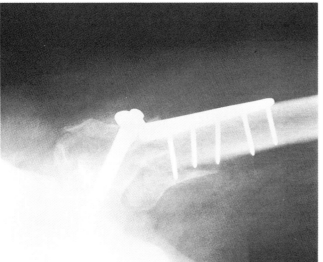

CHECKPOINT

1. Why is a compound fracture more serious than a simple fracture?
2. What factors determine how long it will take for a broken bone to mend?
3. Why must a broken bone be immobilized in a rigid cast or splint?
4. What is meant by the R.I.C.E. treatment of injuries?

2.8 Muscle: Getting Moving

Animals employ many different methods of locomotion. Some have legs and move swiftly across the land, others have wings for flying, or fins for swimming. Whatever the means of locomotion, it is muscles that provide the moving power.

Types of Muscle

Not all of our muscles are large ones used for locomotion. Some muscles allow us to smile or frown, to wink an eye, swallow food, even to wiggle our ears. Muscles account for about 40 percent of the body mass. There are more than 650 muscles in the human body. All muscles are classified into three types. (See Figure 2.31.)

Figure 2.31
The three types of muscle tissues

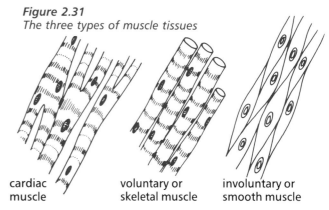

cardiac muscle | voluntary or skeletal muscle | involuntary or smooth muscle

Chapter 2 / The Body in Motion

Although each type has slightly different characteristics at the cell level, they all have the same main function, that is, to contract and, in so doing, perform work.

The following is a summary of the three types of muscle.

- **Skeletal** or **voluntary muscle** is attached to bones and contracts to move the limbs or some group of tissues. (See Figure 2.32.)

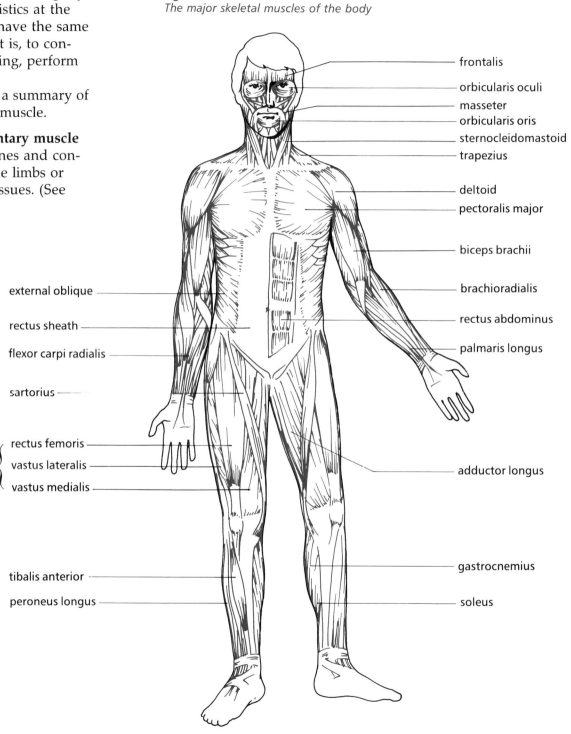

Figure 2.32
The major skeletal muscles of the body

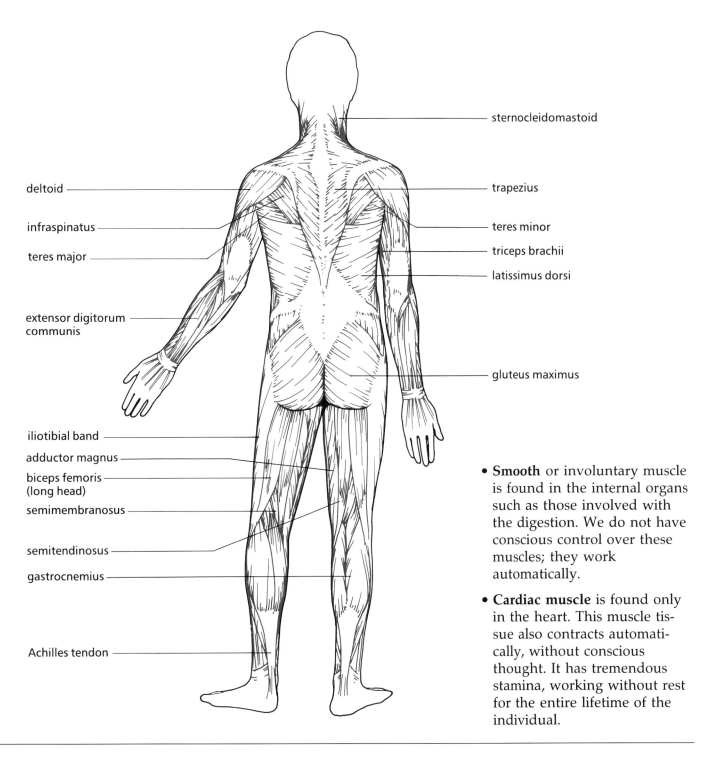

- **Smooth** or involuntary muscle is found in the internal organs such as those involved with the digestion. We do not have conscious control over these muscles; they work automatically.

- **Cardiac muscle** is found only in the heart. This muscle tissue also contracts automatically, without conscious thought. It has tremendous stamina, working without rest for the entire lifetime of the individual.

Activity 2F

The Structure of Skeletal Muscle

What will a close look at skeletal muscle reveal?

Materials

piece of stewing beef
probe or dissection needle
forceps (tweezers)
microscope
two microscope slides and cover slips
methylene blue stain

Procedure

1. Examine the piece of stewing beef on all sides. *Describe the arrangement of the muscle fibres. What is the white tissue?*

2. Use the probe to separate some of the fibres. Use the probe like a rake, and draw it across the side of the fibres.

3. Gently remove one of the fibres and place it on a microscope slide.

4. Add a few drops of water to the slide and place a cover slip over the fibre. Use a piece of paper towel to blot up any excess.

5. Examine the fibre under the microscope using low power. Find a thin, intact section of the fibre and *sketch what you see.*

6. Place another fibre on the second microscope slide. Add a few drops of methylene blue stain to the fibre. After about 2 min, rinse away the stain with a few drops of water.

7. Repeat steps 4 and 5.

8. Observe the fibre under medium and high power. *Sketch what you see under high power. What do you notice that you didn't see before?*

Muscle Contraction and Extension

When muscles are stimulated, they contract and accomplish work. Skeletal muscles are arranged in pairs so that as one muscle contracts upon stimulation, the opposing muscle extends (relaxes). Muscles thus demonstrate both **contraction** and **extension**. Smooth muscles in organs such as the urinary bladder or the stomach are able to relax considerably in order to extend the walls of these structures to carry or store large quantities of liquid or food.

Muscle cells either contract fully or extend fully. They do not work partially. If you lift a heavy load, you simply use more cells than you would for a light load. Cells cannot contract halfway; it is an "all or none" event.

When muscle cells contract, they produce heat. About 80 percent of the energy used in muscle contraction is converted into heat and "lost" to the body. Sometimes when we are cold, the body makes rapid contractions of the muscles, just to produce heat and maintain the body temperature. This is known as shivering.

How Muscle Cells Contract

A muscle is made up of several bundles of cells. The cells of skeletal muscle, called *fibres*, are quite small in diameter, about 10 to 100 μm, but they are very long in comparison with most cells, varying from 3 mm to 7.5 cm in length. Each fibre is enclosed within a sheath (the **sarcolemma**) that is continuous with the tendons at the ends of the muscles. (See Figure 2.33.)

Inside the sheath there are many nuclei, rather than a single nucleus as is found in most cells. These nuclei are scattered around the perimeter of the fibres. The fibres have characteristic cross-banded markings or **striations**. Figure 2.34 shows how these markings appear under a microscope.

Muscle fibres are, in turn, made up of many **myofibrils** bundled together. It is in the myofibrils that the banding or striations are created.

Figure 2.33
The structure of skeletal muscle

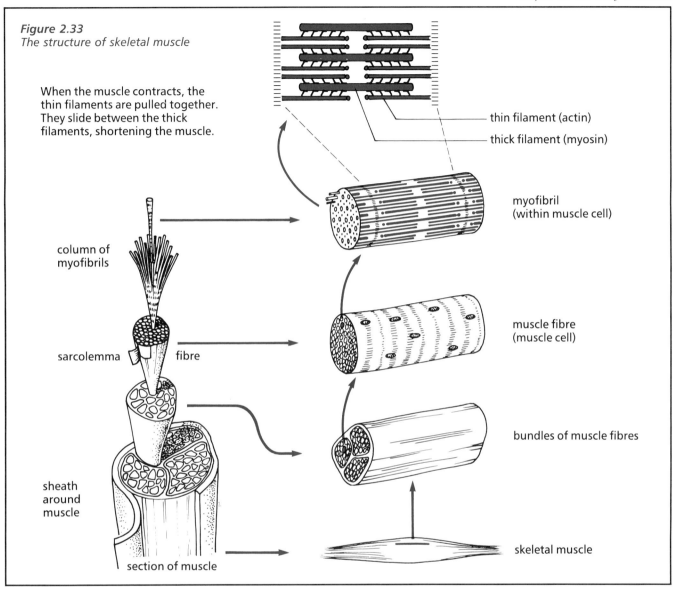

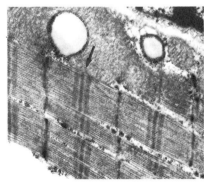

Figure 2.34
Electron micrograph of muscle striations

The striations, which appear when muscles are viewed through a light microscope, are caused by the presence of many tiny protein strands within the myofibrils. (See Figure 2.33.) It is these proteins that shorten the cell as it contracts. Two types of protein strands are suspended in the muscle cell cytoplasm. The *thick* strands are **myosin** filaments which are surrounded by *thin* **actin** filaments.

The thin actin filaments have heavy club-shaped endings. (See Figure 2.33.) These filaments are drawn in between the myosin filaments during a contraction. Although each group of filaments moves only a very small distance, the sum of these tiny movements produces a considerable shortening of the whole muscle. The thick filaments have ridged projections along their length; these attach to the thin filaments during contractions.

The muscle cell cytoplasm also contains many substances important to its function, such as glucose, creatine phosphate, and ATP. All of these help to supply energy for the contracting cell.

The Stimulation of Muscle Contraction
If you wish to bend your arm to scratch the end of your nose, nerve impulses must deliver the order before the muscle will contract to move the arm. This stimulus (a trigger which causes a reaction) arrives at the muscle along a motor nerve which possesses many tiny extensions. Each extension has a small patch buried in the muscle fibres; these are called motor **end plates**. (See Figure 2.35.) A particular muscle may have from a few to many thousands of end plates. As the impulse arrives at the end plate, it causes the instant release of a chemical (acetylcholine), which passes through the muscle fibre membrane and produces the muscle contraction. An enzyme then quickly destroys this chemical so that the muscle can relax and be free to contract when stimulated again.

CHECKPOINT

1. Name the three types of muscle, and tell where each can be found in the body.

2. What purpose does actin play in muscle contraction?

3. (a) What is myosin?
   (b) What is its function?

4. List the order of the events in a muscle contraction from the time the stimulus arrives at the muscle to the muscle's relaxation.

2.9 The Muscle-Bone Connection

Bones provide the body with a system of levers and muscles supply the force required to move the bones. You probably have discussed levers in an earlier science course. Figure 2.36 shows some examples of how the body uses the various classes of levers.

Figure 2.35
Motor nerve end plates and their contact with the myofibrils of skeletal muscle

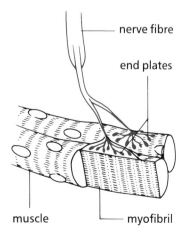

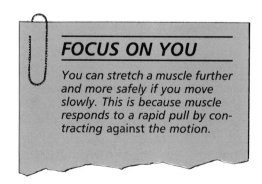

FOCUS ON YOU

You can stretch a muscle further and more safely if you move slowly. This is because muscle responds to a rapid pull by contracting against the motion.

Figure 2.36
The bones of the body act as levers.

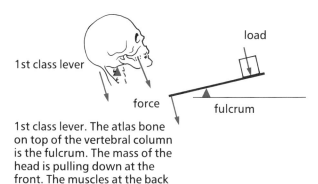

1st class lever. The atlas bone on top of the vertebral column is the fulcrum. The mass of the head is pulling down at the front. The muscles at the back of the neck pull the head down at the back, in a seesaw motion.

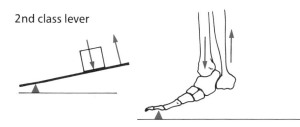

2nd class lever. The ball of the foot acts as the fulcrum. The end of the tibia carries the load and the calf muscle pulls up to raise the body up onto the toes.

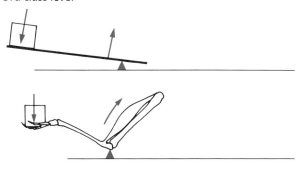

3rd class lever. The elbow acts as the fulcrum. The load rests on the hand and the biceps muscle pulls the forearm upward, raising the hand and the load.

Muscle Attachment

For one bone to move toward another bone, a muscle is required. This muscle will have two attachment points. One end of the muscle must be anchored to a stationary bone, the other to the bone that will move. Then, as this muscle contracts, it will draw the movable bone toward this anchoring structure. The place at which a muscle is attached to the stationary bone is called the **origin** of the muscle. The site of muscle attachment on the movable bone is known as the **insertion**. The insertion of a muscle moves toward the muscle's origin when the muscle contracts. (See Figure 2.37.)

Figure 2.37
Muscle origin and insertion. The bone in which the muscle is inserted moves toward the bone in which the muscle originates.

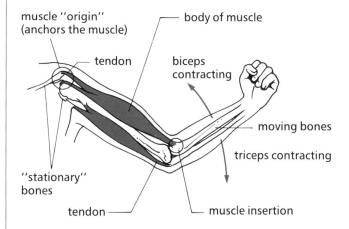

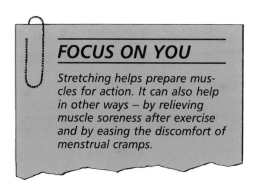

FOCUS ON YOU

Stretching helps prepare muscles for action. It can also help in other ways — by relieving muscle soreness after exercise and by easing the discomfort of menstrual cramps.

Tendons

Tendons are tough, inelastic bands of connective tissue, which anchor the muscle firmly to bone. (See Figure 2.38.) They are so strong that sometimes a bone will break before the tendon tears or can be pulled away from the bone. A tendon as thin as a pencil can take a load of several thousand kilograms.

As the tendons are small, they can pass in groups over a joint or attach to very small areas on the bone, areas too small for the muscle itself to find room for attachment. The tendons are tough, but they are subject to wear and tear as they rub across bony surfaces.

Tendons may become inflamed (tendonitis). This can happen when athletes work out in cold weather without adequate warm clothing, or do vigorous exercise without warm-up exercises.

Figure 2.38
Tendons are tough and flexible. They attach muscle to bone.

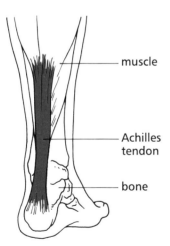

Activity 2G

A Look at Muscles and Tendons

Many features of muscles and tendons can be observed by examining a dissected specimen.

Materials

A freshly killed chicken, with feet attached, may be used for a demonstration dissection, or chicken legs can be used for individual student dissections.

Procedure

This dissection will be arranged by your teacher.

Observations

The dissection will show the following:

- That the thigh muscle is made up of many muscles with different functions
- That the muscles are surrounded by a sheath
- How muscles and tendons are attached. The student should be able to clearly distinguish between the two.
- The flexibility of tendons, their toughness, and that they move in a lubricated sheath
- Smoothness and design of articulating surfaces
- The insertions of tendons into bones
- That the antagonistic actions of flexors and extensors should be examined by pulling on opposing muscles or tendons
- That lean meat is made up primarily of muscle.

Activity 2H

Tendons in Action

What happens to a tendon during muscle contraction?

Procedure

1. Remove your shoes and stand on one foot. Now raise yourself onto your toes. Feel the calf muscles and *describe what has happened to these muscles. Make a simple sketch of the position of the foot and the lower leg when in the raised position. Sketch in the bones, muscles, and tendons. Describe the movement made by the heel. How are the calf muscles attached to the heel?* Feel the Achilles tendon. *What has happened to this tendon during the action taken?*

2. Now point the toes upward, so that the heel is down. Feel the tendon and calf muscles. Feel, also, the muscles at the front of the leg. *Describe what you feel and any changes that pointing the toes up has brought about.* Feel for the tendons in the front of the foot. *How many are there?*

Antagonistic Pairs of Muscles

Many muscles act in pairs. A muscle can only pull (by contracting); it cannot actively push. Thus, once a bone (or other structure) has been moved, movement in the opposite direction can occur only if there is another muscle that can pull the bone in that direction. For example, if the forearm is moved toward the shoulder (flexion), the biceps muscle contracts. Simultaneously, the triceps muscle, which is paired with it, extends. If you wish to move the forearm back into a horizontal position (extension), the biceps extends while the triceps contracts to pull the arm down. To hold the arm in a halfway position, both the biceps and triceps contract slightly to balance each other. Muscles acting in this way are known as **antagonistic pairs**. (See Figures 2.39 and 2.40.)

Figure 2.39
Antagonistic pairs of muscles

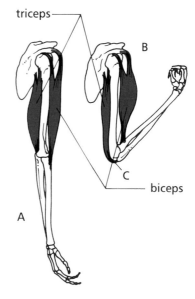

At A the biceps is thin and relaxed while the triceps is fat and tensed. This pulls the forearm back (extension).

At B the biceps is thick and tensed while the triceps is relaxed. The biceps has shortened and pulled the forearm up (flexion).

Note how the tendon of the triceps carries the pull of the muscle around the elbow to attach to the ulna (C).

Figure 2.40
The biceps and triceps muscles

Many antagonistic groups of muscles are required to keep the body upright. (See Figure 2.41.) These muscles contract and act against each other to provide the necessary support for the body. When a position is held and no movement occurs, it is the result of isometric contraction of the muscles.

Figure 2.41
Muscles and posture

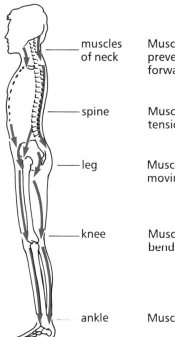

muscles of neck	Muscles keep the head up and prevent it from dropping forward.
spine	Muscles keep the spine in tension.
leg	Muscles keep the leg from moving at the hip.
knee	Muscles keep the knee from bending.
ankle	Muscles keep the ankle tensed.

FOCUS ON YOU

Do you get cramps when you work out?

- Stretch before and after exercise.
- A cramp which occurs repeatedly in the same location may mean a torn muscle. Take it easier for a few days. See your doctor if it persists.
- Cramps also result from dehydration. Drink water before and after exercise.

Activity 2I

Working Against Each Other

Problem

If muscles can only pull by contracting, how can we push a chair away from us?

Procedure

Place your left hand around the upper part of your right arm, with your fingers over the biceps, and your thumb under the triceps. Slowly stretch out your right arm. *What is this action called?* Slowly bend your right arm. *What is this action called? Describe what happens to the biceps and the triceps.* Stretch out your arm again, and *describe how the two muscle groups change. Which muscle is pulling when you do this? Which muscle is pulling when you push a chair away from you?*

In your notebook, *sketch or trace the diagram of the antagonistic pair of muscles in the arm shown in Figure 2.39. Label your diagram with the names of the muscles, and indicate which muscle is contracting or pulling to move the arm into each position.*

CHECKPOINT

1. Explain the difference between the origin and the insertion of a muscle.
2. What is a tendon?
3. What is tendonitis, and when might it occur?
4. Why is a pair of muscles required to move the bones of the body?

2.10 The Body in Action

Your body is like a machine which improves its efficiency with use, and deteriorates rapidly if not used. For this reason, it is important to exercise, and to keep fit. Any student who has trained for some athletic event will be familiar with the term **isometric** exercise. In this type of training, muscles are pitted against each other, with no movement involved. Pressing against a wall or hooking the fingers together and trying to pull the hands apart are examples of isometric exercises. Such exercises have been shown to increase strength and muscle size very rapidly. **Isotonic** exercises involve movement, for example, lifting weights or running. (See Figure 2.42.)

Many of the muscles in the body employ isometric contractions to support the body or some part of it, such as the head. (See Figure 2.41.)

When you are sitting watching television, you might think that you are relaxed, but at least some muscles are working to hold up your head. If you fall asleep while watching, the head will gradually nod forward, or flop sideways. A hand may slip off your lap and hang down while the body gradually relaxes and slumps in the chair. This is caused by the relaxation of the muscles that were acting isometrically to hold the parts of the body in place.

Muscle Fatigue and Energy Needs

Although skeletal muscles are referred to as "voluntary" muscles, there are times when they do not seem to respond to our wishes. When you are extremely tired, your legs often feel rubbery and out of control.

Muscle fatigue is usually caused by lack of energy and a build-up of metabolic waste products in the muscles. Just as a car requires gasoline, the muscle "engines" must be kept supplied with energy or they simply come to a halt. The energy used by the cells is carried by ATP (adenosine triphosphate) molecules. A limited amount of energy is stored in ATP molecules inside each muscle cell. More energy is stored in glycogen, a type of storage sugar molecule. The muscle cells can then convert this to ATP to use as required. The supply of stored energy is quickly used up and the muscles must rely on fresh deliveries from blood circulating through the muscles. During vigorous or prolonged exercise, the body uses reserves of energy kept in the liver and delivers this to the muscles. In order to make use of sugar for energy, oxygen is also required. This is why athletes breathe deeply, gasping for extra air to supply the oxygen needs created by strenuous exercise.

If the body cannot deliver oxygen quickly enough to satisfy the demand, lactic acid (a waste product produced during the chemical breakdown of sugar without oxygen present) is

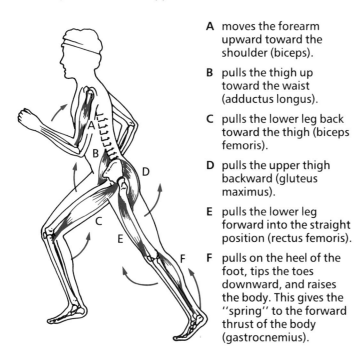

Figure 2.42
Some of the muscles involved in running. Identify the origin and the insertion of each muscle shown. Explain what will happen as each muscle contracts.

- **A** moves the forearm upward toward the shoulder (biceps).
- **B** pulls the thigh up toward the waist (adductus longus).
- **C** pulls the lower leg back toward the thigh (biceps femoris).
- **D** pulls the upper thigh backward (gluteus maximus).
- **E** pulls the lower leg forward into the straight position (rectus femoris).
- **F** pulls on the heel of the foot, tips the toes downward, and raises the body. This gives the "spring" to the forward thrust of the body (gastrocnemius).

built up and this can cause severe muscle pain. You may have seen runners with severe cramps at the end of a race. As lactic acid builds up and the amount of available energy declines, the muscles become fatigued. When the immediate demand for oxygen decreases, the **oxygen deficit** that has built up during exercise can be gradually repaid. Cramps then relax until the pain is no longer felt, and fatigue passes.

Activity 2J

Muscle Fatigue

Problem

How does muscle fatigue affect the muscle's ability to contract?

Materials

one 2-kg dumb-bell (or object which is easily grasped)
stopwatch

Procedure

1. *Prepare a table in your notebook with the following headings to record your observations.*

Trial #	1	2	3	4	5
Time (s)	SAMPLE ONLY				

2. Stand up at the chalkboard, facing the board. Raise your arm to the side, with the palm down, until it is horizontal. Have your partner mark the position of your palm on the board.

3. Face away from the chalkboard. Pick up the dumb-bell and prepare to lift your arm to the side in this horizontal position.

4. Have another student begin timing as your arm is raised.

5. Keep the arm horizontal for as long as possible. The timer will note the time elapsed as soon as your arm drops below the mark on the board, and tell you to rest.

6. *Have the timer record the time elapsed for the trial as you rest for exactly 10 s.*

7. Repeat the trials five times.

8. *Graph your results on a time vs. trial number graph.*

Questions

1. How does repeated use affect your ability to hold your arm up?

2. What do you think would happen if you continued with further trials?

3. How can you explain these results?

Activity 2K

Reaction Times

Problem

Will a tired muscle react as quickly as one that is rested?

Materials

metre stick
rubber ball

Procedure

1. *Prepare a table in your notebook with the following headings to record your observations.*

Trial #	Distance (cm) Before Exercise	Distance (cm) After Exercise
1		
2		
3	SAMPLE ONLY	
4		
5		
average		

2. Place your forearm on a desk so that your hand extends over the end. Use the hand that you do not use to write with.

3. Open your index finger and thumb so they are 3 cm apart.

4. Have your partner hold the metre stick vertically, with the bottom end just above the space between your thumb and finger.

5. Have your partner indicate that the trial is about to begin, and a few seconds later, without warning, drop the metre stick.

6. As soon as you notice that the stick is falling, catch it between your thumb and index finger. Your partner will note the place on the metre stick where you have caught it, and *record this in the observation table.*

7. Repeat steps 3 to 6 to obtain five readings.

8. *Calculate the average distance the metre stick fell, and record it in the table.*

9. Squeeze and release the rubber ball as many times as you can in 1 min. Use the same hand that you have been using for this experiment.

10. Repeat steps 2 to 8.

Questions

1. Was there any noticeable difference in reaction time as the first five trials were conducted? How do you account for this?

2. How did the reaction time change after you squeezed the rubber ball repeatedly? How can you explain this?

Fitness and Exercise

Fitness is the ability of the body to do work, and the body's level of fitness is related to the amount of activity or exercise which is performed. Many parts of the body are affected by exercise, but perhaps the most obvious are the muscles. (See Figure 2.43.)

Figure 2.43
Fitness is related to the amount of exercise you do on a regular basis.

The number of muscle cells in the body remains fairly constant throughout our lifetime. As we exercise, our muscles grow larger. The increase in the size of the muscles is the result of an increase in the amount of cytoplasm in each muscle cell, not an increase in the number of muscle cells. This increase in size is called **hypertrophy**.

In the absence of exercise, the cells gradually shrink, and lose their effectiveness. If you are ill in bed for some time, or have a cast which prevents you from using some of your muscles, parts of the muscle cells get smaller, or **atrophy**.

Exercise affects other parts of the body besides the muscles. Regular, sustained mild exercise, such as brisk walking, can slow the loss of minerals from bones, and the loss of lean body tissue, which naturally occurs as we age. The efficiency of the heart and lungs improves, and the risk of heart disease and blood pressure problems decreases, especially as the amount of body fat decreases. People who exercise feel more optimistic even after a mild workout, and notice an increase in their stamina and capacity for work. Some studies suggest that it may even prevent the occurrence of some diseases.

It is important to prepare or warm up before exercising. A proper warm-up begins with deliberate, slow stretching of the muscles, *never* to the point of discomfort. Then, begin gradually with 5 to 10 min of exercising at low intensity. This increases the blood supply to the muscles, and elevates the heart rate and breathing rate gradually.

After exercising vigorously, it is important to cool down, or slow down gradually. After jogging, for example, you should not sit or lie down, but slow to a walk. This gives the muscles of the body, including the heart, a chance to adjust to the new conditions gradually. Follow with a repeat of the stretches you performed before exercising in order to prevent stiffness.

Activity 2L

How Flexible Are You?

Flexibility, having the ability to move at your joints freely, helps you feel good and prevents injuries. Here are some questions to ask yourself concerning your own level of flexibility.

1. Is it *comfortable* to touch your toes while standing? (You have good flexibility in calves and lower back.)

2. Do you feel *relaxed* when sitting up straight, without slouching? (You have good flexibility in chest and shoulders.)

3. Does playing your favourite sport make you *stiff* the next day? (This is a warning of not enough flexibility. Stretch before and after you play to prevent injuries.)

4. Does your back ache a lot? (Work on the flexibility of your lower back and pelvis, or the problem may become worse.)

5. Have you ever injured your inner thigh muscle playing sports? (Very few sports activities give this muscle enough work, so stretching it before and after exercise is very important to prevent injury.)

How can you improve your flexibility? Just by doing what feels good – proper stretching to lengthen muscles, done slowly and with care.

CHECKPOINT

1. *Explain the difference between isotonic exercises and isometric exercises.*

2. *What causes muscle fatigue?*

3. *What may cause the severe pain in the muscles during a long period of strenuous exercise?*

4. What should you do if you experience severe leg cramps at the end of a race?

5. Distinguish the difference between hypertrophy and atrophy.

6. What are three benefits of regular, sustained, mild exercise?

7. Why should you warm up and cool down after exercising?

BIOLOGY AT WORK

CORRECTIVE THERAPIST
PHYSIOTHERAPIST
OCCUPATIONAL THERAPIST
ATHLETIC TRAINER
MASSEUR/MASSEUSE

HEALTH CONCERNS

shin splints
bursitis
torn ligaments
arthritis
muscular dystrophy
poliomyelitis

CHAPTER FOCUS

Now that you have completed this chapter, you should be able to do the following:

1. List and give an example of each of the four main types of bones.
2. List the functions of the skeletal and muscular systems.
3. Describe the structure of a typical bone.
4. Explain the differences between cartilage and bone.
5. Describe the process of bone formation.
6. Name and locate some of the bones and muscles of the body.
7. Label a diagram of a synovial joint and explain the function of each part.
8. Explain the structural and functional differences between a ligament and a tendon.
9. List the three kinds of joints and give an example of each.
10. Explain the differences between simple, compound, and green-stick fractures.
11. Outline the four steps in the repair of broken bones.
12. Give three examples of skeletal adaptations in other vertebrates which permit them to perform specific activities.
13. Contrast the three types of muscle tissue in terms of their location, appearance, and function.
14. Give the meaning of the terms hypertrophy and atrophy, and explain how these conditions may occur.
15. Briefly describe the structure of a skeletal muscle.
16. Explain the process of muscle contraction.
17. Contrast isometric and isotonic exercises and give an example of each.
18. Explain how pairs of muscles function together to move body parts.
19. Explain the cause of muscle fatigue.

Chapter 2 / The Body in Motion

SOME WORDS TO KNOW

Match each of the descriptions given in the left-hand column with a word shown in the right-hand column. DO NOT WRITE IN THIS BOOK.

1. Point of contact between two bones
2. Individual bones that make up the spine
3. Bone-forming cell
4. Line formed when two bones have fused together
5. Seals the ends of a broken bone
6. Lubricates joints
7. Flexible attachment wrapping the bones of a joint
8. Thin, double membrane covering a bone
9. Joins muscles to bones
10. Smooth muscles along intestines
11. Muscle contractions which produce movement
12. Attachment of a muscle to a "stationary" or less movable bone
13. Thin filaments of protein
14. Muscle contractions which produce no movement
15. Muscles which are attached to bones

Select any three of the unmatched words and, for each, give a proper definition or explanation in your own words.

A periosteum
B vertebrae
C callus
D suture
E osteocyte
F cartilage
G ligament
H synovial fluid
I joint
J voluntary
K involuntary
L antagonistic pair
M isotonic
N isometric
O tendon
P myofibril
Q origin
R actin
S myosin
T sacrum

SOME QUESTIONS TO ANSWER

1. List the four categories of bones and give the main function of each.
2. List the parts of a typical bone and give the function of each part.
3. How is cartilage different from bone tissue?
4. What is an osteocyte?
5. Describe the process of bone formation.
6. Why should you not push down on the top of a baby's head?
7. Is it correct to say that humans have tails? Explain your answer.
8. Why are the curves of the spine important?
9. What should you do if you suspect that someone has suffered a serious back injury? Why?
10. When lifting heavy objects, you should keep your back straight and lift with your legs. Why?
11. What is the function of the pelvis?
12. What may cause the arches of the foot to break down?
13. List four kinds of freely movable joints and give an example of each.
14. Describe the structure of a movable joint.
15. Name three kinds of bone fractures and explain the difference between them.
16. Briefly describe the events which occur at each of the four steps in the repair of a broken bone.
17. Give an example of a skeletal adaptation which allows an animal to perform a task which we are unable to do.
18. List two distinguishing characteristics of each of the three types of muscle tissue.

19. Make a simple sketch of the inside of a skeletal muscle and explain how it contracts.
20. How is a tendon different from a ligament in
    (a) location?
    (b) structure?
    (c) function?
21. Why do you continue to breathe deeply and rapidly for a short time after you stop exercising vigorously?
22. Why do professional baseball pitchers have one arm which is more muscular than the other?
23. Examine the diagram shown in Figure 2.44 and determine what movement or movements would result if the muscles indicated by the letters were to contract.

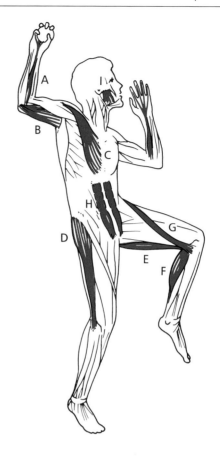

Use this illustration to answer question 23.

SOME THINGS TO FIND OUT

1. How is damage to a joint repaired surgically?
2. (a) What parts of the skeleton can be replaced with artificial parts?
   (b) How is this done?
3. Why does a sharp pain sometimes develop in your side while running?
4. What part does calcium play in muscle contraction?
5. Is there a difference between weight-lifters and body-builders? If so, what is different about their training methods?
6. Investigate a weight-lifting program. Draw up a list of the exercises used, and show which muscles are developed by each exercise.
7. Is it possible for runners to avoid muscle cramps at the end of a race by eating special foods or drinking special liquids before or during the race?

THE NERVOUS SYSTEM: IN CONTROL

During a physical education class, it is easy to see how muscle and bone work together to produce movement. But how do the muscles "know" when and how to move? Why are some movements easy and others difficult to learn? Why do you get better with practice? And why is it so hard to pay attention to your teacher's instructions in the middle of a noisy game?

The answers to these and other questions about your behaviour lie in the nervous system – the control centre of your body.

KEY IDEAS

- *The nervous system receives, sorts, interprets, and responds to information.*
- *The brain is made up of distinct parts, each with a specific set of functions.*
- *The spinal cord links the brain and parts of the body.*
- *The autonomic nervous system maintains the internal environment of the body.*
- *The endocrine system produces chemical messengers which regulate chemical reactions in the body.*
- *The nervous system is vulnerable to the effects of foreign chemicals.*
- *Behaviour, both learned and innate, results from information received and processed by the nervous system.*

3.1 The Nervous System and Behaviour

Can you remember the last time that you were told to behave? Usually, we are told this when our behaviour is not appropriate for the surroundings or circumstances in which we find ourselves. In a scientific sense, we are behaving all the time, if we define **behaviour** as observable muscular activity. Our behaviour or activity is usually a response to a **stimulus**, which is something which prods us into action.

A stimulus can be in any one of a number of forms, such as light or taste. Whatever the stimulus, it first must be received before it can have any effect. Throughout the body, specialized cells of the nervous system act as receptors. These cells receive the stimulus and relay its message. Parts of the nervous system interpret the message being received, and send out other messages in response. The response message is sent to a part of the body, such as a muscle. The muscle contracts, and the body carries out the action, which is a behaviour.

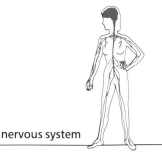

nervous system

Parts of the Nervous System

The nervous system is organized into three major parts:

- **Central Nervous System** (CNS) is made up of the brain and the spinal cord. It co-ordinates and directs the activities of the body.

- **Peripheral Nervous System** (PNS) is made up of the nerves which extend beyond the brain and spinal cord. These nerves bring information in from the sensory and internal organs or carry impulses to effect reactions by the muscles.

- **Autonomic Nervous System** (ANS) controls those parts of the body that act without our thinking about them, for example, the stomach, intestines, and glands. This system helps to prepare the body for emergencies and then returns the body to a normal state after the emergency has passed.

There is only one basic type of nerve cell that transmits all types of impulses. It is called a **neuron**.

The Neuron

Neurons are designed to carry nerve impulses, which are tiny electrical charges, from one point to another. They are living cells, having a nucleus like other cells, but possessing special extensions, the nerve processes, which carry impulses a considerable distance. The cell body and processes of a neuron have a special type of cytoplasm inside called neuroplasm. (See Figure 3.1.)

Figure 3.1
The nucleus, cell body, and processes of a neuron

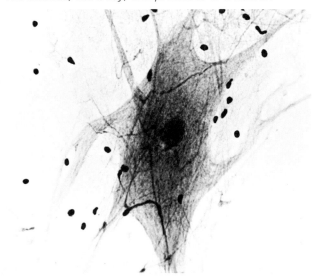

There are two types of nerve processes, **dendrites** and **axons**. Both of these structures contain neuroplasm and are surrounded by a thin membrane. In some locations, axons are very long, and may reach a length of 1 m. Even in such cases, the cell body may be no larger than that of other cells.

Dendrites are shorter than axons and branch extensively. Their function is to pick up impulses and conduct them *toward* the cell body. Axons carry the impulse *away* from the cell body, passing it on to other neurons or cells. (See Figure 3.2.)

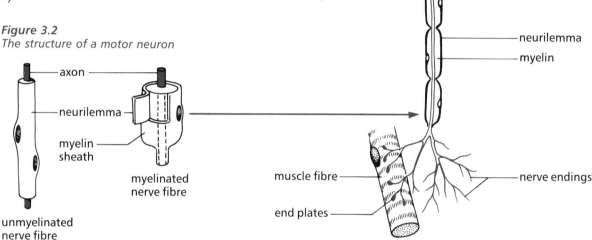

Figure 3.2
The structure of a motor neuron

A close examination of what is usually called a "nerve" shows that it is actually many axons bound together, rather like a trunk telephone cable. The axons of many neurons are covered with a white, fatty protein sheath known as **myelin**. Nerves covered in this way are called myelinated nerves. The main function of this sheath is to insulate the axon, preventing the loss of chemical ions that are present in the nerve. Since these ions are necessary for the transmission of impulses along a nerve cell, the presence of a myelin sheath increases the speed of transmissions. Axons not protected in this way by a myelin sheath are called unmyelinated axons.

Within the brain and spinal cord, the myelinated fibres form areas referred to as **white matter**. The cell bodies of the neurons and unmyelinated fibres make up the **grey matter** of the brain and spinal cord. Myelinated nerve fibres found outside the brain and spinal cord are covered with a delicate membrane known as the **neurilemma** (nur-i-lem-mah). The function of the neurilemma is to protect the fibres and promote the regeneration of damaged axons. Nerves which are not provided with a neurilemma are unable to regenerate. For this reason, damage to the central nervous system is more serious than injury to a nerve of the peripheral system.

Transmission of the Nerve Impulse

When the dendrites of a neuron are stimulated by some sensation (the heat of a flame, for instance) a tiny electrical charge is produced. (See Figure 3.3.) The thin cell membrane which surrounds the nerve fibre is semipermeable, that is, some molecules can pass through, but not others. This membrane can control the passage of certain electrically charged atoms across it, letting them into or out of the cell. Charged atoms are called ions. They have either gained or lost an electron. Chloride ions, for example, have an extra electron and are negatively charged (Cl^-), whereas sodium and potassium ions lack electrons and are therefore positively charged (Na^+, K^+).

When the cell is at rest (has not been stimulated), the membrane allows potassium and chloride ions to move freely across the membrane in either direction. The membrane holds out the sodium ions, which are found in high concentrations in the tissues. These sodium ions build up a positive charge on the outside of the membrane. Inside the membrane, there will be a negative charge, because of the greater number of negative than positive ions (because of the Cl^-).

When an impulse is started, the permeability of the membrane changes; sodium ions then flow into the fibre and some potassium ions are pushed out. The result is a reversal of the charge on the membrane at the point where the impulse is passing (during its trip along the fibre). (See Figure 3.4.)

Reaction Time

The nerve impulse takes some time to travel to its destination in the central nervous system and then on to the muscle where it causes a reaction. This is known as **reaction time**. If you have taken driving instruction, you have probably experienced a test for driver reactions to visual or sound stimuli. An emergency can happen at any time while you are driving a car. Let us use the example of a child running into the road after a ball. The driver sees the child (stimulus); a message is directed to the CNS; motor neurons are

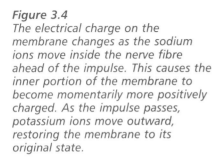

Figure 3.3
The nerve impulse is picked up by the dendrites, passes through the cell body, and is then conducted along the axons to the fine net of nerve endings at the end of the axon.

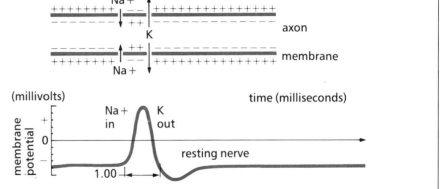

Figure 3.4
The electrical charge on the membrane changes as the sodium ions move inside the nerve fibre ahead of the impulse. This causes the inner portion of the membrane to become momentarily more positively charged. As the impulse passes, potassium ions move outward, restoring the membrane to its original state.

stimulated; and muscles then contract to move the foot operating the brake. The car may travel many metres during this time, as the speed at which an impulse travels is relatively slow – from 1 to 120 m/s. Factors such as fatigue or preoccupation may also have considerable adverse effects on the reaction time.

The Synapse

The impulse eventually reaches the end of the axon and crosses to another neuron. The gap between the end of the axon of one neuron and the dendrite of another neuron is called the **synapse** (sin-aps). (See Figure 3.5.) This gap must be bridged for the impulse to be transmitted farther. This is accomplished by a chemical called *acetylcholine*. When the impulse reaches the end of an axon, it causes acetylcholine to be released. This substance crosses the gap and stimulates the dendrites of the next neuron. Immediately after the acetylcholine has served its function, an enzyme is released at the synapse that destroys acetylcholine, thus preventing it from stimulating the dendrite continuously. The whole nervous system is made up of millions of chains of neurons linked in this way.

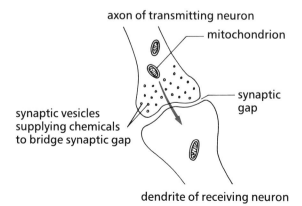

Figure 3.5
The synapse. A chemical such as acetylcholine is released by the axon of one neuron and carries the impulse across the gap to the dendrites of the next neuron. Here it stimulates the production of an impulse in this neuron for further transmission.

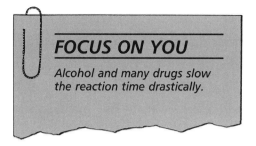

FOCUS ON YOU
Alcohol and many drugs slow the reaction time drastically.

CHECKPOINT

1. What are the four main functions of the nervous system?

2. Compare the activities of the central nervous system and the peripheral nervous system.

3. What is a neuron?

4. Explain how a nerve impulse travels from one neuron to the next.

3.2 The Brain and Its Parts

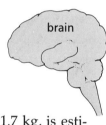

The brain, with a mass of about 1.7 kg, is estimated to contain about 10 billion neurons and about another 9 billion supporting cells. Its size varies with a person's age, sex, and body size. Brain size, like body size, is greater in the average male than in the average female. The greater size, however, is not an indication of greater mental attributes or superiority. The brain continues to grow until we reach about 20 years of age. Reasoning and intellectual ability, however, depend upon pathways, or routes, among the cells of the brain; these pathways continue to develop with use. The brain must be exercised to ensure its fitness and development, just as the muscles require regular exercise to keep them in shape.

The Parts of the Brain

The brain is divided into three distinct parts – the hindbrain, midbrain, and forebrain.

The Hindbrain

The hindbrain consists of two major parts – the **medulla oblongata** and the **cerebellum**.

- The **Medulla Oblongata** (vital body functions, information crossroads)

  The medulla appears as a swollen extension of the spinal cord. Although it is quite small, its functions are vital. Nerve impulses that stimulate the diaphragm and the muscles for breathing originate in the medulla. The heartbeat and regulation of the diameter of blood vessels are also controlled by this part of the brain. The medulla is like a complex telephone exchange, sorting and relaying incoming and outgoing calls. The medulla provides a pathway for impulses moving from higher parts of the brain to motor nerves and muscles and for impulses travelling to the brain from sensory receptors in various parts of the body. (See Figure 3.6.)

- The **Cerebellum** (muscle co-ordination)

  The cerebellum is located just above the medulla. It has curved grooves running all over the surface, giving it a furrowed appearance.

  The cerebellum is responsible for balance, co-ordination of movement, and muscle tone. Its function is to organize impulses which originate in the cerebrum and integrate these with the stream of signals coming in from sensory organs. Some very complex muscular sequences are controlled by the cerebellum. For example, imagine the co-ordination required for a gymnast to make a successful vault. The run up to the horse must be timed so that both feet land together on the springboard. The gymnast must jump up, place the hands upon the horse, and somersault the body over the horse. The knees must be flexed, the feet kept together and ready for landing. The final balance and upright position must be controlled so that the gymnast does not fall forward. This complicated series of movements involves dozens of muscles, all programmed to contract in exactly the right sequence.

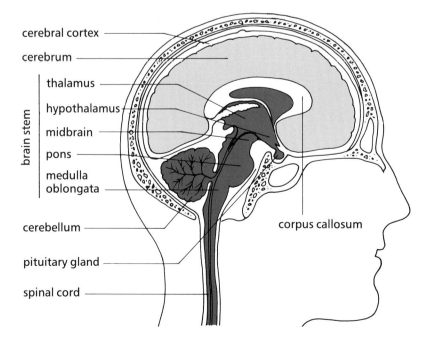

Figure 3.6
A section through a human head showing the major divisions of the brain

BIOTECH

Brain Mapping

A young man lies on the operating table of the Montreal Neurological Institute. Part of his brain is exposed and a surgeon is delicately probing the surface of the brain with a tiny electric probe. The young man is fully conscious, although he feels no pain due to the use of local anesthetics. As the doctor probes, the young man experiences a vivid impression in minute detail of a previous experience. He is acutely aware of the people present on that occasion, what was said, and the details of the room in which it occurred. As the probing continues at another location, his arm moves or his hand clenches without conscious control, although he knows that it has moved.

Dr. Wilder Penfield, a foremost authority on the human brain, pioneered the mapping of much of the human brain. He determined the areas of the brain responsible for many functions previously not known. One of his particular achievements was isolating the portion of the brain responsible for epilepsy. In people with epileptic seizures, certain regions of the cortex of the brain are damaged. As a result, the electrical charges increase and produce violent reactions or seizures. In addition to mapping the cortex of the brain, and establishing the motor, sensory, and psychological areas present, Dr. Penfield discovered how to remove the cells that caused these violent electrical "storms" within the brain and freed many patients from the distress of epilepsy.

Dr. Penfield received worldwide recognition for his contributions to medical science. He was a Companion of the Order of Canada and received the British Order of Merit, an honour conferred on only 25 people, including such individuals as Winston Churchill, Dwight D. Eisenhower, and Lester B. Pearson.

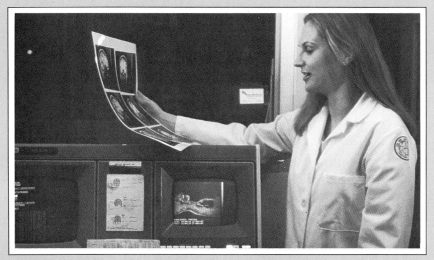

This physician is using the most modern techniques available to produce a visual scan of the living brain. By using computers to interpret electrical and magnetic signals, information on normal function and the presence of disease can be obtained.

The Pons (message transmission)

The word pons means *bridge*, which is a good description of the function of this structure. Lying between the medulla and the midbrain, the **pons** contains fibres which transmit messages within the cerebellum as well as from the cerebellum to the cerebrum, midbrain, and lower centres, such as the medulla. A few special cranial nerves originate in the pons, which also contains part of the system that controls breathing.

The Midbrain (relay for sight and sound)

In terms of size and functions, the midbrain is of only minor importance. It is located just below the centre of the cerebrum and forms part of the brain stem. The midbrain consists of four small spheres of grey matter which act as relay centres for some eye and ear reflexes. Below these spheres of grey matter are some conducting tissues of white matter that connect the higher centres of the cerebrum with the pons, cerebellum, and spinal cord.

The Forebrain

- **Hypothalamus and Thalamus** (internal control)
  There are two small areas of grey matter squeezed in between the midbrain and the cerebrum; these are the **thalamus** and **hypothalamus**. The hypothalamus is directly connected to the most important gland in the body, the **pituitary**. This gland controls all the other glands in the body and is often called the master gland. The pituitary works in conjunction with the nervous system, for example, the sensory nerves bring it "feedback" information that aids its regulatory functions.

  The hypothalamus controls the autonomic nervous system and the internal organs of the body. It directs the production of special secretions, the activities of the intestinal tract, and the blood pressure, as well as behavioural and emotional responses. Another of its important functions is the regulation of water balance and the control of urine production in the kidneys.

  The thalamus forms a sensory relay centre for impulses on their way to the cerebrum. It affects consciousness, temperature, and the degree of awareness of pain.

- **The Cerebrum** (voluntary movement, thought)
  The cerebrum is the largest part of the human brain and consists of billions of neurons and synapses. It is the highest centre of nervous control and is developed to a far greater degree in humans than in any other animals.

  The surface of the cerebrum, known as the **cerebral cortex**, is composed of grey matter (2 to 4 mm thick). It spreads like a coat over the surface of the brain. In order to pack in more cells with specialized functions, the coat is thrown into many folds, which increases the surface area. The folds or creases can easily be seen. The deep folds are known as fissures. The deepest longitudinal fissure divides the cerebrum almost completely into two halves called the **cerebral hemispheres**. These are connected internally by a bundle of fibres.

  Each cerebral hemisphere is further divided into four lobes, each of which is also marked by fissures. The lobes carry the same names as the bones of the skull that cover them – **frontal** at the front, **temporal** at the sides, **parietal** at the top and back, and **occipital**, which is quite low at the back of the skull. (See Figure 3.7.)

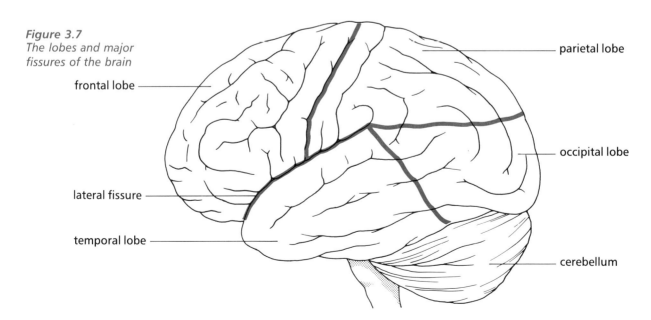

Figure 3.7
The lobes and major fissures of the brain

The frontal lobe is greatly developed in humans. This lobe controls all voluntary movements, for example, regular walking and running action, arm movements, and speech patterns. At the very front of the frontal lobe, there is an area that appears to control our basic intelligence and personality.

In addition to the neurons present in the brain, there are also millions of **glial** cells which support and nourish the neurons. Oxygen and nutrients are delivered by a network of blood vessels and capillaries, which spread over the surface of the brain and surround the tightly packed neurons.

The Motor Activities of the Cerebral Cortex

The motor functions are controlled by a band of nerve tissues located in front of the central fissure. A separate patch of tissue controls each group of motor muscles. (See Figure 3.8.)

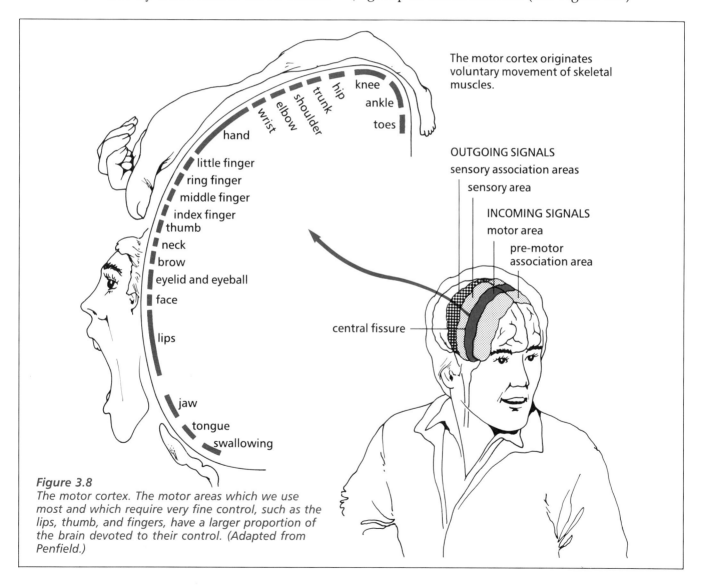

Figure 3.8
The motor cortex. The motor areas which we use most and which require very fine control, such as the lips, thumb, and fingers, have a larger proportion of the brain devoted to their control. (Adapted from Penfield.)

The impulses governing voluntary muscular movements are carried by nerve fibres that pass through the brain stem to the medulla. In the medulla, a large number of these nerve fibres cross over to the opposite side and descend through the spinal cord to eventually stimulate the muscles; thus, the majority of impulses arising in the right side of the cerebral cortex control muscles on the left side of the body and vice versa. (See Figure 3.9.)

Figure 3.9
Follow the path of the motor impulses from the left side of the brain to the muscles on the right side of the body.

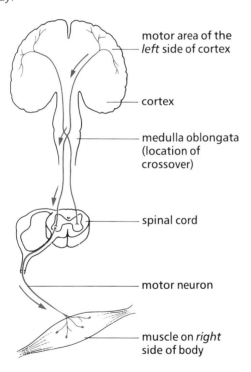

LIFE · SIGNS

SEIZURE

Mary was a happy 14-year-old, friendly with her classmates, and co-operative at school. Mary appeared to be in perfect health, and played on an intramural basketball team. This past year, however, Mary's marks had dropped. Teachers and friends had noticed that her attention seemed to wander at times, and she sometimes gazed off into space, oblivious of others around her.

Mary, at the insistence of her parents, made an appointment with her doctor. On the morning of her visit, while brushing her hair, she gave a loud scream and fell to the floor unconscious. Mary's parents ran upstairs and found her lying stretched out on the floor, her arms and legs making rapid, jerking movements. In a few minutes, Mary opened her eyes and seemed to be all right again.

When Mary and her mother talked to the physician, he asked some questions about Mary's general health and medical history. He asked when they had first noticed any changes and started to make notes in the file.

Mary told the doctor that sometimes she seemed to "drift off" for a few moments so that when people spoke to her, she could not remember what they had said. The doctor then asked Mary's mother about Mary's appearance when she found her on the floor that morning.

The doctor arranged for Mary to have an EEG (**electroencephalogram**) and gave her a general checkup. Eventually, as a result of the tests and reports, the doctor felt confident of his diagnosis and

The Sensory Activities of the Cerebral Cortex

Sensory activities are controlled by a band of nerve tissue located behind the motor area. This band receives the impulses from the sensory organs responsible for such things as touch, pressure, and heat. Some of the major sense organs, such as those for sight and hearing, are controlled by special areas. The area for hearing is found on either side of the brain. Sight is controlled by an area at the back of the cerebrum.

Once information is received, it is transferred to the association areas found beside the sensory explained to Mary and her mother that Mary had **epilepsy**. The doctor immediately assured them that, with treatment, the disease could be controlled so that Mary would be able to lead the life to which she was accustomed.

The medication prescribed for Mary contained anticonvulsive drugs, but it was almost a year before the right combination of drugs brought Mary's symptoms completely under control. During this period, Mary suffered several seizures and her friends learned to help her and look after her while the attacks lasted. At first, the friends were frightened by the convulsions that Mary experienced. (This was to be expected, as our fears are often due to lack of knowledge and not knowing what to do to help someone.) The friends talked with Mary's parents, who themselves had experienced similar fears until they were given special instructions on how to look after their daughter. They told the friends that even if the attack was severe (a *grand mal* seizure), it usually would last only a few minutes, and that a doctor's care would not normally be required. The friends were also told that, during a seizure, an epileptic person is unconscious and feels no pain at all. Even when the body jerks and is thrown about, or the person's skin changes colour, or the body goes quite rigid, there is still very little danger to the person, and there is not very much to be concerned about.

To help someone during an epileptic seizure, the following directions should be followed:

- Keep calm. You cannot stop a seizure once it has started. Let the seizure run its course. Do not try to revive the person.
- Ease the person to the floor and loosen his or her clothing.
- Try to prevent the person from striking his or her head or body against any hard, sharp, or hot objects, but do not interfere otherwise with the progress of the seizure.
- Turn the person on his or her side, so that the saliva may flow freely from the mouth.
- Do not put anything in the person's mouth.
- Do not be frightened if the person having a seizure seems to stop breathing momentarily.
- After the seizure stops and the person is relaxed, allow the victim to sleep or rest if he or she wishes to do so.
- If the person is a child, the parent or guardian should be notified that a seizure has occurred.
- After a seizure, most people can carry on as before. If, after resting, however, the person still seems groggy, weak, or confused, it would be better to accompany the person home.
- If the person undergoes a series of convulsions, with each successive one occurring before the person has fully recovered consciousness, you should immediately seek medical assistance.

What Causes Epilepsy?

The exact mechanism that produces an epileptic seizure is not well-understood. In some way, the normal pattern of electrical activity in the brain is disrupted, and a sudden, violent disturbance of electrical waves develops, rather like a short-lived electrical "storm". The brain tissues are very sensitive to changes in amounts of certain chemicals and quite small changes in acid-base conditions. Certain chemicals in the blood can also promote a seizure. Flashing lights and emotional upsets can induce an attack in some persons.

---

areas. These areas interpret the various sensations. The impulses are then redirected to the motor region for action, or to another area for storage for future reference.

Perhaps one of the most striking, yet least understood, functions of the brain is its ability to retain, change, or modify ideas. Information, some of it received a long time ago, can be recalled to be reused. Known as **associative memory**, this capacity makes reasoning, moral judgment, and moral sense possible. Only a few theories exist to explain how the associative memory operates.

Activity 3A

Dissection of a Sheep Brain

The sheep's brain contains the same structures as a human's, although in different proportions.

Materials

eye/face protection
plastic gloves
protective apron
sheep brain, preserved, soaked overnight in water
dissection tray
dissection kit
hand lens
paper towelling

 CAUTION!
Eye/face protection, plastic gloves, and a protective apron must be worn for this activity.

Procedure

1. Rinse the brain under running water. Carefully blot dry.

2. Place the brain on the dissection tray so that the dorsal surface (surface with many folds) is uppermost.

3. *Locate and describe the appearance of the following structures:*
   (a) cerebrum – largest part of the brain, with many folds; divided into left and right cerebral hemispheres
   (b) cerebellum – smaller part of the brain, located below and at the rear of the cerebrum
   (c) medulla oblongata – the narrow, cylindrical part of the brain
   (d) meninges – the three membranes which cover the brain (Note: Your teacher will tell you if these structures are present on your specimen.)
   (e) cerebral fissure – separates the cerebrum into right and left halves.

4. Turn the brain over and examine its underside. Look for any swellings or cords which protrude from the surface. *What could these be?*

5. Carefully divide the brain in half by cutting though the tissue at the bottom of the cerebral fissure. Continue until you have cut through the medulla and the cerebellum. (See Figure 3.10.)

Figure 3.10
Dissection of a sheep's brain

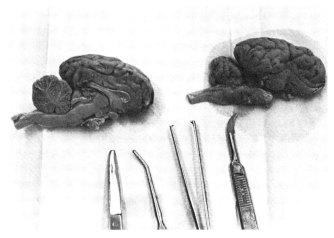

6. *Make a sketch of the inside of one half of the brain. Locate and label the areas of grey matter, white matter, and the corpus callosum, a central band of tissue which lies just below the cerebrum.*

7. Examine a cross-section of the top of the spinal cord (lower part of the medulla). *Sketch a view of the cross-section and label the white and grey matter areas.*

Questions

1. Where does the cerebellum connect to the rest of the brain?
2. How is the outside of the cerebrum different from the inside?
3. Is the left side of the cerebrum different from the right? If so, what differences are there?

CHECKPOINT

1. *What functions of the body are controlled by the medulla oblongata?*
2. *What is the function of the*
   (a) cerebellum?
   (b) hypothalamus?
   (c) thalamus?
3. *The cerebral cortex is spread like a coat with many folds over the surface of the brain. What purpose is served by these many folds?*
4. *What part of the body is controlled by the left side of the cerebral cortex?*

3.3 Protecting the Brain

The brain is one of the best-protected organs of the body. It is entirely enclosed within the hard bones of the skull. Between the bony case and the actual nerve tissue of the brain are three protective membranes called the **meninges** (men-in-jeez).

The brain is further protected by the **cerebrospinal fluid**, which fills a space between the arachnoid and the pia mater. (See Figure 3.11.) This fluid bathes the cells and helps to cushion the brain. It carries nutrients to the cells and removes wastes for transfer to the bloodstream. This fluid and the membranes around it continue down from the brain to surround the spinal cord, giving it the same protection and nourishment. The cerebrospinal fluid helps to provide a buffer against knocks and bumps to the brain and spinal cord.

Serious brain injury may occur if there is a blow to the head, even if the skull is not fractured. Such a blow may cause microscopic damage to the brain. This is called a *concussion*, and may cause dizziness, "seeing stars", or unconsciousness. With a more serious injury, the brain may

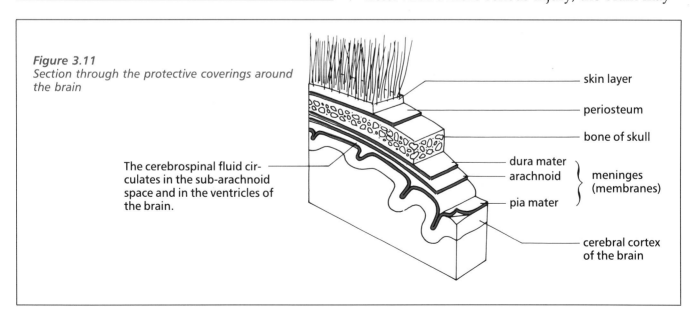

Figure 3.11
Section through the protective coverings around the brain

The cerebrospinal fluid circulates in the sub-arachnoid space and in the ventricles of the brain.

- skin layer
- periosteum
- bone of skull
- dura mater
- arachnoid } meninges (membranes)
- pia mater
- cerebral cortex of the brain

be bruised, or there may be a hemorrhage on the brain surface. This is called a *cerebral contusion*. It is accompanied by unconsciousness, poor circulation, and shallow breathing. Bleeding within the skull causes pressure on the brain. An injury this severe could cause permanent damage to the brain or even be fatal.

In cases where damage to the brain is suspected, have the person lie face down with the head to one side, and get medical help immediately. The best solution is prevention. (See Figure 3.12.)

Figure 3.12
Some serious head injuries can be prevented by wearing an approved helmet.

FOCUS ON YOU

*A bacterial infection may cause the membranes around the brain to become inflamed, a condition called **meningitis**. The early symptoms of this disease are fever, headache, a stiff neck, and vomiting. If not treated, the disease will cause a loss of consciousness, and may be fatal.*

3.4 The Spinal Cord

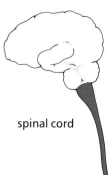

The **spinal cord** extends from the opening at the base of the skull, down through the vertebral canal, to the lower back. Each of the bones of the spine, the vertebrae, contains a large hole through which the spinal cord passes like a thread passing through the holes in a string of beads.

At each vertebra, nerves connect with the spinal cord. Sensory nerves carry impulses to the central nervous system by way of the **dorsal root** which lies at the back of the cord. Motor impulses are carried away from the CNS by the **ventral root**, which is found at the front of the spinal cord. (See Figure 3.13.)

Figure 3.13
A cross-section through the spinal cord. Recall that the grey matter is composed of unmyelinated fibres and cell bodies. The white matter is formed of myelinated fibres.

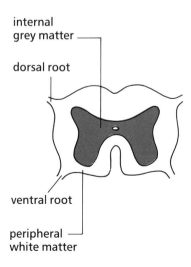

Functions of the Spinal Cord

- It forms a two-way conduction system between the brain and peripheral nerves, both sensory and motor.
- It controls reflex actions that do not require supervision by higher centres in the brain.

The Reflex Arc

The nervous system provides the means of co-ordinating the body's response to its environment. This co-ordination is a network which ties together your senses, brain and spinal cord, and muscle. The **reflex arc** is the simplest basic unit of this network, consisting of the following five parts:

1. **A receptor**. The receptor recognizes some change in the environment, whether of heat, light, sound, or some other factor. The receptor is stimulated to initiate a nervous impulse.

2. **A sensory neuron**. The sensory (afferent) neuron conducts the impulse from the receptor to the central nervous system (the spinal cord).

3. **A central or association neuron**. This nerve cell switches the impulse from the sensory "informing" neuron to the "acting" motor neuron. It allows the impulse to be routed into a number of possible pathways, including information to be sent to the brain.

4. **A motor neuron**. The motor (efferent) neuron carries the impulse to the appropriate organ (usually a muscle) to produce the response.

5. **An effector**. The effector is the muscle or organ that will contract or otherwise respond appropriately to the stimulus. (See Figure 3.14.)

The Reflex Act

A **reflex act** is an automatic or involuntary action which is always the same when a particular stimulus is involved. Drawing your hand back from a source of pain is a reflex act. The closing of the eye's pupil in response to increased light is another example. Reflex acts occur in a fraction of a second, before a person has time to think consciously about what appropriate action is required. Such rapid reaction protects the body from harm.

Figure 3.14
A three-neuron reflex arc. What is the function of each of the five components of the reflex?

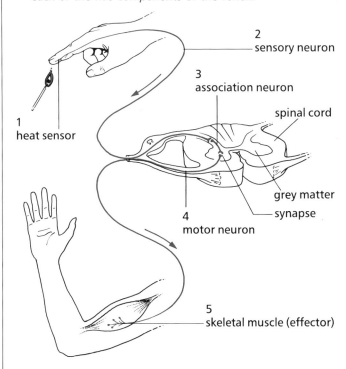

Activity 3B

Human Reflexes

The normal reflex response requires three neurons. Each reaction involves the transmission of an impulse from a receptor, through the sensory neuron, to the spinal cord and back by the motor neuron to the muscles that produce the response. The knee reflex differs in that it involves only two neurons.

Materials

rubber reflex hammer
penlight or microscope light source

Procedure

The Patella Reflex (knee jerk)

1. Have your partner sit on the edge of the bench so that his or her legs hang freely and do not touch the floor. Feel the position of the patella (kneecap) and locate the tendon that is just below it. Strike the tendon sharply with the edge of the hand, or a rubber reflex hammer. The foreleg should extend involuntarily. Try to determine which muscles cause this upward lift of the leg. The receptors, in this case, are stretch receptors present in the tendon below the knee. These receptors have an important role in helping to maintain our upright position. *Record your observations and explanations.*

2. Repeat the experiment, but this time, ask your partner to clasp his or her hands together and, just before striking the tendon, have your partner try to pull the hands apart, so that the muscles are strongly tensed. *Record your observations.*

3. Distract your partner's attention by assigning a task, such as counting the number of individual letters in a sentence on a printed page. Once started on this task, again strike the tendon. *Record your observations.*

The Achilles Reflex

1. Have your partner kneel on a chair with his or her feet hanging over the edge of the seat. (Have your partner remove his or her shoes.) Push the foot forward and then lightly tap the Achilles tendon. *What happens? Which muscles respond?* Repeat the test. *Is the second reflex stronger or weaker? Explain why this might be so.*

Questions

1. In the knee jerk experiment, what differences did you observe between the three steps? Explain your observations.

2. Why do we tightly clench our fists before a fight or under conditions of special stress? What advantage might such action provide?

3. Draw and label the nerve impulse pathway for a knee jerk reflex.

Conditioned Reflexes

Conditioned reflexes are reflexes that have been modified by training or learning. For example, Ivan Pavlov, the Russian physiologist, conditioned animals so that they would respond to certain signals. Dogs produce saliva abundantly when food is placed near them (an unconditioned, reflex act). When Pavlov rang a bell each time the food was offered, the dogs learned to associate the bell with food. Eventually, each time the bell was rung, the dogs produced saliva even if no food was present. A conditioned reflex was thus established.

We are constantly modifying our behaviour. Reflexes allow us to take time-saving, even life-saving, short cuts. The action of our foot on the brake pedal of a car, for example, is often the result of a conditioned or learned reflex response to some visual danger signal.

Activity 3C

A Conditioned Reflex

You have read of Pavlov's experiment on conditioning dog reflexes. Can you design a simple experiment to produce a conditioned reflex in your partner? Shine a light into your partner's eyes at the same time you tap his or her hand.

Repeat this 15 to 20 times, then tap the hand but do not shine the light. Observe the response. If there is no result, continue the conditioning and test again. Once you have achieved success, repeat it several times without the light. This conditioned response will quickly be lost. Note how many times the eye responds to a hand tap before the conditioning is lost. Try to invent a conditioned reflex of your own design.

Injury to the Spinal Cord

Recall that the central nervous system is surrounded by three membranes as well as a fluid-filled space. (See Figure 3.15.) These protective layers cushion and normally prevent injury to the spinal cord, but a severe blow may fully or partially sever the cord. If this happens, a loss of sensation and voluntary control of the motor muscles will occur below the point of injury. Should this occur in the neck, paralysis of almost the entire body results, including both the upper and lower limbs (quadraplegic, from "quad" meaning four). If the cord is permanently damaged near the centre of the spine, only the legs will be affected (paraplegic).

Figure 3.15
A cross-section of the spinal cord through a vertebrae showing the protective meninges of the central nervous system

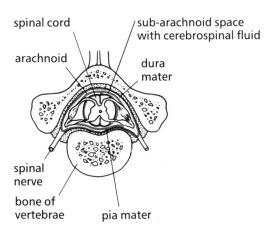

Activity 3D

Getting Around on Wheels

People who must use a wheelchair, due to an injury to their spinal cord or other cause, still need to get to the same places as anyone else. (See Figure 3.16.) How do they manage? What problems might they encounter?

Figure 3.16
What modifications or adjustments were needed so this person could enjoy this activity?

To find out, *make a list of your typical daily travels room by room and place by place. Now, imagine you are going through your typical day in a wheelchair. Note, beside each location and building on your list, where you could move freely without assistance. What locations would be difficult or impossible to enter alone? Why are these areas important to you? Make some suggestions as to how these areas could be made accessible.*

CHECKPOINT

1. Describe the three different ways the brain is protected from injury.

2. What is a concussion?

3. What are the two major functions of the spinal cord?

4. Name and give the functions of the five parts of a reflex arc.

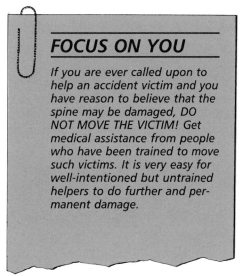

FOCUS ON YOU

If you are ever called upon to help an accident victim and you have reason to believe that the spine may be damaged, DO NOT MOVE THE VICTIM! Get medical assistance from people who have been trained to move such victims. It is very easy for well-intentioned but untrained helpers to do further and permanent damage.

3.5 The Autonomic Nervous System

The name of this part of the nervous system is derived from the same root as the word automatic. This is reasonable because almost all of its functions are performed without thinking. The **autonomic nervous system** controls most of our internal organs and processes – respiration, circulation, digestion, excretion, and reproduction. This system should not, however, be thought of as entirely separate from the central and peripheral nervous systems. The autonomic nervous system is controlled largely by the hypothalamus, which is part of the forebrain. It also involves the medulla, the spinal cord, and the extensive system of peripheral nerves.

The Function of the Autonomic Nervous System

In general, the autonomic system looks after our internal environment, helping to keep the body systems operating normally. It functions to make rapid adjustments in the operation of these systems during an emergency. These activities are automatic; we do not have to think about them.

The Divisions of the Autonomic Nervous System

The autonomic nervous system is divided into two parts – the **sympathetic** and the **parasympathetic** portions. These perform in opposite ways and tend to counteract each other. The sympathetic system prepares the body for emergencies, whereas the parasympathetic system reverses the effect, bringing the body back to its normal condition to conserve energy. (See Figure 3.17.)

The Sympathetic System
If a stressful situation arises, such as when a child runs in front of a car, the body must react rapidly to meet the emergency. The sympathetic nervous system sends messages to the heart to speed it up. Other messages are sent to the skin to shut down the blood vessels there, diverting extra blood to the muscles of the heart and limbs. The sympathetic system also dilates the pupils of the eye to improve peripheral (side) vision, and shuts down digestive processes for the duration of the emergency. Special glands, such as the adrenals, are stimulated to back up the emergency measures already taken. The adrenal glands unlock stores of sugar in the liver to supply muscles with extra energy.

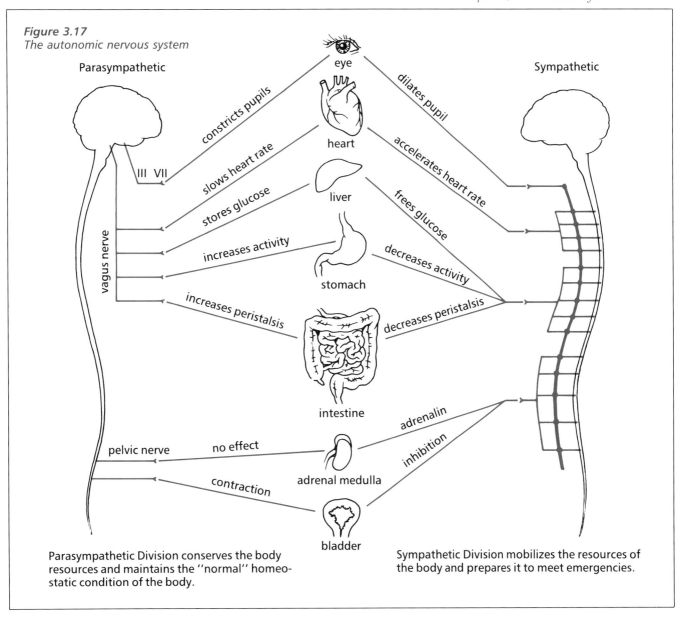

Figure 3.17
The autonomic nervous system

Parasympathetic Division conserves the body resources and maintains the "normal" homeostatic condition of the body.

Sympathetic Division mobilizes the resources of the body and prepares it to meet emergencies.

The Parasympathetic System
The parasympathetic portion of the autonomic system helps to control the rate of operation of organs such as the heart, stomach, intestines, liver, and pancreas. It returns them to a normal working pace after they have been speeded up or shut down during an emergency.

CHECKPOINT

1. What is the function of the autonomic nervous system?

2. Explain the differences between the functions of the parasympathetic and sympathetic divisions of the autonomic nervous system.

3.6 The Endocrine System

During an emergency, the sympathetic nervous system stimulates the production of special chemicals from certain glands. Such messenger chemicals, called **hormones**, serve many other vital functions in your body as well. The glands which produce them are part of the **endocrine system**. (See Figure 3.18.)

Each hormone acts to regulate one or more chemical reactions in the body. Hormones are pumped through the blood to their **target organ**. They are released in response to signals from the nervous system, other glands, or because of the presence of key substances in the body.

An Example of a Gland and Its Function: The Pituitary

The pea-sized pituitary gland, often called the "master gland", lies at the base of the brain and is attached to the hypothalamus. The hypothalamus sets the basic metabolic rate of the body by controlling the secretions of the pituitary. Among its other roles, the pituitary gland regulates growth as well as the secretion of hormones from the adrenal glands and gonads. (See Figures 3.19 and 3.20.)

Only a very small amount of a hormone is required to produce a very large response within the body. The pituitary gland, for example, produces only 0.001 mg of hormones per day.

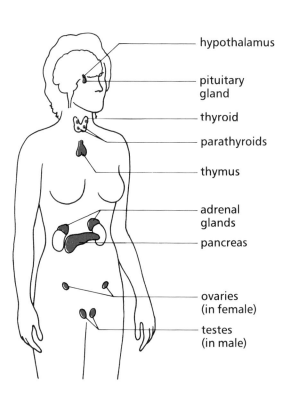

Figure 3.18
These are some of the glands of the endocrine system, shown with the general function of the hormones they produce.

Hypothalamus: sets the basic metabolic rate of the body, controls pituitary secretions

Pituitary gland: the master gland, controls or influences all other endocrine glands

Thyroid gland: responsible for energy metabolism, growth, decreases blood calcium

Parathyroid glands: increase blood calcium

Thymus: aids in immunity reactions in younger people

Adrenal glands: help to prepare the body for stress

Pancreas: regulates sugar in the blood

Gonads (ovaries and testes): responsible for growth and development of the reproductive system

Figure 3.19
Pituitary hormones, their target organs and effects

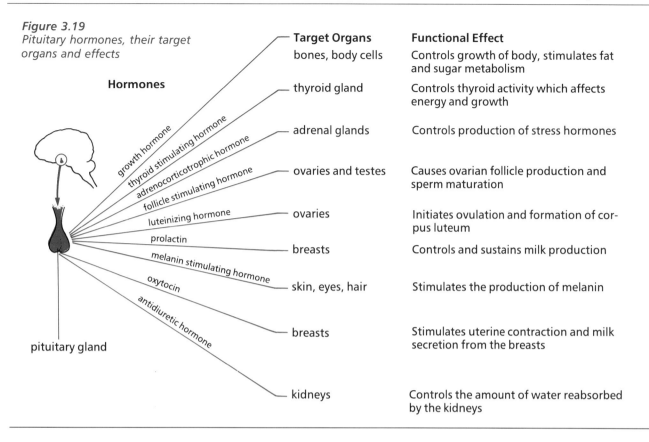

Hormones	Target Organs	Functional Effect
growth hormone	bones, body cells	Controls growth of body, stimulates fat and sugar metabolism
thyroid stimulating hormone	thyroid gland	Controls thyroid activity which affects energy and growth
adrenocorticotrophic hormone	adrenal glands	Controls production of stress hormones
follicle stimulating hormone	ovaries and testes	Causes ovarian follicle production and sperm maturation
luteinizing hormone	ovaries	Initiates ovulation and formation of corpus luteum
prolactin	breasts	Controls and sustains milk production
melanin stimulating hormone	skin, eyes, hair	Stimulates the production of melanin
oxytocin	breasts	Stimulates uterine contraction and milk secretion from the breasts
antidiuretic hormone	kidneys	Controls the amount of water reabsorbed by the kidneys

Increases blood supply to skeletal muscles, heart, lungs, and brain, ensuring maximum delivery of oxygen and sugar energy

Dilates the pupil of the eye to admit more light

Relaxes the smooth muscles of the bronchiole walls to provide a better supply of air to the alveoli of the lungs, stimulates respiration

Increases the heart rate and the amount of blood pumped out by the heart, raises the blood pressure

Contracts muscles in the skin so that hairs stand on end, produces "goose flesh", reduces the supply of blood to the skin

Increases the rate at which blood coagulates, redirects blood to areas where it is most needed during an emergency

Converts liver and muscle stores of glycogen into glucose, quickly raises blood sugar level and delivers it to muscles where needed

Contracts the ureters and sphincter muscles of the bladder

◀ *Figure 3.20*
Responding to emergency. The pituitary gland influences the production of the hormone adrenalin from the adrenal glands. During moments of special excitement or stress, adrenalin is released by the adrenal glands into the bloodstream to prepare the body for the emergency. Shown here are the effects this hormone has on various organs in the body.

3.7 Foreign Chemicals and the Nervous System

Hormones are powerful internal chemical messengers, affecting almost all body tissues. The nervous system is also affected by the presence of foreign chemicals. Some of these chemicals come under the heading of drugs, either medicinal or non-medicinal in nature.

A **drug** is any substance other than food which, when taken into the body, affects the living cells of the body. Medicinal drugs have a variety of uses. A vaccine, for example, is used to prevent the occurrence of a disease. Some drugs aid the diagnosis of an illness. Most commonly, drugs are prescribed by a physician for the treatment of an illness or condition.

Drug Abuse and Addiction

Used properly, drugs can improve people's health and prolong lives. Unfortunately, drug abuse has the opposite effect. (See Table 3.1.) The use of a drug after the medical need for it is over, or when there was never a genuine medical reason for its use, can lead to **drug dependence**, sometimes called **addiction**.

In a *psychological dependency*, a person comes to believe, falsely, that a particular drug is essential to health, performance, or happiness. Withdrawal from the drug may cause personality disorders. In a *physical dependency*, the normal activity of a person's body has been so influenced by the presence of a particular drug that it no longer functions properly. Withdrawal from the drug causes physical symptoms ranging from tremors to painful convulsion and death. Some drugs result in both kinds of dependency. (See Figure 3.21.)

To make matters worse, the body can build up a tolerance to many kinds of drugs. This means that larger and larger doses will be necessary for the user to feel the same result. Drugs such as cocaine will cause very severe effects at high doses, possibly to the point of being fatal, but an addict seeking a "high" may be unable to comprehend the danger.

Figure 3.21
A person suffering from a drug addiction needs professional help. This person can obtain confidential counselling from a physician, support groups, or school counsellors.

Table 3.1 Types of Drugs

Stimulants

Effect
- stimulate sympathetic nervous system
- increase metabolic activity, including heart rate, breathing rate, and blood pressure

Examples
- caffeine (tea, coffee, colas, chocolate), nicotine (tobacco), amphetamines (such as "speed"), cocaine ("crack")

Risks from Abuse
- delusions and hallucinations
- personality changes; bizarre or violent behaviour
- addictive
- burst blood vessels or heart failure

Depressants

Effect
- sedative, cause relaxation and drowsiness
- reduce tension

Examples
- alcohol, barbiturates, tranquillizers, opiates (opium, heroin, morphine, codeine, methadone)

Risks from Abuse
- addiction (withdrawal symptoms are severe)
- reduced brain function, ranging from slurred speech and slow reaction time to coma

Analgesics

Effect
- relieve pain by interfering with prostaglandin production (Prostaglandin is made by most cells after an injury and causes headache, fever, and sensitivity to irritation.)

Examples
- acetylsalicylic acid (aspirin), opiate derivatives, acetaminophen

Risks from Abuse
- irritation of stomach lining
- bleeding ulcers
- dehydration and kidney failure
- to pregnant women: bleeding, low birth weight, stillbirth
- to children: possibly fatal if taken during a viral illness such as chicken pox or flu because of association with Rye's Syndrome

Hallucinogens

Effect
- no proven medicinal value
- small doses cause distortion of senses and emotions

Examples
- lysergic acid diethylamide (LSD), mescaline

Risks from Abuse
- large doses cause hallucinations in which a person sees or hears things which are not there
- addiction
- loss of motivation

It's Up to You

Like all bodily systems, your nervous system is delicately balanced. Upsetting this balance is to risk temporary and possibly permanent damage to the main control of your body.

It is tragic that the abuse of drugs has produced a large number of individuals who can no longer function normally. Their minds are so distorted that rational thinking is no longer possible. How can it happen? The profits available to criminals dealing in illegal drugs make it to the dealer's advantage to promote a drug as "harmless". Why should they care about your health or well-being? The decision to protect yourself – to say no – can only be made by you.

Yet it is important to note that the abuse of two non-medicinal drugs – nicotine and alcohol – is still responsible for most of the health and social problems in Canada. These drugs are legally available to adults. Their use, or abuse, is a matter of choice. And that's up to you as well.

FOCUS ON YOU

Although the amount of an individual drug taken at a time might not be harmful, combinations can be deadly. Taking barbiturates and alcohol, for example, may cause breathing to stop.

FOCUS ON YOU

Marijuana can be considered a hallucinogen, although it also acts as a depressant. It contains a mixture of active ingredients, but the one which produces the major effects is tetrahydrocannabinol (THC). The drug affects the perception of time, colour, and sound. Continued use contributes to psychological dependence.

LIFE · SIGNS

OVER THE LIMIT

It was Jill's birthday and a group of friends had gathered at Jill's home to celebrate the occasion. Beer, spirits, and other beverages were available and many people were having alcoholic drinks in moderation. Ross, however, was setting a faster pace.

By nine o'clock that evening, Ross was talking loudly, dominating the conversation, and rudely interrupting anyone who tried to talk. Some laughed at him as he stumbled against the furniture on his way to the kitchen for another drink. His behaviour was annoying to several other guests.

Between eleven o'clock and midnight, Ross fell asleep in a chair in one corner of the room and his friends were content to leave him alone to sleep off the effects of his drinking. Jill's parents were expected home about 1:30 in the morning and Jill wanted her friends to leave before her parents arrived so that she could put the room in order.

Someone woke Ross up and, after a good deal of encouragement, he went out to the kitchen, drank a cup of coffee, and ate a sandwich. His friends asked him if he was all right and he replied, "Yes, I'm okay." Indeed, he appeared to be a great deal more capable than before he had fallen asleep. The others left in small groups. Ross was the last to leave. Jill talked to him and, since he seemed to have sobered up, she was only mildly concerned that he would be driving home alone.

Ross does not remember too much about driving home and, as far as he was concerned, he had "no problem"; however, a flashing red light and a wailing siren made him pull the car over to the curb. The police officer asked Ross for his driver's license. As the officer got close enough to smell Ross's breath, he asked Ross to get out of the car. Trying hard to appear relaxed and sober, Ross started to get out, but caught his shoulder on the seat belt harness and stumbled against the door. After a few more questions, the officer asked Ross to get into the police car. Ross also was asked to lock his own car, which was then left by the roadside to be collected later.

When Ross arrived at the police station, he was asked a number of questions and then was required to write down his name and address. Several tests followed. First, he was asked to walk along a straight line, turn at the end, and walk back. Ross had some difficulty staying on the line and, on turning at the end of the line, found himself with one foot treading on top of the other. He was asked to touch his finger to his nose and then to pick up some coins. Finally, Ross was asked to take a breathalyzer test and the results showed a blood alcohol content of greater than 80 mg of alcohol per 100 mL of blood. The officer wrote a report stating the test results and describing various physical observations, such as the condition of Ross's eyes.

It was obvious that Ross's responses were impaired by his use of alcohol. His ability to respond to requests and to control and co-ordinate his movements, keep his balance, or even to write his own name and address clearly were all seriously affected by the alcohol he had consumed. He should not have been driving a car. The breathalyzer tests confirmed the officer's observations, and Ross was charged with impaired driving.

When the case came to court, Ross was convicted and fined $300. His license was suspended for three months. The judge warned him that a second offence would result in a minimum two-week jail sentence, and a six-month suspension.

Chapter 3 / The Nervous System: In Control

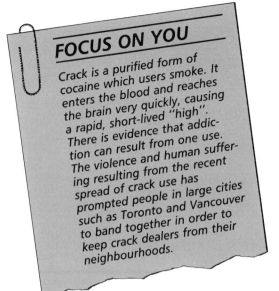

FOCUS ON YOU

Crack is a purified form of cocaine which users smoke. It enters the blood and reaches the brain very quickly, causing a rapid, short-lived "high". There is evidence that addiction can result from one use. The violence and human suffering resulting from the recent spread of crack use has prompted people in large cities such as Toronto and Vancouver to band together in order to keep crack dealers from their neighbourhoods.

FOCUS ON YOU

An overdose of acetylsalicylic acid (aspirin) is one of the most common causes of poisoning.

The appendix at the end of this book lists the drugs most often abused and their effects.

CHECKPOINT

1. *What is a drug?*

2. *What is the difference between a stimulant and a depressant?*

3. *What is an analgesic?*

4. *What are some of the problems associated with the use of aspirin?*

5. *Explain the two forms of drug dependency.*

3.8 Behaviour

Behaviour, simply defined, is how we act. You will recall from the beginning of the chapter that it was defined as observable muscular activity. Behaviour occurs in response to a stimulus.

The stimulus may be *internal*, like hunger or thirst. When we are deprived of basic biological requirements of the body, such as food, water, or sleep, we develop a physiological need. This need becomes a **drive**, which is a condition of the body causing it to become active. If we are hungry, for example, we will begin to look for something to eat. As we eat, our need for food diminishes, and other activities take precedence. (See Figure 3.22.)

A stimulus can also be *external*, such as something heard or seen. Behaviours as varied as moving from sun to shade, or stopping to help someone who has dropped packages are responses to the world around us. In humans, many behaviours are based on an individual's system of values and beliefs. Sometimes an individual's personal beliefs may come into conflict with that of a larger group. If the approval of friends is important, this *peer pressure* may cause a person to participate in behaviour that he or she personally believes to be unacceptable.

Figure 3.22
What need are these students trying to satisfy?

107

Innate Behaviour

There are certain behaviour patterns which are not learned, being already present, or **innate**, in an organism. Such behaviours are called **fixed action patterns**. Birds demonstrate several fixed action patterns when feeding their young. (See Figure 3.23.) Such behaviours seem to be a programmed response, not something an organism does because it realizes the need for the behaviour. The greylag goose, for example, has a fixed action pattern of reaching out with her head and using her beak to retrieve eggs which roll from the nest. This behaviour increases the probability that all her eggs will hatch. The goose will even attempt to roll a nearby volleyball into the nest!

Rhythms are another type of unlearned innate behaviour. Humans operate under several, including the **circadian** (24-hour) **rhythm** during which we eat, are active, and sleep.

Figure 3.23
Fixed action patterns. These small minnows nest under rock surfaces. When a male is put in a tank with an overhanging surface, he will immediately rub his back on the surface to prepare it for a nest.

FOCUS ON YOU

Individuals sometimes behave differently in large groups than they would if they were acting alone. Riots are examples of what has been called "mob behaviour".

Activity 3E

An Example of Innate Behaviour

Many animals exhibit innate behaviours important to their species for survival. How will an earthworm respond to bright light?

Materials

live earthworms
dissection tray
paper towelling
black paper
tape
bright light source

(Note: This activity requires you to work with a living organism. Before you begin, your teacher will explain how you should handle and care for this living specimen.)

Procedure

1. Place wet paper towelling on the bottom of the dissection tray.

2. Place two or more live earthworms on the paper towel and observe for 5 min. Describe the movement of the earthworms. *Which way do they move?*

3. Remove the earthworms from the tray. Cover one-half of the tray with a piece of black paper, and tape it to the side of the tray.

4. Place the earthworms in the middle of the paper towel so that their anterior ends (end closest to the light band of tissue, the clitellum) are in the light. (See Figure 3.24.)

Figure 3.24
Start your earthworms in this position. Handle them gently. What do you predict the earthworms will do?

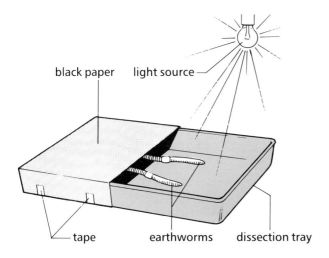

5. Observe the earthworms for 5 min. *Which way do they move?*

6. Repeat steps 4 and 5, but place the earthworms with their posterior ends in the light.

7. Compare your results with those of five other students.

Questions

1. (a) What trend do your observations support?
   (b) Which way do earthworms move?
   (c) Which end of the earthworm is sensitive to light?

2. How would this fixed action pattern increase the earthworm's chances for survival?

Activity 3F

More About Fixed Action Patterns

Many animals have fascinating fixed action patterns. Select one or more of the following to investigate, and report on:

- communication in honey bees
- caribou migration
- the mating behaviour of the stickleback
- nest building in birds
- web construction by spiders
- feeding routine of the digger wasp
- lobster migration.

Personality Disorders

Most people consider themselves to be normal. They believe that their habits, and the other things they do, are done in a normal way. Most of us also observe what we might consider abnormal behaviour during our daily contacts with other people. Most of these behaviours are simply someone else's way of doing things. (See Figure 3.25.)

Truly abnormal behaviours may be the result of a **personality disorder**. For example, a person may hear voices that no one else can hear (a hallucination), or may believe that he or she is a famous person from the past (a delusion). These are some of the symptoms of a disorder called **schizophrenia**.

A person who is suffering from a personality disorder, such as schizophrenia, can receive help through psychotherapy. Some of the symptoms of the disorder can be relieved with the use of medication. The afflicted person can also benefit from the support and structure provided by family and friends. A knowledge of the symptoms and instruction in how best to help is invaluable.

Figure 3.25
How might your behaviour seem to these people?

Activity 3G

Learning About Personality Disorders

Some personality disorders are well-known, although possibly misunderstood. Others may not be as familiar. Select one of the following personality disorders, and learn about its possible causes, symptoms, and accepted treatment. The library and the local branch of the Canadian Mental Health Association can provide you with information about the disorder you select. Prepare a brief report which can be shared with other students.

- phobia
- obsession
- compulsion
- kleptomania
- psychopathy
- drug abuse
- senility
- depression
- schizophrenia
- paranoia.

CHECKPOINT

1. Define drive and give an example.

2. (a) Define fixed action pattern and give an example.
   (b) Why are fixed action patterns important to an animal species?

3.9 Learning

Many behaviours are learned, that is, the behaviour is a result of some experience the individual has had. In animals, behaviours taught by a parent, or self-taught, can increase the individual's chances of survival. Foxes, for example, must learn to hunt for food. Several innate behaviours are improved through learning. For example, birds which weave nests will construct an adequate first nest without experience. Every year, however, the same bird will improve its technique.

Human learning continues for a lifetime. Our capacity to learn seems to be limited only by our desire to make the effort to do so.

FOCUS ON YOU

Learning may begin before birth. Studies have shown that newborn babies have already learned the sound of their mother's voice.

FOCUS ON YOU

Just how much can a person remember? In order to prove that storytellers of ancient Greece used to commit their lengthy epic poems to memory, a 60-year-old man decided to spend 1 h a day memorizing Homer's Iliad — 24 books and 600 pages worth and in the original Greek. After 11 years, he can recall 22 books without error. Soon he will have the entire poem in his mind. It will take 22 h to recite aloud.

Memory and Learning

Memory is the essential component of learning, yet little is known about how we remember things we have learned. Research has shown that in order for something to be stored in our long-term memory, it must have some significance or association to an important event.

Learning can be assisted with the use of mnemonic devices. A **mnemonic** (new-mon-ic) device is any system of coding information to make it easier to remember. The last four digits of a telephone number, for example 1867, are easier to remember if one recalls they also represent the year of Confederation. (See Figure 3.26.)

Figure 3.26
When looking for a book on a particular subject in the library, you use organized information stored in the library's "memory".

Our ability to recall information is also influenced by how the information was presented to us. If we learn the words to a poem or song, we normally do so from start to finish. Later, we may find ourselves reviewing the entire work in order to remember one particular line or phrase. Pictures, on the other hand, are presented all at once. Information presented this way is likely to be recalled in the same manner.

The phrase "practise makes perfect" applies to learning and memory as well. Repetition increases learning, as does frequent review. More importantly, the active involvement of the learner in the learning process contributes to learning.

Activity 3H

Memory and Habit

A habit is a behaviour pattern that is so regularly followed it becomes almost involuntary. Many of our daily activities have become habits. We perform a series of steps to complete the task without needing to give much thought to the sequence of movements which are required.

Drape a necktie around your neck and ask someone who wears a necktie to explain how to tie the knot. Do not allow the person you ask to demonstrate, or do it for you. *What problems are there with the explanation?*

Questions

1. (a) What daily habits do you have?
   (b) List four which are useful or helpful for remembering things.
2. (a) Do you, or does someone in your family always salt food before tasting it?
   (b) What problems might this cause?

Activity 3I

Sense and Nonsense

In order for information to be learned and stored in our long-term memory, it must have some significance to us.

Below you will find a series of numbers, each in a specific location. Have another student begin timing, and attempt to memorize the numbers and their positions. At the end of 5 min (or less if you like), close the book, and write down the numbers in their correct positions. *How many of the numbers did you place in the proper position? Why is this difficult?*

11	7	4	2	1
10	12	8	5	3
6	9	13	9	6
3	5	8	12	10
1	2	4	7	11

After you have completed the exercise, read the hint on page 115 and attempt the exercise again. *Was there an improvement? Explain your answer.*

Learning Disorders

Have you ever tried to learn the rules to a complex game, especially when the other players are quite familiar with them? How do you feel when you can't seem to learn, even when you are trying very hard? For some people, attempting to learn new words or concepts is even more frustrating. Although they have normal intelligence, and do not have any visual, hearing, or physical handicaps, it is difficult for them to listen, to read, to speak, to remember, or to work with numbers. Difficulties such as these are called **learning disorders**. They may include difficulty in remembering what has been said or read, or difficulty in writing, arithmetic, or spelling.

Many who suffer from a learning disorder have developed their own methods of coping with their learning problems. For example, individuals who have difficulty reading may ask someone else to read for them.

There are a variety of treatments which are used for learning disorders. In some cases of hyperactivity, medication is prescribed to reduce the hyperactivity and increase the individual's ability to concentrate on learning. Highly structured learning environments, which reduce distractions, are also used. A student who has difficulty writing may be tested orally.

In other cases, a multisensory approach is used to improve learning. For example, if new words are to be learned, the individual will trace the words with a finger, while looking at the words, and saying them aloud. This multisensory approach provides additional cues for help in learning. Individualized programs allow students to learn using the methods which work best for them.

Activity 3J

Distractions and Learning

Problem

Do distractions affect how you learn?

Materials

cassette tape player with headphones
cassette tape of vocal music and narration
two vocabulary lists of ten words and brief definitions which you do not know (Your teacher will provide you with a list, or have another student select some obscure words from the dictionary.)

Procedure

(Note: For this activity, the class will be divided into two groups. One group will follow the instructions in the order they are listed below. The other group will begin with steps 3 and 4, and then do steps 1 and 2.)

1. Examine the first list of ten words and definitions for 5 min, while listening to the tape. Another student will inform you of the time every 15 s.

2. At the end of 5 min, have another student ask you the meaning of each word. *Record the number of correct responses.*

3. Examine the second list of ten words and definitions for 5 min. Do *not* listen to the tape, and do *not* have another student inform you of the time every 15 s.

4. At the end of 5 min, have another student ask you the meaning of each word. *Record the number of correct responses.*

Questions

1. How did your scores compare?

2. What effect do distractions have on your ability to learn?

3. Considering the results of this experiment, where might be a good place to do homework and study?

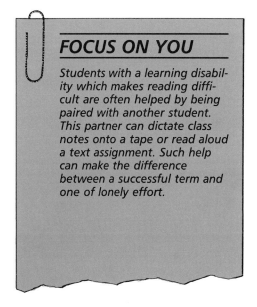

FOCUS ON YOU

Students with a learning disability which makes reading difficult are often helped by being paired with another student. This partner can dictate class notes onto a tape or read aloud a text assignment. Such help can make the difference between a successful term and one of lonely effort.

Activity 3K

Trial and Error Learning

Trial and error is a method of learning which can require a great deal of patience. When a problem is presented, a trial method of solving the problem is developed. This method is applied to the problem. If it does not solve the problem, then a different approach must be sought. This process continues until the problem is solved.

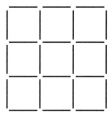

Using the trial and error method, take away eight of the lines so that there are only two squares left.

Questions

1. How many trials did it take to get the right answer?
2. Did you repeat any of the trials?
3. What would improve the rate of learning using this method?

See page 153 for the solution to this problem.

Activity 3L

Organizing Information

We live in a world which bombards us with information or data. No one can learn everything there is to know, let alone remember everything they ever see, hear, or read. It is important to be able to organize this information so that it is easy to use, to store, and to retrieve.

Design a table which will allow you to collect and organize information about student preferences in music. After collecting the information, how will you know what music is the most popular? How can you identify those students who have similar interests? How else could this information be organized so that it is easy to use? How would you organize a music store which has a large selection of LP's, cassette tapes, and compact discs? How are most music stores organized?

Activity 3M

Seeing, Hearing, Touching: How Best to Learn?

Our eyes, ears, and fingertips (sense of touch) are constantly sending information to our brain. Most of the time we take them for granted, and are not even aware that they are informing us of our surroundings. When we learn new things, the information enters through our senses. Is one route better than another for learning? To investigate this, work with a partner.

Procedure

1. Assemble a group of ten common objects and place them in a container so that your partner cannot see them. (For Part A)
2. Write the names of ten other objects on a piece of paper. (For Part B)
3. Assemble another group of ten objects and place them in a separate container so that your partner cannot see them. (For Part C)

Part A: Learning by Seeing (Visual Learning)

1. Inform your partner that you will remove ten objects, one at a time, from the container. Your partner will be allowed to look at each one carefully, then it will be returned to the container. After all ten have been viewed, you will ask your partner to name all ten objects (not necessarily in order).

2. Remove one object from the container and allow your partner to look at it for 1 min. Return the object to the container and repeat with the next one.

3. After all ten objects have been viewed, ask your partner to name all ten objects. *Record the responses, and determine the number of correct ones.*

Part B: Learning by Hearing (Aural Learning)

1. Inform your partner that you will read a list of ten objects, as many times as your partner wishes in 5 min. After the 5 min have passed, you will ask your partner to name all ten objects (not necessarily in order).

2. Note the time, and begin reading the list of objects. Read slowly and clearly. After completing the list, ask your partner to indicate when you should begin again.

3. After the 5 min have passed, ask your partner to name all ten objects. *Record the responses, and determine the number of correct ones.*

Part C: Learning by Touching (Tactile Learning)

1. Blindfold your partner. Inform your partner that you will place ten objects, one at a time in his or her hands, to be handled for 1 min. After all ten have been handled, you will ask your partner to name all ten objects (not necessarily in order).

2. Place one object in your partner's hands and allow your partner to handle it for 1 min. Remove the object and repeat with the next one.

3. After all ten objects have been handled, ask your partner to name all ten objects. *Record the responses, and determine the number of correct ones.*

Questions

1. *Which of the three trials produced the most correct responses?*

2. (a) *Where did any errors or omissions occur?*
   (b) *Is this surprising?*

3. (a) *Which of the three methods do you prefer for learning?*
   (b) *How could you use this to improve your learning and study habits?*

4. *Survey other members of your class to see which of the three methods produced the most correct responses.*

Hint for Activity 3I (page 112):
Look for a pattern in the numbers, beginning with the numbers 1, 2, and 3 in the top right-hand corner, or in the bottom left-hand corner. Once you determine the pattern, spend a moment to learn the positions of the numbers and repeat the exercise.

CHECKPOINT

1. *Explain how each of the following are involved in learning:*
   (a) experience
   (b) memory
   (c) habit.

2. *List three ways to improve learning.*

3. *Describe three different types of learning disorders.*

Chapter 3 / The Nervous System: In Control

BIOLOGY AT WORK

- PSYCHOLOGIST
- PSYCHIATRIST
- NEUROLOGIST
- PUBLIC HEALTH EDUCATOR
- PHARMACIST
- PROSTHETIST-ORTHOTIST

HEALTH CONCERNS

meningitis	amylotrophic lateral
cerebral palsy	sclerosis
spina bifida	hydroencephaly
multiple sclerosis	Alzheimer's disease

CHAPTER FOCUS

Now that you have completed this chapter, you should be able to do the following:

1. State the four general functions of the nervous system.
2. State the three major parts of the nervous system and give the major function of each part.
3. Sketch and label a diagram of the basic nerve cell, a neuron.
4. Describe how a nervous impulse is transmitted.
5. Explain how a nervous impulse crosses a synapse.
6. Identify the major parts of the brain and give at least one function of each part.
7. Describe how the brain and spinal cord are protected from injury.
8. List the two major functions of the spinal cord.
9. Explain how a reflex arc operates to reduce reaction time.
10. Contrast the functions of the two divisions of the autonomic nervous system.
11. List three of the organs of the endocrine system and describe the effect that they have on the body.
12. Name four general categories of drugs and describe, in general terms, the effect that each has on the body.
13. Explain how an organism's chances of survival are improved because of drives, fixed action patterns, and learned behaviours.
14. List three ways to improve learning.

SOME QUESTIONS TO ANSWER

1. Name the three main parts of the nervous system and give the main function of each.
2. (a) Draw a neuron and label its various parts.
   (b) Give a simple explanation of the function of each part.
3. Explain how a nerve impulse is transmitted along the membrane of a neuron.
4. (a) What is myelin?
   (b) Explain how myelin affects the passage of a nervous impulse.
   (c) What is the difference between white matter and grey matter?
5. Explain the meaning of the following terms:
   (a) reaction time   (d) concussion
   (b) synapse   (e) reflex act
   (c) meninges   (f) target organ.
6. List the major divisions of the brain and give an important function of each part.
7. How is the brain protected?
8. What changes are produced in your autonomic nervous system when a child unexpectedly runs into the road in front of your bicycle?

9. Which portion of the brain looks after each of the following activities?
   (a) arm movements
   (b) vision
   (c) co-ordination
   (d) heart rate
   (e) respiration rate
   (f) hearing
   (g) memory
   (h) sleep.

10. How do the glands of the endocrine system cause changes in body activities?

11. List six effects resulting from the release of adrenalin into the bloodstream during an emergency.

12. Why is the pituitary gland sometimes called the master gland of the body?

13. Name the two main parts of the pituitary gland and list the secretions of each.

14. Select one of the drugs listed in this chapter and describe how it affects the body.

15. Explain how a biological drive may improve the chances of survival for a species, giving an example.

16. (a) What is a mnemonic device?
    (b) Explain why it is an aid to learning.

SOME WORDS TO KNOW

Match each of the descriptions given in the left-hand column with a word shown in the right-hand column. DO NOT WRITE IN THIS BOOK.

1. Small gap at the junction between two neurons
2. Carries the nerve impulse away from the cell body
3. Protective membranes around the CNS
4. Part of autonomic nervous system that reacts to stress
5. Carries nerve impulses from receptors to CNS
6. Involuntary response
7. Co-ordinates muscular activities, balance, etc.
8. Largest portion of the brain, controls motor and sensory functions
9. Composed of the brain and spinal cord
10. Controls autonomic nervous system and the internal organs
11. Nerves outside of the brain and spinal cord
12. The chemical messengers of the body
13. A condition of the body causing it to become active
14. Promotes regeneration of damaged axons
15. A muscle or organ which responds to a stimulus

A hypothalamus
B drive
C axon
D central nervous system
E autonomic nervous system
F peripheral nervous system
G hormones
H dendrites
I neurilemma
J receptor
K sensory neuron
L motor neuron
M cerebrum
N cerebellum
O meninges
P reflex
Q synapse
R sympathetic nervous system
S effector

Select any two of the unmatched words and, for each, give a proper definition or explanation in your own words.

SOME THINGS TO FIND OUT

1. What happens when we sleep? Are there several stages in sleep? Investigate current theories.
2. What are biorhythms? Can they predict behaviour? As part of your research, try to work out your own biorhythm.
3. What is jet lag? What explanations are there to explain the human biological clock?
4. Find out what is meant by "body language". Do you use it? Can you determine what your friends are thinking from this process?
5. How does stress affect the body?
6. What progress has been made in the treatment of injuries to the brain or spinal cord?

SENSING THE EXTERNAL ENVIRONMENT

An air traffic controller must decide on the safest path for each and every aircraft. The controller relies upon a constant supply of information concerning the position, speed, and direction of approaching and departing planes, as well as other factors such as the weather. Failure to receive information could mean a collision and loss of life.

You, too, need a constant supply of information concerning your external environment. To know what to do, and when to do it, your body receives and processes signals from several different sources at once. This sensory network monitors everything from temperature (whether to shiver or sweat) to the approach of danger (whether to run or remain still). Quite simply, you couldn't survive without it.

KEY IDEAS

- *Control systems receive a stimulus, transmit and determine appropriate responses.*
- *The eye is the most important of our senses, receiving more than 80 percent of our information.*
- *The ear is responsible for our sense of hearing and balance.*
- *Chemoreceptors detect the presence of chemicals dissolved in body fluids.*
- *The skin contains receptors sensitive to touch, pressure, pain, and temperature.*

4.1 Processing Information

Our senses give us information about the world around us. The light entering the eye, sound waves entering the ear, a pinprick on the fingertip, or salt on the tongue, are examples of the information that our senses receive. These physical processes produce a **stimulus**. For a stimulus to be sensed, it must be received by the appropriate **receptor**. A receptor responds to a particular type of stimulus. For example, when light is shone into the ear, we don't see or hear anything.

Once the receptor has been stimulated, it converts the stimulus into an electrical message, a nerve impulse. This message is carried to the central nervous system along a sensory nerve, which is one form of a **conductor**. When the message reaches the CNS, it is delivered to an **interpreter**. The interpreter determines the meaning of the message. It is at this stage that we finally see, hear, touch, or taste.

Within the brain are specialized areas responsible for interpreting sensory information. (Refer to Chapter 3.) These areas also contain nerve cells which initiate any required response. If someone called your name, you might respond by turning around. Whatever the appropriate reaction, instructions are sent in the form of nerve impulses. This message travels along another conductor, a motor nerve, until it reaches the appropriate organ. This organ, which is usually a muscle, is called an **effector**. When it receives the message, it acts to produce a response. The response may cause other stimuli, which starts the whole process again. Together, the receptor, conductors, interpreter, and effector make up a **control system**. The many control systems of the body process information and act to maintain the best possible conditions for the body.

Activity 4A

Investigating Reaction Time

Using the procedure in Activity 2K on page 76 as a guide, design an experiment to investigate the effect of the dominant hand (right- or left-handedness) on how quickly a person reacts to a stimulus, their *reaction time*. What other factors affecting reaction time could you test? Here are some suggestions: whether the person is given a visual cue or a sound cue only; whether the person has performed the activity before (learning); whether the person is distracted by being asked to concentrate on more than one task at a time, etc. *State your hypothesis for each experiment. Record your observations in chart form.*

Questions

1. What effect does the dominant hand have on reaction time?

2. What generalizations can be made about reaction times?

3. What is the difference between the action you investigated in this activity and a true reflex? (Refer to Chapter 3.)

4.2 Vision: The External Structure of the Eye

The eye is, undoubtedly, the most important of our senses. A larger portion of the brain is devoted to vision than to any of the other senses. The eye responds to light energy, converting stimuli into impulses which are conveyed to the brain for interpretation. It is important to realize that unless impulses from the eye reach the brain, we do not really see. The eye may receive light, the receptors respond and send impulses, but if these impulses do not reach the visual centre of the brain, we are completely blind.

The Cavities That Contain the Eye

The eyes are contained within two hollow depressions in the skull, called the **orbital cavities**. The bones above the eye, as well as the nose and cheek bones, project forward and offer the eye some protection from large objects, such as basketballs and doorposts. Openings in the bones of the skull at the back of the orbital cavity provide a passageway for the arteries and veins that supply the eye as well as the optic nerve, which carries the nerve impulses to the brain. The orbital cavity also contains several other structures, including the muscles which move the eyeball, the tear-producing apparatus, nerves, blood vessels, and some fatty material. This fat helps to absorb shocks and provides a soft seating for the eyeball within the hard bone. The nutrient reserves stored in this fat may be needed during a severe illness. If such a need arises, the eyes take on a sunken appearance because the reserve of fat has been used.

Muscles of the Eye

The muscles found in the eye may be divided into two groups. Muscles on the outside of the eyeball serve to move the ball in its socket; muscles located on the inside of the eyeball control the lens and the iris. (See Figure 4.1.)

When the eyes are focussed on an object to one side of the body, both will move together in that direction. If the object is to the left, for example, the left eye will move toward the left side of the face away from the nose, and the right eye will move toward the nose. If you hold a finger at arm's length, then slowly move the finger in toward your nose, gradually both eyes will turn further and further inward until you appear "cross-eyed". Watch your partner's eyes as he or she performs this exercise.

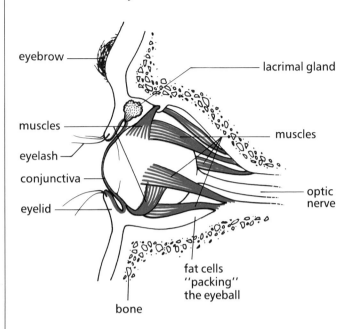

Figure 4.1
The external features of the eye and the muscles which move the eyeball in its socket

Protecting the Eye

The eyes are well-protected because they are so vital for survival. Above the eyes, on the crests of two bony ridges, are the **eyebrows**. The hairs of the eyebrows help to prevent dust and other falling particles from entering the eye. The ridge and eyebrow also act as a shading device in very bright light. We sometimes extend this ridge by raising the flattened hand above the eyes, to give added protection from very intense sunlight.

The **eyelids** are two very thin folds of skin which close to cover the eye completely, shutting out light or protecting the eyes from damage. The eyelids can be voluntarily controlled, as when we close the eyes or wink, but they may also operate automatically. They can close reflexively, faster than we can think, to protect the eye from an approaching object; however, the eye is not always able to sense very small objects such as fragments of metal. For this reason, you should always wear eye protection when working or playing where there is a chance of something getting into the eye. (See Figure 4.2.) What are some activities for which you should always wear eye protection?

The **eyelashes** are attached to the edges of the eyelids. These short hairs are strong enough to remain extended in a gentle curve. They protect the eyes by forming a fine screen to trap dust and other particles.

Figure 4.2
Always wear eye protection when there is a chance that something may strike the eye.

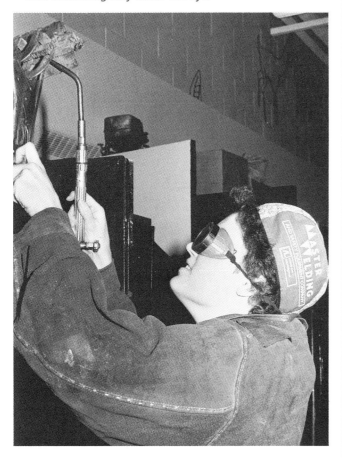

The Conjunctiva

An opening in the body must be protected from bacteria, and the eye is no exception. A thin, transparent membrane called the **conjunctiva** (con-junk-ti-vah) is attached to the edge of the upper lid. It covers the inside of the upper eyelid and then folds down to completely cover the eyeball. The membrane then doubles back to cover the inside of the lower eyelid. (See Figure 4.1.) It completely seals the eye from any bacterial agent trying to enter the body through the eye. The disorder conjunctivitis occurs when this membrane becomes infected and inflamed. One type of conjunctivitis, known as "pink eye", is very contagious.

The Lacrimal Glands

Tears are produced by the **lacrimal glands**, located above and to the outer edges of the eye socket. They produce a watery, salty, slightly germicidal fluid. Secretions of the lacrimal glands flow onto the surface of the eye through several tiny ducts to wash away dust particles and to lubricate the eye. Much of this fluid evaporates; the rest drains into the nasal chamber through the **lacrimal ducts** on the inner edge of the lower eyelids. (See Figure 4.3).

Figure 4.3
The lacrimal glands and the ducts that drain surplus fluid into the nasal cavity

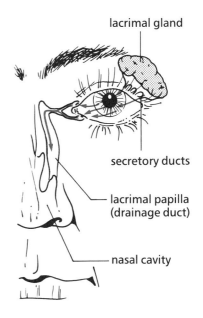

During periods of emotional stress, pain, or irritation, the flow of fluid increases under the direction of the autonomic nervous system. If this increased supply is greater than the tear ducts can drain away, the fluid spills over the lids as tears.

Artificial Tears and Care of Contact Lenses
Commercial solutions described as artificial tears are available. These products are intended to soothe the discomfort of dry, irritated eyes. Other products claim to reduce the redness in "bloodshot" or irritated eyes. Neither product is designed for long-term use, so if a problem persists, you should consult with your ophthalmologist.

Contact lenses are worn directly on the surface of the eye, separated only by the fluid coating the eye. Contact lens wearers must be careful to clean their lenses and follow the instructions they have been given for cleaning, storage, and handling. Lenses and lens cases which are not properly maintained can be a continual source of infection.

Some contact lenses, called extended wear lenses, are designed to be worn for up to a month without removing the lenses from the eyes. Regular lenses can only be worn during waking hours, usually because of problems associated with cleanliness and the amount of oxygen reaching the cornea. Extended wear lenses are not for everyone, and may increase the risk of corneal infections. An ophthalmologist should be consulted if a decision is made to purchase extended wear lenses.

Students wearing contact lenses still require safety goggles in the laboratory. The contact lens can hold a splashed substance against the eyeball, and cause serious damage.

CHECKPOINT

1. *List the five parts of a control system and give the function of each part.*

2. *What is the function of the muscles attached to the outside of the eye?*

3. *What structures protect the eye?*

4. *What is the purpose of the conjunctiva?*

5. *How do the lacrimal glands help to cleanse the eye?*

6. *How can improper use of contact lenses be harmful to the eye?*

FOCUS ON YOU

If something does get into your eye, do not rub it. This may cause permanent damage to the eye by scratching the eyeball, as well as making the particle more difficult to remove. Instead, close both eyes for a few minutes, and allow the natural flow of tears to wash out the particle. If this method does not bring relief, fill a clean medicine dropper with warm water and release several drops into the affected eye. This should flush out the particle. If the particle still remains, or if you feel pain in the eye, see a physician immediately.

FOCUS ON YOU

If a chemical is splashed into your eye, it must be rinsed away immediately. Hold your head with your eyes open directly under a gently running stream of warm tap water. Continue rinsing for several minutes. (In a laboratory, and in many workplaces, an eye wash station is available.) After rinsing, see a physician at once.

4.3 Vision: The Internal Structure of the Eye

The Iris

The **iris**, which is the coloured portion of the eye, ranges from blue to dark brown, including mixtures of hazel and green. The colours, which result from the presence of various pigments, are determined by genes inherited from our parents.

The iris contains two sets of smooth muscles. These act antagonistically, one set to open the iris, the other set to reduce the size of the opening. The inner muscles are circular; when they contract, they close like a ring, reducing the amount of light entering the eye. (See Figure 4.4.) The outer muscles radiate outward; as they contract, they draw the edge of the iris back, increasing the diameter of the opening (dilation), thus allowing more light to enter. If you are familiar with the workings of a camera, you will recognize the similarity behind the action of the iris and the diaphragm of a camera. (See Figure 4.5.)

The black centre of the eye, known as the **pupil**, is simply the opening in the iris. The eye is dark inside the opening, and it shows up as black, like the dark opening of a deep cave or tunnel. The size of the black opening changes as the muscles of the iris react to the change in the amount of light shining into the eye.

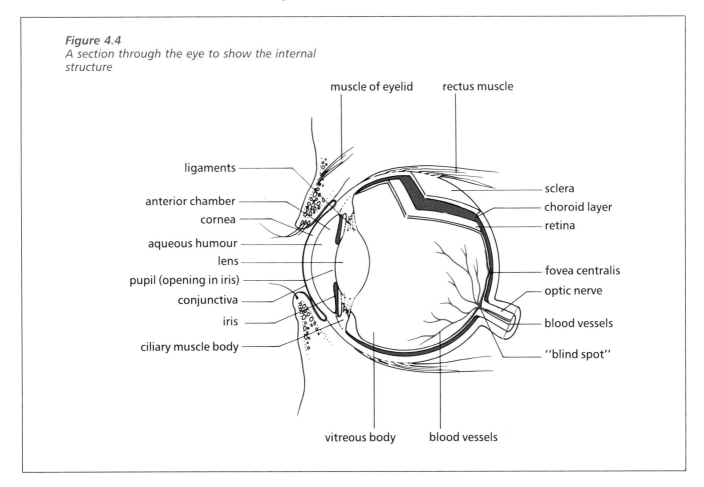

Figure 4.4
A section through the eye to show the internal structure

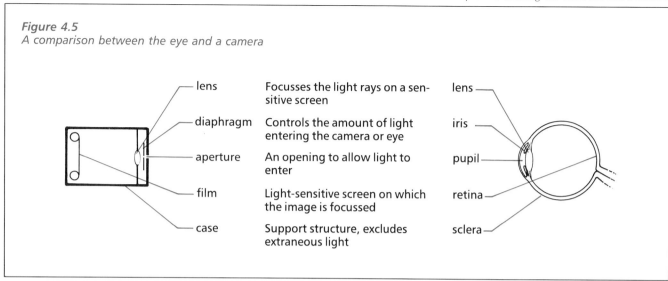

Figure 4.5
A comparison between the eye and a camera

The Lens

The **lens** is located immediately behind the opening of the pupil. It is a transparent disc-shaped structure, which is elastic and therefore capable of changing its shape. The interior is filled with a clear, jellylike fluid. The lens is held in place by tiny ligaments that are attached to the ciliary muscles. If these muscles were totally relaxed, the lens would be more spherical in shape. It is, however, held under tension by the suspensory ligaments and is thus pulled into a flattened disc with two convex surfaces. When the ciliary muscles contract, they release tension on the ligaments and allow the lens to assume its thickest, most convex shape. When the muscles relax, the ligaments pull the lens into a flattened shape with less convex surfaces. (Review the comparison of a camera and the eye in Figure 4.5.)

The Layers of the Eyeball

The wall of each eyeball is composed of three layers. The outer coat, called the **sclera** (sk-lair-ah), is a white, tough layer forming a case which helps to maintain the shape of the eye. Muscles are attached to this layer. At the front of the eye, the sclera becomes transparent and bulges forward to form the **cornea**. The cornea lacks blood vessels and is completely clear; its cells obtain the necessary nutrients from interstitial fluid rather than blood. The cornea contains very sensitive touch receptors.

The middle layer is the **choroid** (core-oid) **layer**. It contains many blood vessels and pigmented granules which prevent light from being reflected within the eye, much as the black paint inside a camera prevents stray rays of reflected light from spoiling the film. The inside layer of the eye wall is the retina. It lines the interior of the eye and is made up of special light-sensitive neurons. (See Figure 4.4.)

The Retina
The innermost layer of the eyeball is the **retina** (ret-i-nah). It contains two main types of cells – a layer of pigmented cells, and several layers of different kinds of nerve cells. The light-sensitive nerve cells respond to light energy entering the eye, and convert it into nerve impulses for transmission to the brain. The other nerve cells are primarily connector neurons which accept the nerve impulses and carry them to the optic nerve. (See Figure 4.6.) There are two types of light-sensitive nerve cells – the **rods** and the **cones**.

Figure 4.6
A section through the retina to show the rods and cones (the light-sensitive nerve endings of the eye)

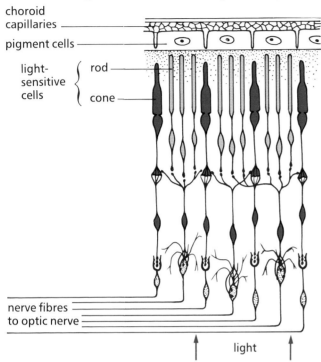

wavelengths, are usually the result of inherited disorders, such as red-green colour blindness.

Cones operate best under bright (daylight) conditions. They are much less sensitive than rods to low light intensity. Have you noticed your ability to distinguish colour is reduced at night or in a dimly lighted place? Witnesses describing car accidents that occur at night are frequently unable to accurately recall the colour of a vehicle.

At the back and very close to the centre of the retina is a small yellowish spot in which there are cones but no rods. This spot, called the **fovea** (foe-vee-ah) **centralis**, is an area of special sensitivity to detail. As you read this page, your eyes move in small jumps to keep the print focussed on the fovea. When you look closely at a scale, or the entry point of a sliver into the skin, you use this specialized location on the retina.

All the nerve fibres from the rods and cones collect at a point near the fovea and leave the eye as the **optic nerve**, which carries impulses to the brain. At the point where these fibres leave the retina, there are no rods or cones and hence no capacity for vision. This is therefore called the **blind spot** of the retina.

The **rods** are sensitive to dim light and are especially useful for night vision. There are more than 125 million rods in each retina; that should give you some idea of how small these neurons are. The rods contain a substance called **rhodopsin**, which requires vitamin A for its production. If the body lacks vitamin A, a condition called night blindness develops, in which the eye recovers more slowly than usual from changes in light intensity. For example, the oncoming glare of car headlights at night temporarily wipes out the visual capacity. The eyes of a person suffering from night blindness readjust to darkness very slowly.

The **cones** of the eye are responsible for detecting colour. The cones are less numerous than the rods and number about 7 to 10 million in each eye. Defective cones, which prevent a person from recognizing some combinations of light

Activity 4B

Dissection of an Eye

The eye of a cow or sheep is very similar to the human eye. This dissection will allow you to see the internal and external parts of the eye which are normally hidden from view.

Materials

cow or sheep eye (soaked in water overnight)
dissection tray
laboratory dissection kit
scalpel or single edge razor blade
plastic gloves
goggles

CAUTION! For this activity, you must wear eye protection and plastic gloves.

Procedure

1. Rinse the preserved eye thoroughly under running water and blot it dry with paper towelling. Place the eye in the dissection tray.

2. Examine and *describe the material on the outside of the eye.*

3. Use the forceps and scissors to carefully remove as much of the yellowish-white material that surrounds the eye as possible. *What is this material?*

4. Locate and *describe the following external parts of the eye*:
   (a) the *muscles* along the top, bottom, and side of the eye
   (b) the *optic nerve*, located at the back of the eye, about as thick as a pencil
   (c) the *sclera*, or outer covering, most easily seen at the front of the eye
   (d) the *cornea*, covering at the front of the eye (Note: May be cloudy in a preserved specimen.)

5. Use the scalpel to cut through the eye. (Do not push down on the eyeball.) Use light pressure and make repeated cuts until you penetrate the eye. Cut in a circle about half way between the cornea and the optic nerve. (See Figure 4.7.) Keep the front and back parts of the eye together until you complete the circular cut.

Figure 4.7
Use light pressure to make the first cut, and make repeated cuts until you penetrate the eye.

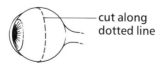

6. *Describe the liquid material found in the eyeball. What is this?*

7. *Describe the texture and thickness of the sclera.*

8. Examine the back half of the eye. Locate and *describe the following parts*:
   (a) *retina*, a thin film that may be folded over itself (Use the forceps to gently lift this layer.)
   (b) the *choroid coat*, a layer beneath the retina (Note: In a human eye, this layer is black.)
   (c) the *optic nerve* (blind spot), where the retina seems held to back of the eye.

9. Examine the front half of the eye. Locate and *describe the following parts*:
   (a) *ciliary muscles* and *ligaments*, found attached to the lens
   (b) *lens*, located in the centre of the eye. Remove the lens and look through it. Try to read through it. Squeeze it very gently. *What happens? Sketch the shape of the lens from the front and from the side.*
   (c) *aqueous humour*, fluid that fills the front chamber
   (d) *pupil*, the opening in the iris
   (e) *iris*, tissue in front of lens
   (f) *cornea*, clear layer at the front of the eye.

Activity 4C

Peripheral Vision

If you look straight ahead, you will find that you can still notice objects, especially moving ones, out of the corner of your eye. This is your **peripheral vision**.

Materials

large protractor or Vision Disk
coloured paper, 5 cm × 5 cm (black, white, red, green, yellow, etc.)
tongue depressors (or similar pieces of wood)
tape

Procedure

1. This activity requires that you work with a small group. Prepare the materials by taping a piece of coloured paper to the end of each tongue depressor. Decide who will serve as the subject for the experiment.

2. Have the subject sit in a chair, looking straight ahead. It is very important that the subject looks straight ahead. This is most easily done if the subject focusses on a target directly in front of her or him. One member of the team will act as the recorder. The recorder should stand to the subject's left side and place the large protractor above the subject's head, so that the mark indicating 0° is in line with the subject's nose, and 90° is approximately over the right ear.

3. Ask the subject to indicate when she or he first notices something moving in her or his peripheral vision, and also to indicate its colour.

4. Another member of the team should stand behind the subject and begin by selecting a coloured piece of paper. The paper should be held by its "handle". The piece of paper should be held at arm's length directly behind the subject's head, at eye level.

5. Move the paper slowly toward the front of the subject, keeping it approximately an arm's length from her or his head at all times. Move the paper up and down *slightly* as you go.

6. When the subject indicates that she or he can see the paper, the recorder should sight along the protractor to determine the position of the piece of paper, and *record its position*. The protractor should be read with 0° starting at the subject's nose. *Record the colour the subject states for the paper in this location.*

7. Continue moving the paper until the subject can determine the correct colour of the paper. *Record this position.*

8. Repeat steps 2 to 7 using at least three other colours of paper. *Record your observations.*

9. Repeat the experiment with the same subject, but reverse sides. *Record your observations.*

Questions

1. How large is the subject's field of vision for movement? (Add the number of degrees for the left eye and right eye results together.) Compare your answer to those from other subjects.

2. What is the field of vision for each colour that was used?

3. Why is the field of vision different for movement and for colours?

4. What problems might be experienced by someone who has limited peripheral vision?

Activity 4D

Discovery of the Blind Spot

Where the optic nerve leaves the eye, there are no light-sensitive cells and each eye has a "blind spot" at this location. (Refer to the diagram of the eye.)

Procedure

1. Hold your textbook about 45 cm away from your face so that the cross is directly in front of your right eye.

Close your left eye, then slowly move the text toward you. Keep your right eye on the cross;

you will be aware of the dot, but do not look at it directly. At some point, the dot will disappear. *Measure the distance from your eye to the book at the point where the dot disappears. Repeat with the other eye. In your notebook, explain what the blind spot is and record your results.*

2. Another method of discerning the location of the blind spot is to use the line of letters below. Hold the book as described in step 1.

   + A B C D E F G H I J K L M

   Close the right eye. Look at the cross, then slowly read the letters from left to right. At some point, the cross will disappear. *Note which letter you have reached when the cross disappears. Is this consistent with the results of others in the class? Did each student keep the book the same distance from the eye?*

CHECKPOINT

1. Prepare a table in your notebook with the following headings:

Structure	Location	Description	Function

   SAMPLE ONLY

   Complete the table for the following structures: iris, pupil, lens, sclera, cornea, choroid layer, retina, fovea centralis, optic nerve.

2. Explain the difference in location, structure, and function of the rods and the cones of the retina.

3. (a) What is rhodopsin?
   (b) Why is it important?

4.4 Vision: The Characteristics of Sight

The Reflexes of the Eye

There are two reflexes involving the pupil of the eye. The **light reflex** is well-known and can easily be demonstrated by shining a small light on the eye and then removing it. The pupil will change in size, growing smaller in brighter light and increasing in diameter as the light dims.

The second pupillary reflex, **accommodation**, also results in a change in the size of the pupil but occurs in response to changes in the distance from the object being viewed. If the eyes are fixed on a distant object, then shifted to view a nearby object, the pupils will decrease in size. It has been estimated that the eyes make about 100 000 accommodation adjustments every day.

Adaptation

If you walk into a dark theatre, your eyes require some time to adjust to the dim light, so that you can find your way to a seat. If you go out of a dimly lit cabin into a bright snowy landscape, you will find that your eyes also require time to adjust to the intensity of light. The eye must shift the active reception of light from rods to cones or vice versa. This ability is known as **adaptation**.

Activity 4E

The Pupillary Reflex

Have your partner close his or her eyes for about 2 min. Observe the pupils when you ask your partner to open his or her eyes. *Describe what happens and explain this reflex.* Shine a soft light into the eye for just a few moments. *Observe and record the results.*

Activity 4F

The Accommodation Reflex

Have your partner look at a distant object, perhaps across the room or out of the window. Examine the size of the pupil. Now ask your partner to look at a book placed about 20 to 30 cm in front of his or her face. Examine the pupil size. *Explain any differences in pupil size that you observe.* Repeat the experiment, but this time observe any movements of the eyeballs within their sockets. *Record your observations and explain why this action takes place.*

Binocular Vision

If you look at a photograph taken with a camera, it appears flat, in two dimensions only. A skillful photographer can give an impression of depth, but in order to make accurate judgments about distance and depth, we require two eyes producing two images. We need **binocular vision**. (See Figure 4.8.)

When you close one eye (**monocular vision**), the brain receives only one set of impulses. When both eyes are open, two sets of impulses are sent to the brain. Because the eyes are set a short distance apart, two slightly different images are perceived. Perhaps you have seen stereoscopic slides projected or have used stereoscopic lenses in a geography class to look at **stereoscopic** aerial photographs. The two pictures taken from slightly different positions give a very realistic, three-dimensional effect.

Look carefully at Figure 4.9. You will see that nerve fibres from the side of each eye nearest to the nose cross over to the visual centre on the opposite side of the brain. Light striking the lateral side of each retina is carried to visual centres on the same side of the brain. This means that the objects seen out of the right side of each eye are interpreted by the same side of the brain and create a single visual impression, in a way similar to the images of stereoscopic pictures. The recognition and interpretation of these two sets of visual stimuli enables the brain to establish depth and distance relationships.

If you use a telescope to sight an object, you must use one eye. The "line of sight" will allow

Figure 4.8
Convergence. In order to see objects very close to us, it is necessary for the eyes to turn inward (converge). Objects at a great distance are viewed with the eyes held approximately at right angles to the facial plane of the body.

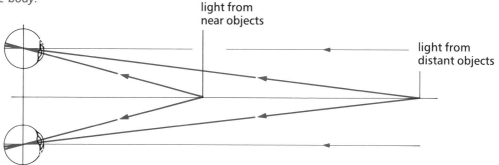

Figure 4.9
This diagram illustrates that light coming from a source on the right-hand side (black) will be viewed by portions of the left side of each retina. Nerve impulses from these areas of both eyes are then co-ordinated in the left visual centre of the brain.

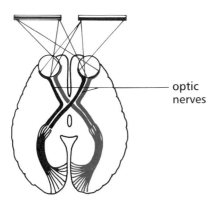

you to see the object, but you will not have a very good indication of exactly how far away the object really is. The other eye would give you a second line of sight. The point at which these two lines meet and cross gives an excellent indication of the relative position of the object. Stand a test tube up in a beaker and place it at arm's length. With both eyes open, try to place the pencil in the test tube. Now with one eye closed, try the same test. You will note how much more accurate distance perception becomes when two eyes are used rather than only one.

CHECKPOINT

1. How are the light reflex and the accommodation reflex similar? How are they different?
2. Why does it take a few minutes before you can see when you walk into a darkened room?
3. Why do we need binocular vision? Give an example.

4.5 Vision: Measuring the Ability to See

The lens adjusts to allow us to focus on nearby or distant objects. It is "elastic" and can be stretched or relaxed by the ciliary muscles that surround it. When we look at nearby objects, the ciliary muscles contract and the lens bulges to become more convex. This causes greater bending of the light rays and brings the image into sharp focus on the retina. To view distant objects, the ciliary muscles relax and the lens assumes a flatter, less convex shape, which does not bend the light rays as sharply. (See Figure 4.10.)

The light rays, whether from nearby or distant objects, must be focussed on the retina. If the eyeball loses its round shape and becomes too long, the light rays will focus before they reach the retina, resulting in *near-sightedness*. Sometimes the eyeball is misshapen so that it is too short and the light rays will focus behind the eyeball, resulting in *far-sightedness*. The use of an artificial lens in front of the eye can make adjustments for those small differences and enable people to achieve normal sight. (See Figure 4.10.)

FOCUS ON YOU

After treating an accident victim who had suffered cuts to the cornea, a physician in the Soviet Union noticed that his patient's vision improved. It was discovered that making tiny cuts to the surface of the cornea changes the shape of the eyeball. Surgery to make such cuts can cause light entering the eye to focus on the retina without the aid of corrective lenses. This operation can cause permanent damage to the eye; however, the success rate is quite high.

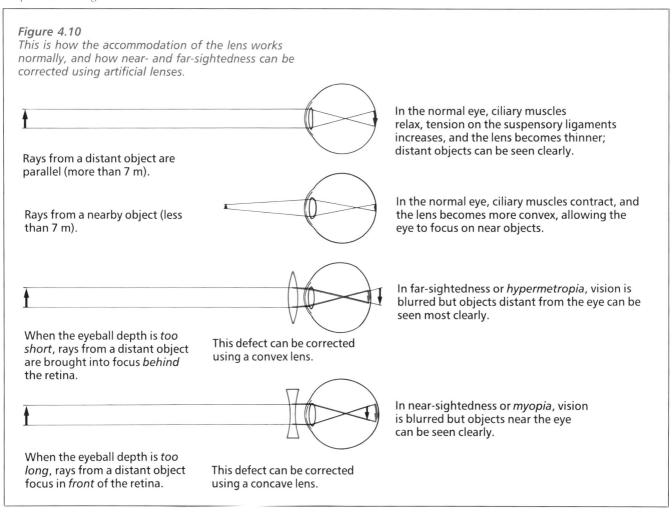

Figure 4.10
This is how the accommodation of the lens works normally, and how near- and far-sightedness can be corrected using artificial lenses.

Other lenses are available which alter the light passing through them. The light reflected from metal, glass, water, or snow contains scattered light waves. Polarized lenses (such as ski goggles or sunglasses) allow only some of these waves to pass, thus reducing the glare of reflected light.

Coloured lenses absorb some of the visible wavelengths of light, and allow others to pass through. For example, a red lens will filter out most visible wavelengths, but allow the red wavelengths to pass through. Coloured contact lenses can be used to enhance or change the apparent colour of a person's eyes.

Ultraviolet light is an invisible, yet powerful component of natural light. Ultraviolet wavelengths cause sunburn, and can damage the tissues of the eye. Sunglasses with lenses which filter out the ultraviolet wavelengths should be purchased to protect the eyes.

FOCUS ON YOU

Until recently, visual acuity was measured in feet and referred to as 20/20 vision. With the introduction of metrication, 6/6 vision is gradually being adopted.

BIOTECH

School and the Visually Impaired Student

Try this simple experiment. Hold a single layer of thin, white cloth, like a T-shirt, over your eyes. Look around and try to describe nearby objects. Try to read a book. While this is only a simulation, it will give you the feeling of what it can be like to be visually impaired.

Some children are born with a serious vision problem, for example, as a result of their mother contracting Rhubella (German measles) while pregnant. Others inherit degenerative diseases which affect the retina. The most common is retinis pigmentosa, a condition in which night vision, then peripheral vision, and eventually central vision is greatly decreased. In many cases, vision, even with corrective lenses, is limited to 6/60, which means that the person sees at 6 m what the average person sees at 60 m. What problems are faced by those who are visually impaired? What special problems would a visually impaired student have?

One way in which a person can adapt to a visual impairment is by increasing the use and sensitivity of the other senses, such as hearing and touch. Listening and memory skills also become more important.

Modern technology can also help. Visually impaired students can be provided with tape recorders for "audio" note tak-

ing. Books printed in large type editions are available for those students who can read the large print, some with the aid of special lighting or magnifiers. For others, special viewing systems magnify printed materials such as books, newspapers and pictures up to 60 times. A closed-circuit television camera is used to scan the material. The magnified image is then projected onto a screen for viewing.

Braille is an alphabet composed of raised dots a person "reads" by touch. Printed materials, such as textbooks, must first be translated into Braille before a visually impaired student can use them. Fortunately, new computer software programs can transfer whatever text appears on a computer monitor to a printout in Braille for the student. The student can do her or his work on a keyboard with Braille keys and instruct the computer to print a copy in type for the teacher to mark.

A visually impaired student may need assistance to become familiar with new surroundings. In some cases, a teacher with special training may work with the student to help with the student's orientation to the school and new classroom situations. Other students may assist by translating visual cues and information into verbal messages. What other kinds of assistance might be necessary?

A visual impairment may cause someone to make some adjustments in the way that they learn, and in the way that they live; however, like any other handicap, it never stands in the way of someone with a desire to achieve.

Visual Acuity

Most people have used a Snellen Eye Chart, which has a series of letters with the largest letters at the top and lines of successively smaller letters in rows beneath. **Visual acuity** refers to the sharpness with which detail is seen. It is measured by demonstrating the size of type that you can read at a standard distance of 6 m. If you are able to read the line of letters that a person of normal vision can read at 6 m, you are rated at 6/6. If you have poor visual acuity, you may only be able to read at 6 m what a person with normal vision can read at 30 m, and you would be rated as 6/30. It is not uncommon to find students with visual acuity better than 6/6. Because there is frequently a difference between the left and right eyes, the test is taken using each eye alone, by covering one while the other is used for reading the chart.

Astigmatism

Another test given is for **astigmatism** (a-stig-ma-tizm), a condition caused by an uneven curvature of the cornea or lens. Light waves passing through the lens or cornea are distorted, giving an image that may have parts in focus and other parts out of focus. A diagram, like the one shown in Figure 4.11, is used to determine what part of the cornea is affected.

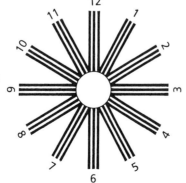

Figure 4.11
A diagram like this is used to help determine if a person has astigmatism. If they do, some lines will appear to be in sharp focus while others will be blurred.

Colour-blindness

Colour-blindness affects males more frequently than females. About six to eight percent of males are colour-blind, whereas only 0.4 to 0.6 percent of females have this defect. Total colour-blindness is usually caused by a lack of colour receptors or a defect in the visual centre of the brain. Red-green colour-blindness is more common and may be due to a defect in the colour receptor cones for red and green. Such persons are unable to take jobs where these colours are important, for example, as train engineers or airplane pilots.

Activity 4G

The Snellen Eye Chart: Visual Acuity

This test determines the sharpness or visual acuity of the eyes. The chart uses a sequence of black letters of varying size. Beside each line is a number that represents the distance at which that letter can be read by the normal eye. If you read the line marked 6 at a distance of 6 m, then you have 6/6 vision in that eye. Your vision is within the normal range. If, however, you can only read the line marked 12 at a distance of 6 m, then you have 6/12 vision. Many young people can read line 4 at 6 m, thus giving them a visual acuity of 6/4 (better than average vision).

Procedure

Have your partner stand 6 m from the Snellen Eye Chart. Cover one of her or his eyes with a piece of card and determine the smallest line that she or he can read clearly. Repeat using the other eye. *Record your results.*

Students who use glasses should complete the experiment with and without their glasses and *record both results.*

Questions

1. Explain the meaning of visual acuity and the meaning of 6/6 vision.
2. What is the minimum acceptable standard of visual acuity for driving without using glasses?

Activity 4H

Colour-Blindness

Colour-blindness is a genetically inherited abnormality resulting from the deficiency of a particular gene on the X-chromosome. The most common type of disorder is red-green colour-blindness. If the red cones are lacking, wavelengths of red light stimulate the green cones and little differentiation between red and green is possible.

Procedure

Ishara Test (using Ishara Test booklet)
Use the colour-blindness test booklet and follow the directions provided at the front of the book. *Record the numbers that you see and explain the significance of these numbers. The numbers are interpreted in the booklet.*

Holmgren Test
In this test, a set of coloured wool strands is used. The subject attempts to match each of the strands with an identical strand in the folder. Mismatching (or hesitation in matching) usually indicates a colour abnormality.

Activity 4I

Near Point Accommodation

This test determines the near point accommodation of the eye.

Materials

metre stick
10-cm ruler

Procedure

Hold a metre stick horizontally, with one end just under one of your eyes. Close the other eye. Have your partner slide a 10-cm ruler along the metre stick toward you, starting at a distance of about 60 cm. Keep the top edge of the ruler, which should be raised just above the edge of the metre stick, in focus and note the distance at which it first becomes blurred. Move the ruler away, until the ruler's edge is sharply in focus again. *Record this distance.* Repeat with the other eye.

The results of this test depend on the resiliency of the lens and its ability to change its shape. This ability decreases with age.

Age in Years	Near Point Accommodation
10	7.5 cm
20	9.0
30	11.5
40	17.2
50	52.5
60	83.3

Examine the chart and *note the age at which a dramatic change occurs.* Have you noticed how middle-aged people have to hold books or papers at arm's length in order to be able to read them? *Explain in your notebook what is meant by near point accommodation, how it can be tested, and your results for each eye. Is one eye stronger than the other?* Test your parents, brothers and sisters, or grandparents and see if your results conform to those in the table.

The term "accommodation" is also used to describe the ability of the eye to focus on near and distant objects. *What muscles are used for each of the two kinds of accommodation?*

Optical Illusions

You have probably heard the saying "seeing is believing". Most of the time, we can believe what we see; however, there are times when our eyes seem to play tricks on us. For example, if you look at the rising full moon, it seems incredibly large. But if you look at it a few hours later when it is overhead, it seems to be smaller. Now you know that the moon can't change its size, so how does this happen? When the moon is low on the horizon, there are other objects, such as trees and buildings, which can be seen in the same view. We see the moon with this frame of reference, and it appears quite large. As the moon rises overhead, we see it without the other objects to affect our frame of reference. This is an example of an **optical** or **visual illusion**. Careful examination will usually show that what we perceive in such illusions is different from reality. Figure 4.12 presents some optical illusions. Examine them carefully, and then read the caption for each. Does your perception change?

Figure 4.12
Optical illusions. (a) The three cylinders are the same size. (b) Distances ab, bc, de, and ef are all equal. (c) The two figures in A are identical. The centre circles of B are the same size. (d) The right-hand circle is the same shape as the one on the left. (e) The two diagonals are the same length.

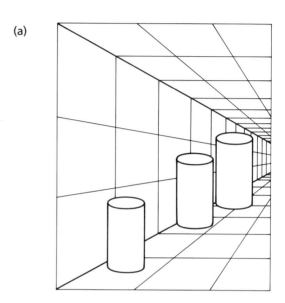

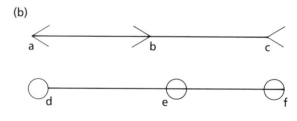

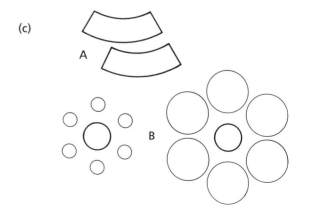

CHECKPOINT

1. *Explain how the lens is able to focus on nearby or distant objects.*

2. *What is the cause of near-sightedness?*

3. *What should you remember when shopping for sunglasses?*

4. *Explain what is meant if a person has visual acuity of 6/6.*

5. *What is the cause of colour-blindness?*

Chapter 4 / Sensing the External Environment

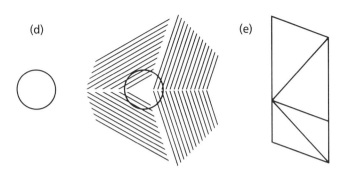

4.6 The Ear and Hearing

Hearing is an important part of our early warning system. It is sometimes called the "watch dog" of the senses, because it is the last to relax before we go to sleep and the first to come on duty, even before we are really awake. It is through the ear that we learn to communicate and develop our skills of speech, by listening to word sounds made by others. The ear enables us to judge distance and location of sounds, whether we hear thunder at a distance or a car horn close behind us.

The ear not only recognizes sound vibrations, converting them to nerve impulses for interpretation by the brain, but also functions as the organ of balance. It enables us to recognize our position, whether vertical or horizontal, and to maintain a steady, walking pace without falling. The ear is organized into three distinct sections – the external, middle, and inner ear. (See Figure 4.13.)

The External Ear

The **external ear** is the part that can be seen on the side of the head. It consists of a flap of skin and cartilage called the **auricle** (o-re-kul) or **pinna**. The shape of this structure is designed to collect sounds and funnel them into the ear canal. Sometimes if we have difficulty hearing some quiet sound, we use the hand to "cup" the ear. This extends the effective size of the pinna and increases the amount of sound energy entering the ear. The channel leading into the head from the pinna is the **auditory canal**. It passes through the temporal bone of the skull. The auditory canal is lined with a thin layer of skin that contains fine hairs as well as cells which secrete a waxy substance (cerumen glands). This wax helps to protect the eardrum, although sometimes it can accumulate in such large amounts that it partially prevents the sound waves from reaching the eardrum.

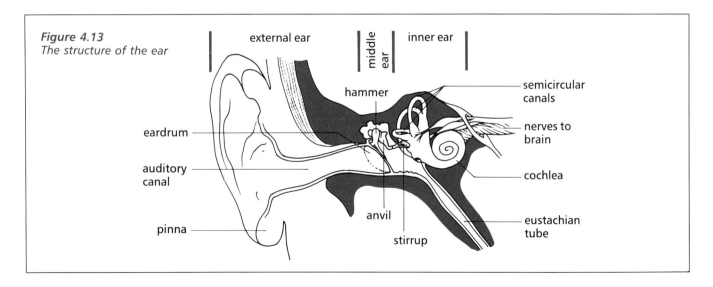

Figure 4.13 The structure of the ear

The Middle Ear

The auditory canal is closed at the inner end by a membrane which completely prevents entry into the head. The membrane is known as the **eardrum** or **tympanic** (tim-pan-ik) **membrane** and it separates the external ear from the **middle ear**.

How Sound Travels

Before explaining the operation of the middle ear, we need to review a few simple ideas about what sound is and how it travels. Sound is produced when something vibrates. It may be from a ruler, extended beyond the edge of the desk, which is plucked to produce a sound, or it may be the vibration of a fan or motor. In order for this vibration to be conducted from its source to our ear, it must pass through a medium, something that transmits the energy of the vibrations to the ear. The medium could be materials such as air, water, or metal. Usually, this energy is carried by air molecules. The wave of energy, which passes from molecule to molecule, finally causes the molecule resting against the eardrum to vibrate. This sets the eardrum in motion, thereby passing the vibrations on to the middle ear. (See Figure 4.14.)

Figure 4.14
How sound reaches the ear. When the ball at A is allowed to fall, the energy is passed from one to another of the suspended balls until the ball at B, which is free to move, is pushed away from the others. In a similar way, the energy of a vibrating source is transferred to the molecules of the air and passed from one molecule to another until it reaches the eardrum and causes it to vibrate.

Activity 4J

Sound and Hearing

In this activity, you will investigate some sound wave phenomena.

Materials

tuning fork
oscilloscope

Procedure

1. Strike the tuning fork on your knee or against the heel of your hand. Hold it at a reasonable distance from your ear, so that you can hear it easily, then rotate it. *What happens to the volume of the sound? Explain why this happens.* Move the sounding tuning fork in a circle around your head and *decide where you can hear it best and where you do not hear it as well. Explain why this is so.*

2. Use an oscilloscope to represent the wave form of the tuning fork. Compare the sound and wave forms of several different tuning forks. *What is the relationship between pitch and the shape of the wave form?*

The chamber of the middle ear is filled with air. If the tympanic membrane is to vibrate freely with an incoming sound, the air pressure must be equal on both sides of the membrane, that is, the pressure in the chamber of the middle ear must equal that in the outer ear. You may have experienced a "popping" of the ears while you were riding in a subway train or aircraft, or even during a car ride when you were climbing steep inclines in a hilly part of the country. This popping is caused by an increase or decrease in pressure on the outside of the eardrum, which means that the membrane will bend inward or outward but cannot move freely. To equalize the pressure, you may swallow or yawn. When you swallow, you allow air to move in or out of the **eustachian** (yoo-stae-shun) **tube**, which connects the middle ear with the throat. The air pressure in the throat is equal to that outside the body and in the outer ear. Swallowing thus tends to equalize the pressure in the middle ear.

Within the middle ear are three tiny bones (ossicles) which are held in place by tiny ligaments. (See Figure 4.15.) These bones conduct the vibrations from the tympanic membrane across the air space of the middle ear to the **oval window** which communicates with the inner ear. They also act to amplify the sound energy by 20 times before it strikes the oval window.

The Inner Ear

The **inner ear** has two functions. It converts sound vibrations into nerve impulses which are transmitted to the brain. It also aids in the maintenance of balance. The **utricle**, **saccule**, and **semicircular canals** are responsible for maintaining balance and body position. The **cochlea**, which performs the hearing functions of the inner ear, is responsible for transforming vibrations to nerve impulses.

The Cochlea

The cochlea is shaped like the shell of a snail. It is like a long, fluid-filled funnel that has been rolled up from its pointed end. In the middle of this funnel is a membrane which holds tiny hairlike receptor cells. This is the **organ of Corti**. Its hairlike cells connect with small nerves which join together and leave the ear as the **auditory nerve**. At the larger end of the cochlea are the **oval** and **round windows**. (See Figure 4.16.)

Hearing Sound

As the oval window receives the vibrations from the middle ear, the window pushes on the fluid inside the cochlea. A fluid wave moves along one side of the cochlea, reaches the end, and then bounces back along the other side. As the wave reaches the end, it meets the round window, where the movement of the wave is absorbed.

The movement of the fluid wave causes the tiny hairlike projections of the organ of Corti to vibrate. The hairs vary in length, and are stimulated by vibrations of different wavelengths. Low notes cause the longer hairs to vibrate, and high notes cause the shorter hairs to vibrate. This stimulates the nerve cells to which they are attached. A nerve impulse is then sent along the auditory nerve to the auditory centre in the brain. It is at this point that the vibration is "heard" as sound.

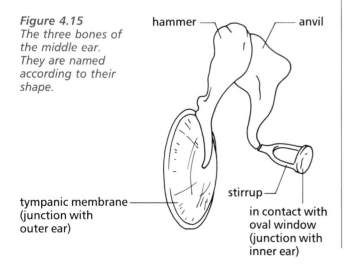

Figure 4.15
The three bones of the middle ear. They are named according to their shape.

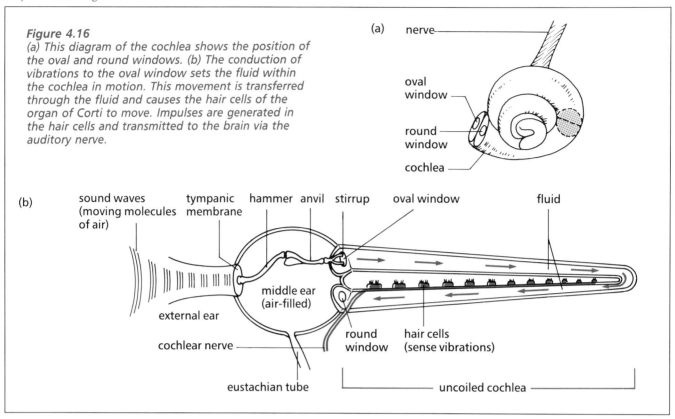

Figure 4.16
(a) This diagram of the cochlea shows the position of the oval and round windows. (b) The conduction of vibrations to the oval window sets the fluid within the cochlea in motion. This movement is transferred through the fluid and causes the hair cells of the organ of Corti to move. Impulses are generated in the hair cells and transmitted to the brain via the auditory nerve.

Humans have an approximate hearing range of 16 Hz (vibrations per second) to 20 000 Hz. A piano can produce sounds ranging from 16 Hz (lowest note) to 4186 Hz (highest note). Most human speech is pitched at about 1000 Hz.

CHECKPOINT

1. Prepare a table in your notebook with the following headings:

Structure	Location	Description	Function

Complete the table for the following structures: auricle (pinna), auditory canal, tympanic membrane (eardrum), eustachian tube, ossicles, oval window, cochlea, round window.

2. What are the two functions of the inner ear?

3. Name the organs which are responsible for maintaining balance and body position.

4. Describe what happens as a sound vibration pushes on the oval window.

5. What is the function of the hair cells in the organ of Corti?

4.7 Hearing Loss

There are several causes of hearing loss or **deafness**. **Conduction deafness** is caused by some interference in the transfer of sound waves to the middle ear. It may be the result of something as simple as a buildup of wax. It also could be caused by damaged or scarred tissue on the tympanic membrane, or by some defect in the action of the bones of the middle ear. **Nerve deafness** is caused by some problem of the nerve cells, either of the sensory cells in the inner ear or the nerve pathway to the brain. This is the type of deafness which occurs as a result of exposure to very loud sound.

Loud Sounds and Hearing

The frequency of a sound determines its pitch, as heard by the ear. The magnitude, or intensity, of a sound determines its loudness. Loudness is recorded in decibels (dB), with 0 dB being the threshold of audible sound for humans. Each increase of 10 dB on this scale represents an increase in actual loudness of 100 times.

Ranges in the 80+ dB range can be uncomfortable, and higher intensities cause physical pain and may damage the ear. (See Figure 4.17.) Musicians depend upon their sense of hearing, yet those who use powerful electronic amplification frequently suffer ear damage that affects their range of hearing, as well as their ability to pick up the fine overtones.

People who have some auditory impairment frequently increase the volume on their sound equipment to compensate for their hearing loss, and are not aware of the discomfort this brings

Figure 4.17
A comparison of some common sound levels experienced in everyday life. (Note: The bel is the unit of sound intensity in recognition of the work of Dr. Alexander Graham Bell, the inventor of the telephone and a man who devoted much time and effort to those suffering hearing defects.)

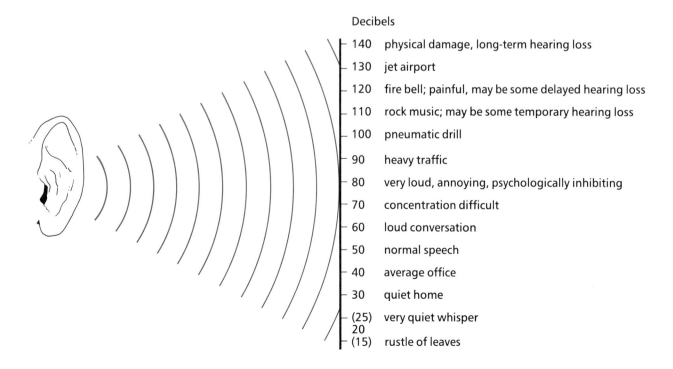

Decibels
- 140   physical damage, long-term hearing loss
- 130   jet airport
- 120   fire bell; painful, may be some delayed hearing loss
- 110   rock music; may be some temporary hearing loss
- 100   pneumatic drill
- 90   heavy traffic
- 80   very loud, annoying, psychologically inhibiting
- 70   concentration difficult
- 60   loud conversation
- 50   normal speech
- 40   average office
- 30   quiet home
- (25)   very quiet whisper
- 20
- (15)   rustle of leaves

to others. It has been well-established that rock music, with its constant high volume of sound, has already caused irreparable damage to many of its fans. A study of 1410 students found that 60 percent (about 850 students) had significant hearing loss in the high-frequency range.

The use of stereo headphones can contribute to hearing problems. They are capable of producing sounds above 100 dB. The average user listens at 88 dB, the volume level of rush hour traffic. Canadian workplace standards require hearing *protection* if a person is exposed to similar levels. To protect your hearing and continue to be able to hear *all* the music, keep the volume adjusted to about the level of a normal conversation.

Activity 4K

How Well Do You Hear: Auditory Acuity

A quiet room is required for these experiments. Perhaps a laboratory preparation room can be made available for these tests.

PART A

CAUTION!
Be careful not to insert the cotton plug too deeply within the ear. Always use a fresh, clean piece.

Your partner will plug one of her or his ears with clean cotton batting. Have your partner sit down with eyes closed. Hold a watch, with a distinct tick, near the unplugged ear and move it slowly away from the ear until she or he can no longer hear it. *Measure and record the distance of the watch from the ear.* Now hold the watch beyond the point at which your partner could hear the tick, and bring it closer and closer until she or he can hear the watch again. *Measure and record the distance. Are these distances the same? What might account for any differences? Repeat the experiment testing your partner's other ear, using* clean *cotton batting. Record your observations.*

PART B

Use an audio generator and a set of headphones to test your auditory acuity in the high- and low-frequency ranges. Before beginning this activity, check the volume level at a frequency of 1000 Hz. Adjust the volume to the level of a normal conversation.

Place the headphones on your head and sit facing away from the audio generator. Your teacher will begin with a frequency of 10 Hz and increase the frequency gradually. Raise your hand when you first hear a sound. *Record the frequency.*

Repeat this test beginning with a frequency above 20 000 Hz, and decrease the frequency gradually. *Record the first frequency you are able to hear.*

Questions

1. How did your hearing acuity (as measured by this activity) compare to that of your classmates?

2. How would you account for the results for your class? (To answer this, refer to the sources of loud sounds, if any, they routinely encounter.)

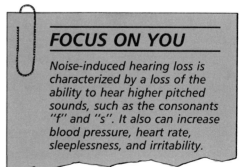

FOCUS ON YOU

Noise-induced hearing loss is characterized by a loss of the ability to hear higher pitched sounds, such as the consonants "f" and "s". It also can increase blood pressure, heart rate, sleeplessness, and irritability.

CHECKPOINT

1. (a) What is the unit used to measure the loudness of sound?
   (b) List three sounds you commonly hear on the way to school and estimate their loudness using Figure 4.17.
2. Explain the difference between conduction deafness and nerve deafness.
3. What precautions can you take to prevent nerve deafness?

4.8 The Organs of Balance

If you have ever been on a rapidly revolving ride at a fairground, or just spun around quickly, you will know that, for a moment or two after stopping, your sense of balance has been thrown off and you become giddy or disoriented. Two separate organs of the inner ear are responsible for maintaining balance. One is a fluid-filled structure that is divided into two parts called the saccule and the utricle. (See Figure 4.18.) The other is formed by a combination of three tubes at right angles to one another, called the semicircular canals.

The Utricle and Saccule

Just inside the vestibule of the inner ear, floating in fluid, are two membranous sacs, the **utricle** and **saccule**. Each of these is itself filled with fluid. (See Figure 4.18.) On the floor of the utricle is a small patch of hair cells, and around them is a jellylike fluid, which contains some tiny particles of calcium carbonate, the **otoliths**. When the head is tilted, these particles move down in response to gravity and trigger hair cells to send impulses to the brain indicating the new position of the head. (See Figure 4.19.)

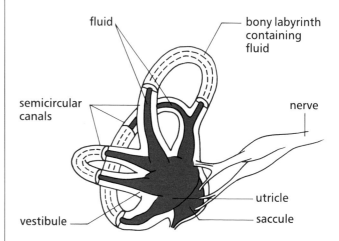

Figure 4.18
Structures associated with balance in the inner ear. The vestibule is filled with a fluid. "Floating" in this fluid are two membranous sacs called the utricle and the saccule.

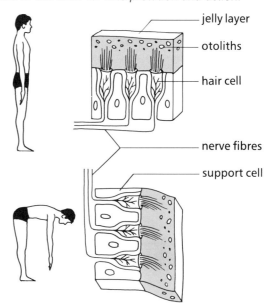

Figure 4.19
Inside the saccule are fine hairs which project into a jellylike substance containing small particles of calcium carbonate (otoliths). When the head is bent forward, gravity affects the otoliths and the jelly that contains them. This movement stimulates the nerve fibres in the hair cells and a "position" message is sent to the brain for interpretation and action.

The Semicircular Canals

While the utricle responds to position, the **semicircular canals** respond to changes in movement such as stopping, starting, or turning. The three tubes of the semicircular canals form part of a circle and each is set at right angles to one another in three different planes. (See Figure 4.20.) There are hair cells present in the semicircular canals and these respond as the fluid is set in motion.

If the head is swung forward and down, the fluid will flow in one of the semicircular canals, in another if the body is bent to the left or right, and the fluid will flow in the third tube if the head, or body, is turned on the body axis. The brain recognizes these impulses and sends out motor impulses to correct, or adapt to, the motion by stimulating the appropriate skeletal muscles.

FOCUS ON YOU

If the flow of impulses from the inner ear is affected by the intake of alcohol or other substances, the body's reactions to changes in position slow down and the response becomes exaggerated. The body swings too far to one side or another before the motor neurons cause the skeletal muscles to react. It becomes difficult to maintain normal balance or to walk a straight line.

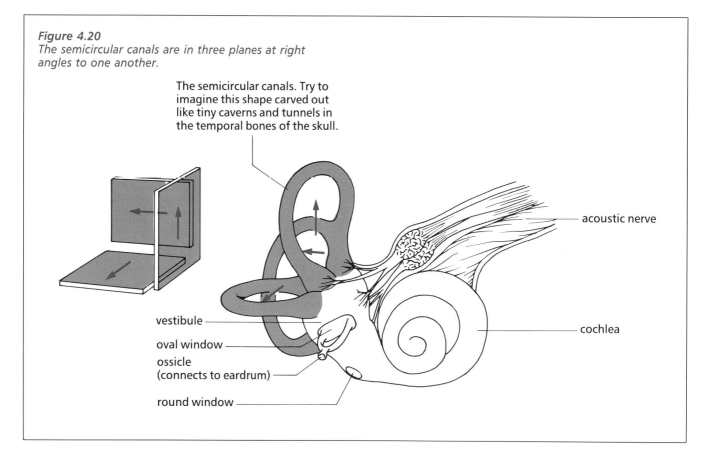

Figure 4.20
The semicircular canals are in three planes at right angles to one another.

Activity 4L

The Romberg Test

The Romberg test allows you to test the sense of balance.

Materials

slide or filmstrip projector

Procedure

1. Have your partner stand close to the chalkboard and place the projector or lamp so that a sharp shadow of your partner is cast on the board. Have your partner remove his or her shoes and stand with feet close together. The eyes must be closed or a blindfold used. You will find that the shadow moves as the body sways and you can mark the degree of sway by making a chalk mark at the edge of the shadow.

2. First have your partner stand facing the board and mark the side-to-side sway. Then have your partner turn at right angles and mark the front-to-back sway. *Record this amount of sway in centimetres in each case.*

3. Try having your partner stand on one foot and repeat the experiment. Ask him or her to repeat the test, but with eyes open. *Record your observations.*

Questions

1. Describe the role of the eyes in maintaining balance.

2. Compare your results with those of other students. How do the results compare?

CHECKPOINT

1. *Explain how the otoliths help to inform the brain of the head's position.*

2. *What is the purpose of the semicircular canals?*

3. *Why are there three semicircular canals?*

4.9 Other Senses: Taste, Smell, and Touch

The Sense of Taste

If you open your mouth and look into a mirror, you can see that the surface of the tongue is covered with tiny bumps (*papillae*). Around these small bumps, buried in the tissues of the tongue, are the **taste buds**. On the top of each taste bud is a small opening, or pore, through which solutions enter to stimulate the receptors. Solids cannot be tasted; the substances must first be dissolved. One of the functions of the saliva is to dissolve substances in the mouth, so that they can flow into the taste buds for identification. (See Figure 4.21.)

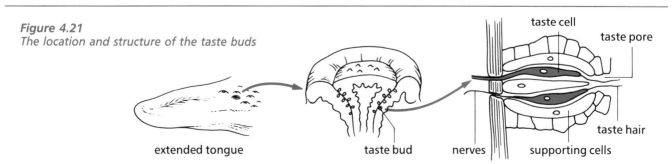

Figure 4.21
The location and structure of the taste buds

There are four main tastes – sweet, sour, salt, and bitter. The tongue is not equally sensitive to these four tastes in all areas. The back of the tongue is more sensitive to bitter tastes, whereas the tip of the tongue quickly responds to sweetness. (See Figure 4.22.) Many flavours are combinations of these four taste types.

Figure 4.22
All four tastes can be detected to a limited extent in all parts of the tongue, but are most readily recognized in these regions.

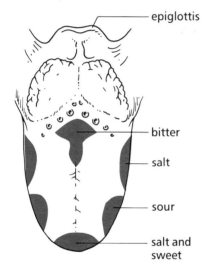

Taste buds are easily damaged by heat or chemical burns, but they regenerate quickly. Many of the cells found in the tongue have a short life span and are frequently replaced. Taste buds usually decrease in number with age, making it harder to detect the flavour of some foods. Most people have experienced the taste changes that are associated with a bad cold. This lack of sensitivity is not only due to the coating that appears on the tongue, but also the mucus that blocks the smell receptors in the nose.

The Sense of Smell

Smell is a subtle aid to our taste buds, refining the first impressions. If you taste an orange or a lemon with the nose pinched shut, you will find that your taste impression is quite different compared with that obtained when tasting the fruit with your nostrils open.

The **olfactory cells** (ol-fak-tore-ee), which are sensitive to odour, are found in a small patch in the upper part of the nasal cavity. (See Figure 4.23.) The cells are above the normal stream of air which passes through the nose. In order to detect faint odours, it is necessary to sniff strongly. When food is in the mouth, odours pass up the passage at the back of the throat into the nose and add to the flavour.

The olfactory cells in the nasal cavity are simple columnar cells with sensitive hairlike cilia protruding from them. Nerve fibres come from the base of these receptor cells and pass through an opening in the bony roof of the nasal cavity, to reach the olfactory centre in the brain.

Figure 4.23
The location of the olfactory nerve endings in the nasal cavity

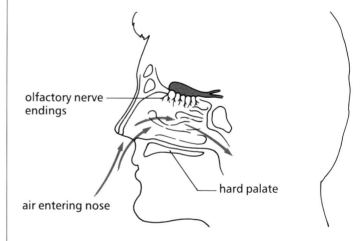

The cells which detect smell and taste are types of **chemoreceptors**. Chemoreceptors respond to individual kinds of molecules, and detect the presence of substances when they are dissolved in body fluids. The stimulus molecule reacts with a special spot on the membrane of a particular receptor cell. This produces a change which causes a nervous impulse to be sent to the appropriate area of the brain, where the message is interpreted. There are also receptors which measure the concentration of solutes and gases in the blood.

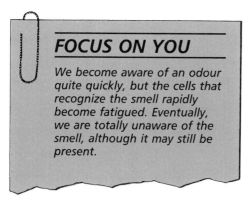

FOCUS ON YOU

We become aware of an odour quite quickly, but the cells that recognize the smell rapidly become fatigued. Eventually, we are totally unaware of the smell, although it may still be present.

Activity 4M

The Sense of Taste

Problem

How is the sense of taste distributed over the surface of the tongue?

Materials

cotton swabs
sterile wipes
10 percent sucrose solution
20 percent salt solution
1 percent acetic acid solution or vinegar
0.1 percent quinine solution
distilled water

Procedure

1. Work in pairs for this activity, with one student recording the observations and the other performing the activity. Rinse the tongue with distilled water and dry it with a sterile wipe.

2. Using a clean cotton swab, apply a small amount of sugar solution to the tip of the tongue, then try the swab on the sides and back of the tongue. Finally apply the swab to the centre of the tongue. *Grade the taste as strong, moderate, mild, or negative. Record the sensations experienced by the subject at each location.*

3. The recording partner will pass the tasting partner a swab dipped in each of the other solutions in turn, without telling what solution is being supplied. Rinse the tongue with distilled water from time to time to clear residual amounts of the solutions. The applications should be in small amounts to a limited area and not spread over the tongue.

4. *Draw a simple map of the tongue indicating which areas of the tongue are most sensitive to each of the four main tastes used in the experiment.*

Questions

1. Are the four tastes located in different areas of the tongue?

2. Which substances are most readily detected?

3. Which tastes seem to last the longest?

Activity 4N

Taste vs. Smell

Do you taste onions or smell them? Try this experiment. Prepare pieces of onion and apple about the same size. Close your eyes and pinch your nose. Have another student choose one of the pieces and place it on the tongue. Do not move the piece of food around in your mouth. Can you tell which piece you were given? Try this with pieces of orange and lemon.

Touch: Sense Receptors in the Skin

You will recall that there are specialized nerve endings in the skin. (See Figure 4.24.) These nerve endings are sensitive to heat, cold, touch, pressure, and pain. These receptors are spread over the skin in varying concentrations. The lips and fingertips, for example, have a great many more touch receptors than does the back. The forehead is also well supplied with pain receptors, but the nose and ear lobes are not. Where on the body would you expect to find the greatest concentration of heat receptors? Why? How could you test this? Would you also expect to find the same concentration of cold receptors here? Why or why not?

Figure 4.24
Special sensory nerve endings in the skin

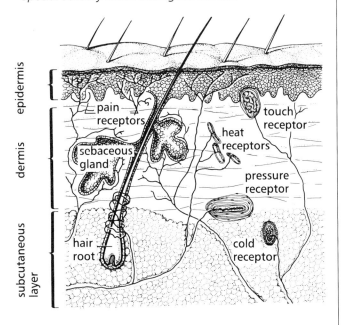

Activity 40

Two-Point Discrimination Test

Problem

Which areas of the skin are more sensitive to touch than others?

Materials

divider
centimetre ruler

Procedure

1. Extend the legs of the dividers until they are about 5 cm apart. (Dividers found in mathematical sets are quite suitable. Sterilize the tips in alcohol.)

2. Work in pairs for this experiment and have your partner close his or her eyes. You will be touching your partner with the dividers, and he or she will try to tell you if you are touching the surface of the skin with one point or both points at the same time. Steady the hand as much as possible so that the points make contact at the same time, rather than one after the other. Try to trick your partner occasionally by using only one point.

3. Gradually decrease the distance between the points until your partner can no longer accurately distinguish whether you are using one point or two. *Measure the distance between the points in centimetres and record this distance in your notebook.*

4. Test the following areas of the body: arm, back of the hand, fingertip, and lips. If you want to pull your shirt out and try the middle of your back, you may have to make the points of the dividers even wider than 5 cm!

Questions

1. Examine the results that you have recorded and try to explain why the areas differ in sensitivity.

Activity 4P

Other Skin Senses: Pain, Pressure, and Temperature

Problem

In which areas of the skin are the senses pain, pressure, and temperature (hot and cold) located?

Materials

straight pin
small finishing nail (or blunt dissection probe)
beaker of hot water (60°C)
beaker with an ice cube
rubber stamp divided into a grid of 100 small squares
non-toxic ink

Procedure

Temperature

1. *Stamp two grids in your lab notebook with the heading "Wrist". Label one grid "hot" and one "cold". Beside them, under the heading "Calf", place two more grids labelled "hot" and "cold". Stamp one grid on the hairless inner surface of your wrist and another on the back of your calf.*

2. *Heat the head of a small finishing nail in the beaker of hot water. Remember to reheat the nail frequently. First try touching the hot nail to several parts of your arm. You will be able to find a "hot" reaction in some spots, but little more than an awareness of pressure in others. Now methodically touch the centre of each of the squares in the grid on your wrist. Mark an X in the appropriate square in your lab notebook every time you feel a "hot" sensation.*

3. Using a small beaker with an ice cube, cool down the nail as much as possible. Then proceed as before, testing each square of the grid on your arm and *recording the "cold" spots in the grid in your notebook.*

4. Repeat steps 2 and 3 on the calf of your leg.

5. *Count and record the number of X's in each grid.*

6. *Examine the grids and decide if the sensory endings form a regular pattern or are quite randomly arranged.*

Pain and Pressure

1. *Stamp four more grids into your notebook. Two under the heading "Wrist" will be marked "pain" and "pressure" and two more grids under the heading "Calf" will also be marked "pain" and "pressure". You can use the same grids already stamped on your wrist and calf for these experiments, or you can stamp two new ones.*

2. Wash the wrist with soap and water. Do not dry your arm. Keep the area moist during the activity.

CAUTION!
There is no need for the skin to be broken in this activity. Touch the skin lightly. Pain is a relative term. You are looking for a response, not trying to hurt your partner.

3. Use an ordinary straight pin and, holding it upright, press it lightly against the skin. The pin should make a small depression but not pierce the skin! You will feel the pressure easily, but in some squares there should be a much

sharper "pain" response. *Mark X in each square of the grid in your notebook to correspond to the "sharp" responses.*

4. Repeat, using the head of the pin. Try to distinguish between dull responses and distinct sensations. *Record your results.*

5. Repeat the activity on the calf of your leg. *Record your results in the grids of your notebook.*

Questions

1. Compare the number of X's in the pain and pressure grids with those for heat and cold for both wrist and calf. What patterns can you find? What differences are there between the sensitivity of the wrist and the calf?

2. What is the approximate mean (average) distance between the sensitive points? (You may find that they are in small groups.)

3. Write a series of generalized statements to summarize what you have found out about the sensitivity of your skin to heat, cold, pain, and pressure.

4. Did you get the same results when you tested other parts of your arm for heat and cold sensitivity? Were your results similar to your partner's?

5. Why should the temperature of the bath water of infants be tested with the elbow rather than the hand?

Activity 4Q

Sensing Temperature

Problem

Can the hot and cold receptors of the skin be used to determine the temperature of the body's external environment?

Materials

one beaker of warm water (30°C)
one beaker of cold water (10°C)

Procedure

1. Immerse the index finger of one hand in the warm water for 2 min. Now dip the other index finger into the same container of water. *What differences in sensations can you distinguish between the two fingers? Record your observations.*

2. Immerse one index finger in the beaker of warm water and the other index finger into the beaker of cold water. After 2 min, switch the beakers and dip your fingers into the water of the opposite temperature. *Record the sensations received.*

3. Again, place the finger of one hand in the warm water and the finger of the other hand in cold water. After 2 min, immerse both fingers in another beaker containing cool water. *Explain the sensations experienced.*

4. You may also wish to try placing your elbow in a tray of ice water. Keep it there until you feel sensations in the fingers. *Describe the progression of sensations and exactly where these sensations occur.*

Questions

1. Could you use your hand to measure temperature? Explain your answer.

CHECKPOINT

1. *What are the four main tastes?*

2. *Why do you experience a taste change when you have a bad cold?*

3. *Where are the olfactory cells located?*

4. *Explain how chemoreceptors function.*

5. *Name the five kinds of receptors in the skin.*

BIOLOGY AT WORK

OPTICIAN
OPTOMETRIST
OPHTHALMOLOGIST
AUDIOLOGIST
HEARING CLINICIAN
SPEECH PATHOLOGIST
ACOUSTICAL ENGINEER
CINEMATOGRAPHER

HEALTH CONCERNS

cataracts
detached retina
glaucoma
"lazy" eye
conduction deafness
nerve deafness
tone deafness
seasickness or motion sickness

CHAPTER FOCUS

Now that you have completed this chapter, you should be able to do the following:

1. List the five parts of a control system and describe the function of each part.
2. Explain how the eye is protected by its position and by the structures which surround it.
3. Explain the function of the conjunctiva and the lacrimal glands.
4. Label a diagram of the internal structures of the eye and give the function of each part.
5. State the difference between rods and cones in structure, location, and function.
6. Explain the meaning of the following terms: light reflex, accommodation reflex, adaptation, after-image, convergence, visual acuity, astigmatism, colour-blindness.
7. Label a diagram of the eye to show the cause of near-sightedness and far-sightedness and explain how each condition can be corrected.
8. Give an example of an optical illusion.
9. Label a diagram of the parts of the ear and give the function of each part.
10. Explain how sound waves are converted to nervous impulses.
11. Suggest ways to reduce hearing loss.
12. Name the organs of balance and explain how they function.
13. List the four main tastes and indicate the areas of the tongue which are most sensitive to each one.
14. Explain how chemoreceptors, such as olfactory cells, function.
15. List the five kinds of receptors found in the skin.

Chapter 4 / *Sensing the External Environment*

SOME WORDS TO KNOW

Match each of the descriptions given in the left-hand column with a word shown in the right-hand column. DO NOT WRITE IN THIS BOOK.

1. Controls the amount of light entering the eye
2. The inner, light-sensitive lining of the eye
3. Responsible for colour reception in the retina
4. Changes thickness to focus the image of nearby or distant objects on the retina
5. A result of distortion of the cornea
6. A super-sensitive portion of the retina
7. Thin membrane covering the outer surface of the eyeball
8. Transparent protective coating in front of the eye
9. The outer coat of the eyeball
10. Converts vibrations to nerve impulses in the inner ear
11. Snail-shaped structure in the inner ear
12. Responsible for balance
13. Permits changes of pressure to take place in the middle ear

A semicircular canals
B oval window
C round window
D organ of Corti
E otoliths
F iris
G cornea
H conjunctiva
I retina
J lens
K rods
L cones
M fovea
N astigmatism
O sclera
P cochlea
Q eustachian tube
R auditory canal

Select any two of the unmatched words and, for each, give a proper definition or explanation in your own words.

SOME QUESTIONS TO ANSWER

1. Explain the difference between a receptor and an effector and give an example of each.
2. What should you do if a chemical is splashed into your eyes?
3. Why must contact lens wearers be careful to keep their lenses clean?
4. What problems are associated with the use of extended wear contact lenses?
5. What is the composition of tears?
6. (a) What are the three layers of the eyeball?
   (b) Give the function of each layer.
7. What condition might develop if your diet was deficient in vitamin A?
8. Why is it difficult to see colours at dusk?
9. Explain why you have a blind spot in each eye.
10. What is the special advantage of having two eyes?
11. What happens to the eye when a person looks first at a nearby object and then looks at a distant object?
12. (a) Label a diagram of the eye of a person who suffers from near-sightedness, showing the position of light rays which enter the eye.
    (b) Label a diagram of the eye of a person who suffers from far-sightedness, showing the position of light rays which enter the eye.
    (c) What kind of lens is used in each case to correct the problem? How do they help?
13. (a) What is astigmatism?
    (b) How is it corrected?
14. What are the functions of the three bones of the middle ear?
15. Draw a diagram of the human ear and label its parts.
16. Explain how the sound of a ringing bell reaches the eardrum.

17. (a) What is the range of human hearing?
    (b) How does this change with age?
18. What effect does prolonged loud music or noise have on the ear?
19. Explain the working of the semicircular canals.
20. Why do we feel "giddy" or disoriented when we suddenly stop after spinning around?
21. Draw a diagram to show where each of the four taste sensations are recognized.
22. (a) Where exactly are the taste buds located?
    (b) Can we taste dry food? Why or why not?
23. What influence does the sense of smell have on our ability to taste foods in the mouth?

SOME THINGS TO FIND OUT

1. (a) What kind of visual tests and regulations apply to drivers?
   (b) Is there a difference for drivers of commercial vehicles?
2. What effects do the following have on the pupil of the eye?
   (a) adrenalin
   (b) cocaine
   (c) morphine
   (d) marijuana.
3. Investigate how lasers can be used to help people with eye problems.
4. Investigate other optical illusions, such as distorted rooms, reversible figures, and impossible objects. There are a wide variety of these illusions, some very deceptive. Try to explain why our minds are tricked by such images into seeing the illusions.
5. (a) Investigate what agencies are involved with helping people with hearing disabilities.
   (b) How are children with hearing problems taught?
   (c) What is closed captioned television?
6. How does weightlessness affect the senses of astronauts?
7. (a) One theory suggests that there are seven different basic types of scent. What are they?
   (b) Try to design an experiment that could be used by students that would support the theory.
8. (a) What is acoustic fatigue?
   (b) Find out what the law says about requirements for protective devices to avoid hearing loss or acoustic fatigue at work.
9. Find out how the different kinds of hearing aids work.
10. Why is the sense of smell more pronounced than the sense of taste in humans?
11. The antibiotic streptomycin may be given to treat certain bacterial infections, but it can sometimes cause deafness. How is this possible?
12. Most people are familiar with the use of seeing-eye dogs by people with a visual handicap. Find out how animals are used by people with other kinds of physical disabilities.

Solution for Activity 3K (page 114):
The removal of eight lines will produce two squares.

BLOOD AND CIRCULATION

In order to keep cells alive and growing, they must be bathed in a fluid solution of chemicals which provides nutrients, oxygen, and moisture. Attempts to produce an artificial substitute for this fluid have increased our understanding of the complex mixture of free moving cells, proteins, and dissolved substances called blood.

KEY IDEAS

- Blood and its circulatory system transport nutrients, wastes, gases, and other vital substances throughout the body.

- Blood contains water, dissolved substances, cells, and other solids.

- Blood cells transport oxygen and fight infection. Platelets prevent blood loss from wounds.

- Heart rate and blood pressure are affected by many factors.

- Knowledge of how the circulatory system and the heart function has saved many lives.

5.1 What is Blood?

Our senses tell us a few things about blood; it is red, it is sticky, and it has a distinctive taste and smell. It also dries much more quickly than other common liquids. Blood is sometimes called the "life stream". It flows through the tiny "rivers" and tributaries of the body, carrying the necessities of life – oxygen, nutrients, chemical messengers – to the cells, as well as transporting away the wastes and carbon dioxide that these cells produce.

Plasma

The average body contains about 5 L of blood. Of this, 55 percent is composed of a straw-coloured fluid called **plasma**. (See Figure 5.1.) Plasma is approximately 90 percent water and carries many dissolved substances, including, for example, waste carbon dioxide from cellular respiration, calcium ions for bone construction, and special blood proteins such as hormones.

Figure 5.1
By centrifuging a sample of whole blood, the liquid plasma can be separated from the heavier blood cells.

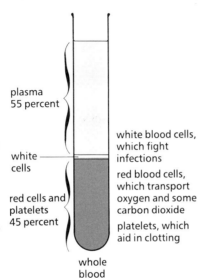

Plasma contains
- hormones, antibodies, and enzymes
- inorganic substances, such as sodium, chlorides, iron, and calcium
- digested food, including simple sugars, fats, and amino acids
- vitamins
- gases (oxygen and carbon dioxide)
- plasma proteins, such as fibrinogen and globulin

The Blood Cells and Platelets

The remaining 45 percent of blood volume is made up of solids – red blood cells, white blood cells, and cell fragments called platelets. Figure 5.2 summarizes some important features of these components.

Figure 5.2
The blood cells and platelets

The Red Blood Cells
- form about 45 percent of whole blood
- contain hemoglobin
- carry oxygen and carbon dioxide

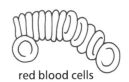
red blood cells

The White Blood Cells (WBC)
The granular white blood cells
- Neutrophils make up 65 percent of the WBC. These cells are actively phagocytic and engulf bacteria and other cells.
- Eosinophils comprise 2 to 4 percent of the WBC. They are moderately phagocytic.
- Basophils make up 0.5 percent of the WBC.

neutrophil   eosinophil

basophil

The non-granular white blood cells
- Lymphocytes make up 20 to 25 percent of the WBC. They aid in the development of immunity against infections.
- Monocytes make up 3 to 8 percent of the WBC. The monocytes are actively phagocytic.

lymphocyte   monocyte

The platelets contain prothrombin. They are responsible for initiating the clotting sequence.

platelets

Red Blood Cells

Red blood cells, or **erythrocytes** (e-rith-row-sights) are round with a hollow domed depression in each side. (See Figure 5.3.) This shape gives the cell a large surface area with only small bulk, helping it pass through even the tiniest blood vessels. One very unusual characteristic of red blood cells is that they have no nucleus. They are produced in the marrow of bones, such as the sternum and vertebrae. During the development of these cells, a normal nucleus is present, but, as the cells mature, the nucleus gradually shrinks and disintegrates.

Red blood cells are the oxygen carriers of the body. They contain **hemoglobin** (hee-ma-glo-bin), a molecule composed of a protein called *globin*, a compound containing iron. The hemoglobin molecule can combine with four molecules of oxygen. When oxygen associates with hemoglobin, a bright red compound, called **oxyhemoglobin**, is formed. When oxyhemoglobin is in an area low in oxygen, such as near active muscle cells, the oxygen is released.

Red blood cells are required to do a great deal of work; therefore, they wear out fairly quickly, lasting only about three to four months. Much of the wear and tear occurs on the very thin membrane surrounding each cell. During its normal life of about 120 days, each cell travels more than 1100 km. Old red blood cells are broken down in the liver and spleen. The iron within them is returned to the bone marrow for use in new cells. The replacement rate is 1 or 2 million cells per second!

FOCUS ON YOU

There are about 5.5 million red blood cells per cubic millimetre of blood in a healthy adult male (4.8 million in a healthy female). Low cell counts can indicate anemia.

White Blood Cells

White blood cells, or **leukocytes** (loo-ko-sights) differ from red blood cells in many ways. They contain a nucleus. They are colourless and, although they are usually spherical, possess the ability to change shape. White blood cells are larger than red blood cells but are far less numerous. There is about one white blood cell for every 600 red blood cells. (See Figure 5.3.) Most white blood cells are produced in the bone marrow, but some are also made in the lymph tissue at various sites around the body. There are several types of white blood cells. They function to protect the body from infection and disease-causing organisms (**pathogens**). Figure 5.4 shows a typical response by white blood cells to an infection.

Figure 5.3
Human blood cells – red blood cells (a), white blood cells (b), and platelets (c)

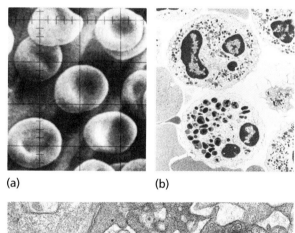

(a)   (b)

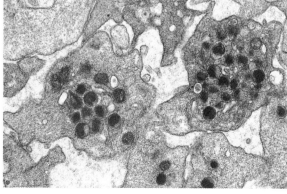

(c)

Figure 5.4
Inflammation. The response of the body to infection involves the circulatory system and white blood cells.

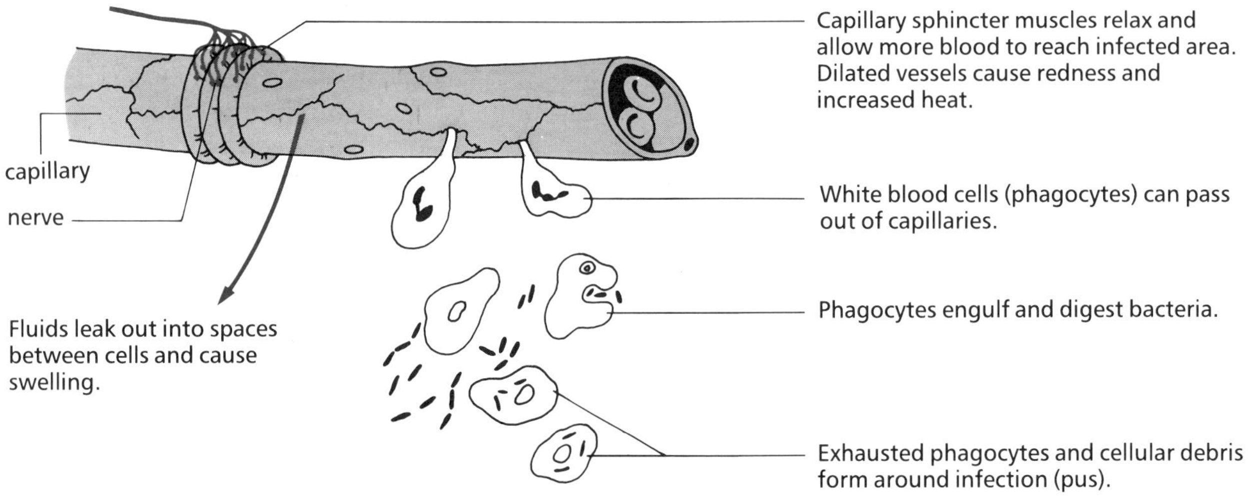

- capillary
- nerve
- Fluids leak out into spaces between cells and cause swelling.
- Capillary sphincter muscles relax and allow more blood to reach infected area. Dilated vessels cause redness and increased heat.
- White blood cells (phagocytes) can pass out of capillaries.
- Phagocytes engulf and digest bacteria.
- Exhausted phagocytes and cellular debris form around infection (pus).

All white blood cells possess, to a varying degree, these main characteristics which help them perform their function:

- **Amoeboid action**. White blood cells can move independently. They can also change shape and squeeze through tiny pores to move into and out of body tissues.

- **Phagocytosis** (phag-o-sigh-toe-sis). White blood cells can surround or engulf bacteria and digest their toxic proteins. They can also join together in order to surround quite large products (for example, small splinters) which must be removed from the body.

- **Chemical properties**. White blood cells react to charged particles, proteins, and other chemicals which may be present in areas of injury and inflammation. Some cells produce special chemicals called *antibodies* to neutralize the effects of certain foreign chemicals.

Each pathogen, and body cell for that matter, has a particular chemical structure on its surface. These chemicals, usually proteins and carbohydrates, are unique to each cell. White blood cells "recognize" these chemicals, called **antigens**, and respond immediately.

One way in which white blood cells respond is to produce **antibodies**. An antibody is a protein which will react with a specific foreign antigen. When the appropriate antibody reacts to an antigen, it may cause the pathogens to stick together in large masses, thus immobilizing them. Alternatively, it may help phagocytes to recognize the pathogens more easily and thus destroy them more rapidly.

FOCUS ON YOU

Sick of blood tests? The analysis of blood provides a physician with information ranging from what was in your last meal to your susceptibility to serious illness – all for the price of a moment's discomfort.

LIFE · SIGNS

ANEMIA

Renata was a likeable 17-year-old, with an exuberant personality. One morning, to everyone's surprise, Renata found herself on the floor looking up at a circle of worried faces. Renata had fainted without any warning that she could remember. Rather shaken by this event, she made an appointment with a physician.

Dr. Wallis questioned her closely about her daily activities and feelings. Had she been feeling tired lately? Did she tire easily? Did she get aches and pains in her muscles? Renata admitted that she had felt rather tired lately and didn't seem to have her old get-up-and-go. She had thought it was caused by working rather hard and getting home late.

The doctor questioned Renata about her breathing. Was she ever short of breath? Did she have any difficulty in breathing, especially after doing strenuous exercise? Dr. Wallis looked at her fingernails and also noted that Renata looked rather pale.

The next questions were about her blood. If she cut herself, did it seem to take a long time before the bleeding stopped? Did she have a heavy menstrual flow?

When it came to the next questions about her eating habits, Renata was embarrassed. She made excuses about being busy and not having time for breakfast, not liking vegetables, and rarely eating fruit. She

Some Iron-containing Foods
liver, dried fruits, kidneys, leafy green vegetables, red meats, enriched grain products, egg yolks, peas, molasses, beans, nuts

said she couldn't stand liver and rarely ate meat except hamburger.

The doctor took a sample of Renata's blood from her arm, to be sent away for analysis. She questioned Renata further about her general health and about the medical background of her parents. Dr. Wallis finally gave Renata a complete physical checkup and asked her to return in a few days when the blood analysis report would be available.

When Renata returned to see the physician, she was not surprised to hear that she was anemic. Dr. Wallis quickly reassured her that, in her case, it was not one of the more serious forms of anemia and that her problem could be remedied by changes in diet and by taking a prescribed amount of iron tablets.

On a final visit, Dr. Wallis told Renata to discontinue taking the iron tablets as, with a suitable diet, the supplement of iron was unnecessary.

FOCUS ON YOU

To assist the activity of white blood cells, histamines and certain other chemicals in the blood cause blood vessels to expand in diameter as much as ten times. In allergic reactions, histamines are released in large amounts even though no infection is present, causing symptoms ranging from congestion to hives.

FOCUS ON YOU

Vaccinations work by triggering your body's own immune system. Injecting a weak version of a disease-causing organism into your body causes the production of antibodies against that organism. This means your blood will be ready to take action should you come into contact with that disease later in life.

Platelets

Platelets are much smaller than red blood cells, numbering about 250 000 to 300 000 per cubic millimetre. They are responsible for the initial stages of *blood clotting*, the process which prevents loss of blood from a wound. (See Figure 5.5.) If blood did not clot, wounds would continue to bleed until the loss of blood was sufficient to cause death. Platelets have no nuclei and are not cells, but rather the tiny fragments of cells containing a special enzyme that initiates clotting.

Figure 5.5
The clotting of blood closes wounds and prevents blood loss.

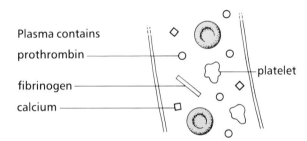

1. Platelets break open at wound site and release thromboplastin.
2. Thromboplastin causes calcium and prothrombin to unite and form thrombin.
3. Thrombin joins with fibrinogen to form fibrin.
4. A network of fibrin threads traps red blood cells and forms a clot that plugs the wound.

Although all the components of this reaction are present in the blood plasma, the starter enzyme, thromboplastin, is kept in separate little containers (platelets) until required. Once the platelets are broken open and thromboplastin is released, the reaction starts.

Activity 5A

Examining Blood Cells

The prepared blood smear that you will use has been stained. The dyes used will make the nuclei of the white blood cells stand out more clearly, but it will also distort the colour of the red cells, making them appear a pinkish purple rather than bright red.

Materials

microscope
prepared slide of human blood smear

Procedure

1. Examine the slide under low-power magnification. The screen will be filled with many oval or round red blood cells. Once you have these sharply focussed, change to the highest magnification your microscope provides. (For good results, you must also adjust the diaphragm and the condenser lens beneath the microscope stage.)

2. Examine the red blood cells carefully. Try to think of them in three-dimensional form rather than flat circles. *Draw four or five of these cells. Draw them about the size of a quarter.*

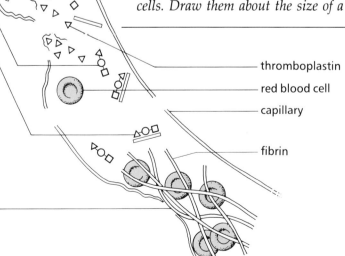

3. In the background are the platelets, which are not much bigger than small particles. *Add some of these to your drawing.* Be sure that you keep them in proportion to the size of your red blood cells.

4. Because there is only one white blood cell for every 600 red blood cells, white blood cells are rather hard to find. The white blood cells can be recognized by their nuclei, which are usually stained a light purple colour. White blood cells are also larger than red blood cells, but their outline is very faint and hard to see.

5. Switch back to low power and make a systematic search for white cells. Move the slide across the stage from left to right, then make another pass across the smear from right to left a little lower down the slide. When you find a white blood cell, centre it in the field of view and return to high-power magnification.

6. *Carefully draw this cell. Make sure that the outline is in the proper size proportion to the red blood cells you have already drawn. Examine and copy the shape of the nucleus.* Many white blood cells have lobed nuclei. Look at Figure 5.2 of your text, and try to *identify the type of white blood cell you have found.*

7. Return to low power and find as many white cells as you can. *If the cell has a different size and shape of nucleus, draw it and try to identify it. If it is the same as the one you have drawn, make a record of this and produce a count of each type of white blood cell that you find.*

Questions

1. How many different types of white blood cells did you find?

2. What type of white blood cell was most common in your slide?

3. Were there any nuclei present in the red blood cells?

Some Blood Cell Disorders

- **Anemia**. When there are insufficient red blood cells to carry all the oxygen that the body requires, a condition called **anemia** (ah-nee-mee-ah) occurs. People who have this condition appear tired and lack the energy to work or play efficiently. They may also catch other illnesses easily. There are several types of anemia, including one very common type that is caused by insufficient iron in the diet. Because iron is a basic component of hemoglobin, a lack of iron results in insufficient hemoglobin being formed. This reduces the oxygen-carrying capacity of the blood. Whole wheat bread, nuts, raisins, spinach, liver, and other meats are good sources of iron in the diet.

- **Hemophilia**. This is a rare blood disorder that is inherited as a sex-linked trait, being most common in males. (See Chapter 10 for details of sex-linkage.) All types of hemophiliac disorders prevent normal blood clotting; the blood may clot either very slowly or not at all.

- **Leukemia** (loo-kee-mee-ah). This is a cancerous disease of the blood-forming organs which affects the leukocytes. Although there are several kinds of leukemia, most types involve a considerable increase in the number of white blood cells and a decrease in the number of red blood cells. As the disease progresses, the number of mature white blood cells decreases and large numbers of immature white blood cells crowd the bloodstream. The body then fails to cope adequately with infections. The reduction in the number of red blood cells and platelets also decreases the amount of oxygen available, and the body becomes unable to cope with the many internal hemorrhages. Excessive exposure to X-rays and to radioactive elements can cause increased production of white blood cells. These factors are recognized as among the major determiners of the disease.

- **Mononucleosis**. This is not really a disease of the blood but it does result in the production of large numbers of white blood cells. Mononucleosis is most commonly transmitted during a transfer of saliva. It is therefore often known as the "kissing disease". It occurs most commonly among young people and in the more affluent countries of the world. Mononucleosis is thought to be caused by a particularly stubborn virus which is not easily eradicated. It increases the production of large numbers of a certain type of white blood cell. Symptoms include fatigue, swollen glands, fever, and sore throat.

CHECKPOINT

1. List four substances found in blood plasma.
2. (a) Describe the shape of a red blood cell.
   (b) What are the advantages of having this shape?
3. (a) What is hemoglobin?
   (b) What is its function?
4. Name three ways in which white blood cells differ from red blood cells.
5. What three characteristics are common to all white blood cells?
6. List two ways in which white blood cells fight infections.
7. Describe one cause of anemia.

5.2 Blood Types

Blood type is determined by the presence or absence of certain antigens on the red blood cells. There are two kinds of antigens, A and B. If antigen A is present, then the individual is said to have type A blood. Type B blood contains the B antigen. Type AB blood contains both A and B antigens. Type O blood has neither of these antigens.

The antibodies found in the blood plasma of a person with a particular blood type are always the opposite of their antigen. For example, type A blood contains the A antigen but the B antibody. Type B blood has the B antigen and the A antibody. Type AB blood contains both antigens, but neither antibody A nor B, while type O blood possesses no antigens, but contains antibodies A and B. (See Table 5.1.)

Table 5.1 Human Blood Types

Type of Blood	Antigens Present on Red Blood Cells	Antibodies Present in Blood Plasma
A	A	B
B	B	A
AB	A and B	Neither A nor B
O	Neither A nor B	A and B

If antigens and antibodies of the same type come together, the antibodies cause the red blood cells to clump together or agglutinate; thus, if two samples of blood of the same type are mixed together, no clumping occurs, whereas if different blood types are mixed, the cells may clump. If this should happen as a result of incorrect matching of blood during a transfusion, the result could be fatal. This happens very rarely, as some of the recipient's blood is always tested with the donor's blood before being transfused. This is called **typing**.

Blood Transfusions

Before a transfusion takes place, it must be established that the antigens in the donor's blood will not cause clumping of the recipient's blood, which might block important arteries. A sample of blood is mixed with a drop of concentrated serum containing A antibodies. Another drop of the sample blood is mixed with a drop of serum containing B antibodies. Whether or not the red blood cells of the sample clump together will quickly serve to identify the type of blood contained in the sample because clumping would indicate the presence of A or B antigens. Blood from a donor is always cross-matched with a recipient's blood to be sure the two types are compatible. (See Figure 5.6.)

Figure 5.6
This figure shows the results of cross-matching the main blood groups. Even distribution of blood cells indicates safe or compatible blood for transfusion. Clusters indicate the clumping of incompatible blood cells. Type O donors cause no clumping and are known as universal donors. People with type AB blood are known as universal recipients.

Table 5.2 Reactions When Blood is Mixed with Serums Containing A and B Antibodies

Sample Mixed with Anti A	Sample Mixed with Anti B	Antigens Present in Red Blood Cells	Blood Type
No clumping	No clumping	None	O
No clumping	Clumping	B	B
Clumping	No clumping	A	A
Clumping	Clumping	A and B	AB

Table 5.3 The Percentage of Blood Types in Several Groups of People

Population Sector	Frequency (percent)			
	A	B	AB	O
North American White	41	10	4	45
U.S. Black	26	21	4	49
Inuit	55	5	4	36
Chinese	25	34	10	31

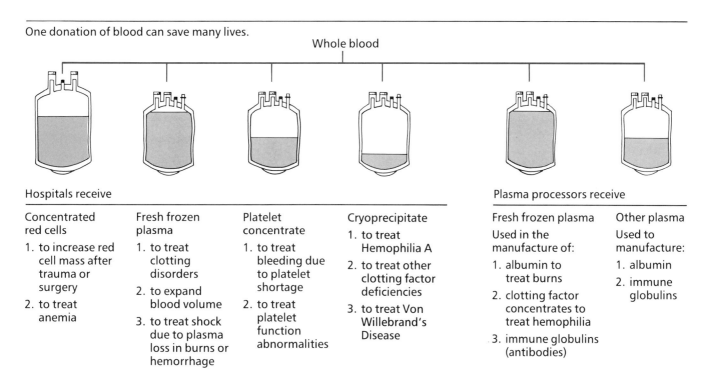

One donation of blood can save many lives.

Whole blood

Hospitals receive

Concentrated red cells
1. to increase red cell mass after trauma or surgery
2. to treat anemia

Fresh frozen plasma
1. to treat clotting disorders
2. to expand blood volume
3. to treat shock due to plasma loss in burns or hemorrhage

Platelet concentrate
1. to treat bleeding due to platelet shortage
2. to treat platelet function abnormalities

Cryoprecipitate
1. to treat Hemophilia A
2. to treat other clotting factor deficiencies
3. to treat Von Willebrand's Disease

Plasma processors receive

Fresh frozen plasma
Used in the manufacture of:
1. albumin to treat burns
2. clotting factor concentrates to treat hemophilia
3. immune globulins (antibodies)

Other plasma
Used to manufacture:
1. albumin
2. immune globulins

The Rh Factor

The **Rh factor** is an antigen found on the red blood cells of most persons. This inherited blood type factor is present in about 85 percent of the Canadian population. Individuals with this factor are said to be Rh positive, and those who do not have it are said to be Rh negative. The disease associated with the Rh factor is "Hemolytic Disease of the Newborn". This disease may occur when there is an incompatibility between the Rh blood type of the mother and that of her baby. This happens when the father is Rh positive, the mother is Rh negative, and the baby is Rh positive.

The danger arises when, during pregnancy or at delivery, some of the baby's Rh positive blood enters the mother's circulatory system through the placenta. When this happens, an antibody is formed in the mother's blood against the Rh factor. In a future pregnancy, the mother's antibodies attack the Rh positive cells in the baby's blood. Depending on the amount of antibody in the mother's blood, the baby's cells may clump, red cells may lose hemoglobin and be unable to carry adequate amounts of oxygen, or the baby may be born with massive swelling of the body, or even be stillborn.

Fortunately, this disease can be prevented with a simple injection of Rh immune globulin. Rh immune globulin is a special blood component produced from blood plasma, collected from persons whose blood already contains a high level of Rh antibodies. When introduced into the bloodstream of the mother, Rh immune globulin will prevent the formation of Rh antibodies.

CHECKPOINT

1. *What are the four blood types and what antigens are found in each type?*
2. *If you know the antigens present in a blood type, what antibodies can you predict will be present, or lacking, in that particular blood type?*
3. *What is the Rh factor?*

5.3 Moving Blood Throughout the Body: The Circulatory System

The Arteries and Other Vessels

An **artery** is a vessel that carries blood *away* from the heart; **arterioles** are simply small arteries. **Veins** carry blood *to* the heart; **venules** are smaller veins. **Capillaries** link arterioles and venules.

The walls of arteries are thick and contain a layer of muscle. (See Figure 5.7.) When blood is pumped out of the heart, it is forced out under pressure. The muscular artery walls are elastic and stretch as this wave of pressure is pushed along. This is the little "bulge" that we feel under our fingers when we take our pulse. The veins also have a muscular coat in the wall, but the pressure of blood flow is greatly reduced and this coat is much thinner than that found in arteries.

Both arteries and veins have walls far too thick to allow any blood plasma, nutrients, or gases to pass out into the surrounding tissues. The **capillaries**, however, have very thin walls, composed of single layers of endothelial cells. The capillary is so narrow that often only one tiny red blood cell can pass through at a time. It is through the walls of the capillaries, which link the arterioles to the venules, that the exchange of oxygen, carbon dioxide, nutrients, wastes, and other substances takes place. (See Figure 5.8.)

Figure 5.7
A comparison of the structure of an artery, a vein, and a capillary. Note that the muscle wall in an artery is much thicker than that in a vein. The capillary has been greatly enlarged in order to show detail.

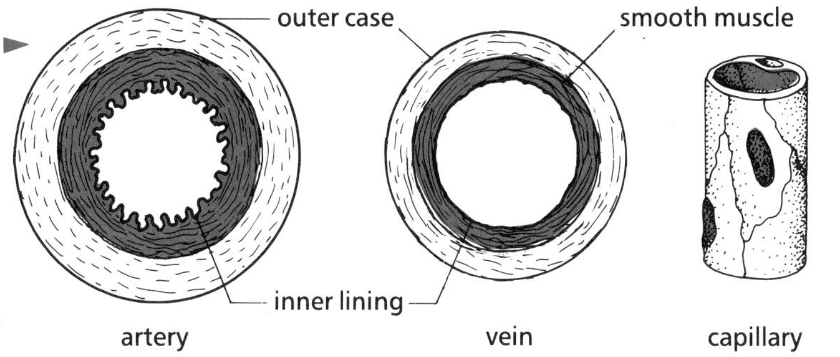

Figure 5.8
The action of blood cells and the diffusion of materials between a capillary and the surrounding tissues

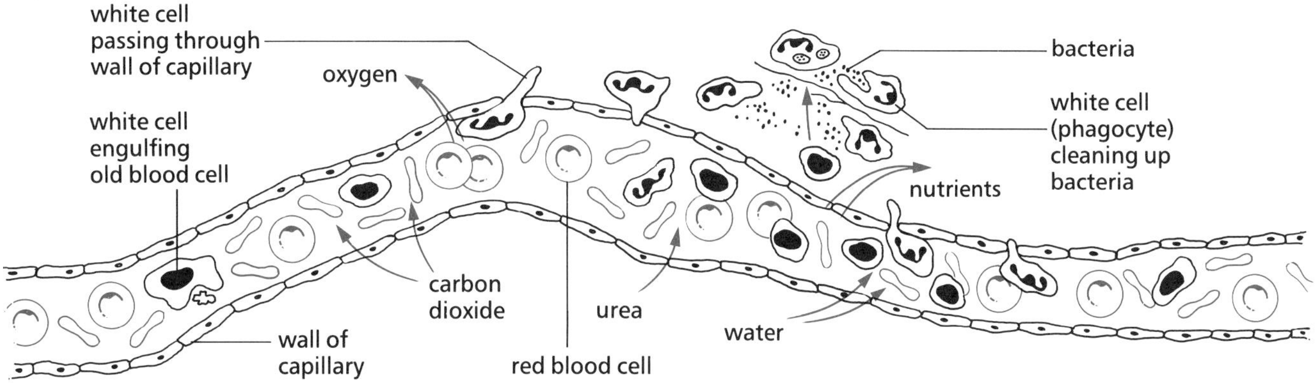

Some of the fluid in the plasma passes through the capillary walls to bathe the cells of the body tissues. A portion of this plasma re-enters the capillaries. Any additional fluid that is not reabsorbed is later collected and returned via the lymphatic system which is related to, but separate from, the blood circulatory system.

Although the blood is forced into the arteries under pressure, by the time it reaches the capillaries this pressure is very low. (See Figure 5.9.) There must be another mechanism for getting blood back to the heart.

Figure 5.9
Blood pressure in the different vessels of the circulatory system

As blood leaves the heart, it flows "downhill" until the pressure in the capillaries and venous system becomes very low. Blood flow in the capillaries is also very much slower than that in the larger vessels, which allows time for the diffusion of many substances between blood and the surrounding tissues.

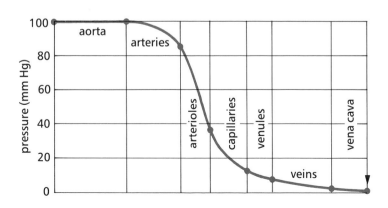

The Veins and Their Valves

If you could see inside a vein, you would find many tiny **valves** regularly spaced along its length. These valves point in the direction in which the blood is flowing. The valves can be pushed open to allow blood to flow toward the heart, but they close to prevent any back flow. (See Figure 5.10.)

Figure 5.10
Valve function in the veins

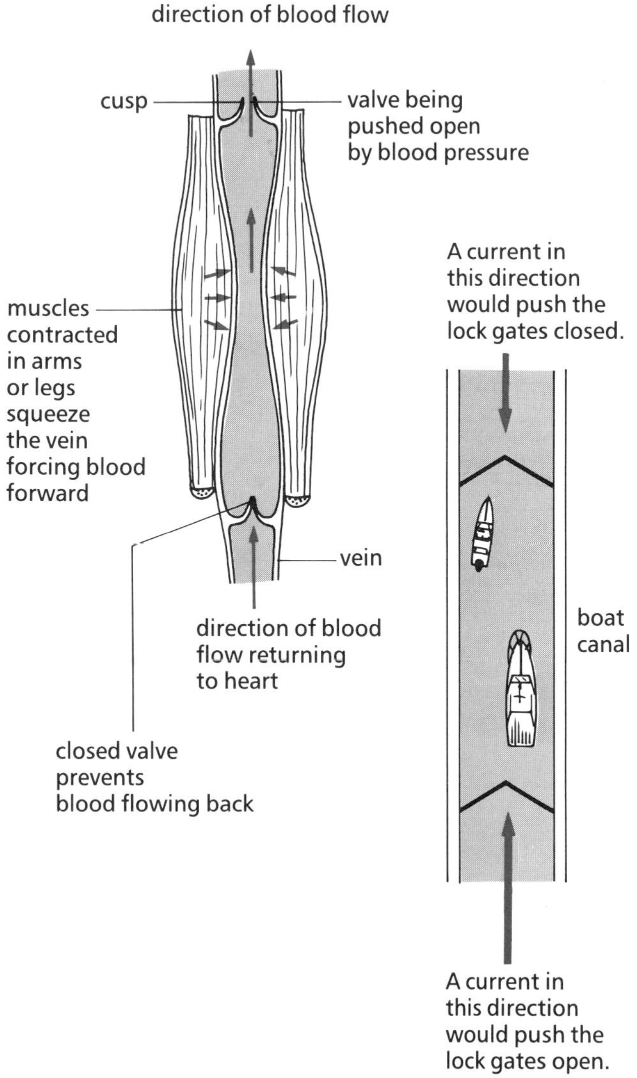

Perhaps you have seen lock gates along a boat canal and have noticed that they meet at a slight angle. If the current in the canal is flowing in one direction, it tends to push against the gates and keep them closed. If the current were to flow in the other direction, it would tend to push the gates open. The valves in the veins act in a similar manner. When the body muscles surrounding these veins contract, they bulge outward, thereby squeezing the veins and pushing the blood forward through the valves. When the valve flaps move back into place, the blood cannot flow back into the vein behind the valve. The blood is thus squeezed back to the heart step by step, by the contraction of the muscles that surround the veins.

Table 5.4 A Comparison of Arteries and Veins

Artery	Vein
Takes blood away from the heart	Takes blood to the heart
Blood travels in small spurts	Blood travels more smoothly
No valves present	Valves present
Thick muscle walls	Thin muscle walls
Blood rich in oxygen, bright red in colour	Low in oxygen, high in carbon dioxide, more like maroon in colour
Pressure very high	Pressure very low

FOCUS ON YOU

In the condition known as varicose veins, portions of the veins lose elasticity and become extended, preventing their valves from closing completely. This allows the blood to form bulging pools. This swelling, particularly in the legs, causes the veins to appear knotted and blue. Varicose veins can be very painful and are aggravated by long periods of standing in one place.

Interstitial Fluid

All of the cells and tissues of the body must be continuously bathed by fluids. These fluids enable nutrients to pass from the capillaries across the spaces between the cells to reach the cell membranes. They also prevent the cells from drying out.

The fluid that surrounds the cells is called **interstitial fluid** (*interstitial* refers to spaces between things). It is very abundant, making up about 15 percent of the body mass. You have probably seen this fluid as the clear, colourless liquid that appears when you graze the skin or break a blister.

Interstitial fluid is very similar to the plasma of the blood. It contains approximately the same substances as plasma, but in different proportions. The differences in these substances are due to the movement of materials and exchanges that take place between cells, interstitial fluids, and the blood plasma.

The blood is laden with substances needed by the cells. Interstitial fluid bridges the gap, carrying materials from capillaries to cells. Some of this fluid then diffuses back into the bloodstream, but much of it enters small tubes which are part of the **lymphatic system**. (See Figure 5.11.)

CHECKPOINT

1. *How do arteries, veins, and capillaries differ in structure and function?*

2. *How does blood pressure differ in arteries, capillaries, and veins?*

3. *Explain how blood is returned to the heart, when there is almost no pressure in the veins to pump it back.*

4. *(a) Where would interstitial fluid be found in your body?*
   (b) What functions are served by this fluid?

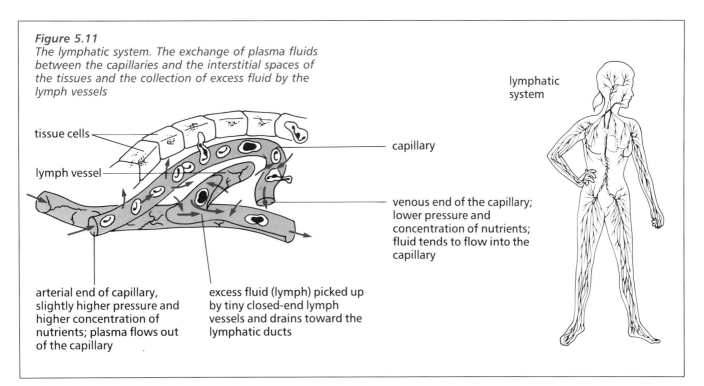

Figure 5.11
The lymphatic system. The exchange of plasma fluids between the capillaries and the interstitial spaces of the tissues and the collection of excess fluid by the lymph vessels

5.4 The System in Action: The Heart

The heart is located in the thoracic cavity, well-protected by the rib cage. It nestles between the lungs with its lower end slightly toward the left side. The adult heart is about the size of a large fist and has a mass of approximately 300 g. The heart is not rigidly attached to any of the surrounding tissues, but is suspended by the large blood vessels that are attached to it. (See Figure 5.12.) Because of this, it is able to move loosely in place as it contracts.

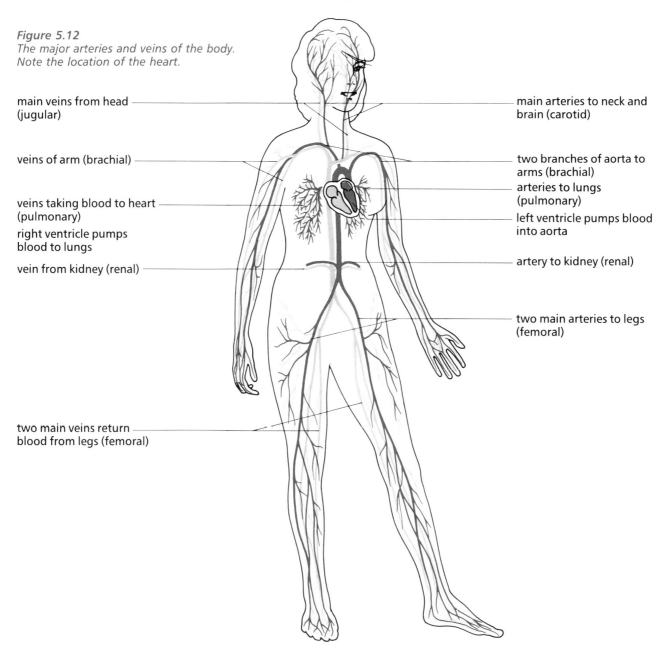

Figure 5.12
The major arteries and veins of the body. Note the location of the heart.

- main veins from head (jugular)
- veins of arm (brachial)
- veins taking blood to heart (pulmonary)
- right ventricle pumps blood to lungs
- vein from kidney (renal)
- two main veins return blood from legs (femoral)
- main arteries to neck and brain (carotid)
- two branches of aorta to arms (brachial)
- arteries to lungs (pulmonary)
- left ventricle pumps blood into aorta
- artery to kidney (renal)
- two main arteries to legs (femoral)

Heart Muscle

The heart is not one pump, but two separate pumps, each complete in itself. The heart is composed of very special muscle tissue, which is like striated muscle in some ways. (See Figure 5.13.) It has incredible stamina, for it must beat continuously without rest from the early weeks of life (about 18 days after conception) until death. Each cardiac muscle cell is attached to, and pulls against, other cardiac muscle cells during contractions; the muscle is not anchored to bones or other tissues. The muscle cells of the heart have an innate ability to contract and they do so spontaneously. Even fragments of heart muscle placed in a plasma solution will continue to contract rhythmically on their own. If they touch, they have the ability to co-ordinate their contractions into a common rhythm.

Figure 5.13
The fine structure of heart muscle

The Parts of the Heart

The right side of the heart is responsible for collecting blood from the body and pumping it to the lungs. (See Figures 5.14 and 5.15.) The blood flows back from the head and arms to this side of the heart by way of a large vein called the **superior vena cava**. It leads into the upper right-hand chamber, the **right atrium** (ae-tree-um) of the heart. (See Figure 5.14.) Blood from the trunk and legs enters this same chamber via the **inferior vena cava**. Both the right and left atria (plural of atrium) are thin-walled chambers which lie above the ventricles. Their function is to collect blood and pass it into the main contracting vessels, the **ventricles** (ven-tri-kul).

The **right ventricle** has thicker, more muscular walls than the right atrium and is much larger. It is connected to the right atrium via the **tricuspid valve** which prevents the blood from flowing back into the atrium when the ventricle contracts. When the right ventricle contracts, it pushes blood out through a semilunar valve into the **pulmonary** (pull-mon-air-ee) **artery**. This vessel carries the blood to the lungs where its load of carbon dioxide wastes is released and a fresh load of oxygen is absorbed. This newly oxygenated blood then flows into the **left atrium** through the four **pulmonary veins**. This portion of the circulatory system, the heart to lungs and back, is called the **pulmonary circulation**. Contractions of the left atrium push the blood through the **bicuspid valve** into the **left ventricle**. This chamber is the largest and most heavily muscled of the heart. As it contracts, it must force the blood to every part of the body, from the brain to the smallest toe. The blood is pushed out from the left ventricle via the aortic semilunar valve, into the largest blood vessel in the body, the **aorta**. This portion of the circulatory system is called the **systemic system**, since it supplies blood to the remaining body systems.

Chapter 5 / Blood and Circulation

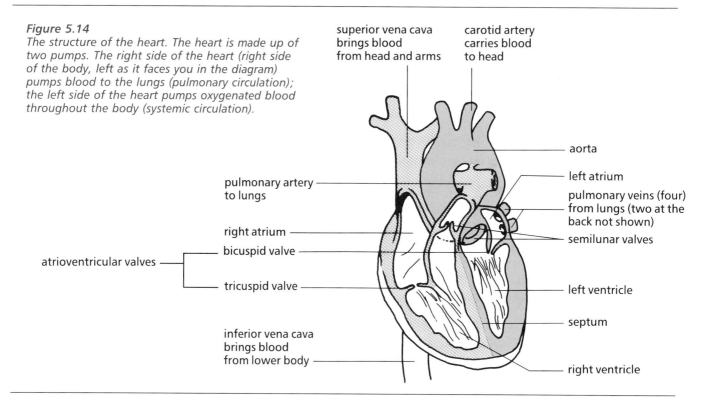

Figure 5.14
The structure of the heart. The heart is made up of two pumps. The right side of the heart (right side of the body, left as it faces you in the diagram) pumps blood to the lungs (pulmonary circulation); the left side of the heart pumps oxygenated blood throughout the body (systemic circulation).

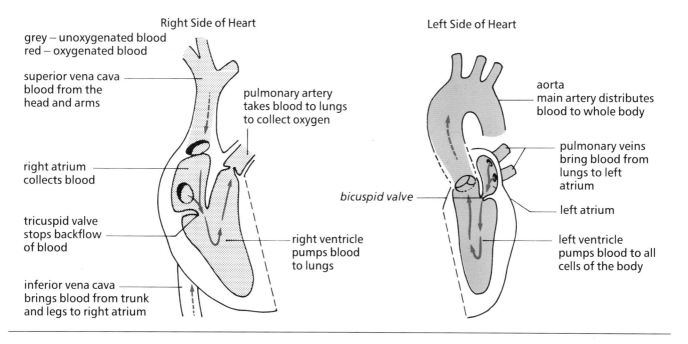

Figure 5.15
The two pumps of the heart

The Heart Valves

In the right side of the heart, the tricuspid valve connects the atrium and ventricle. (The tricuspid and bicuspid valves are also known as the *atrioventricular* valves after the two chambers that they separate, the atrium and ventricle.) The tricuspid valve is closed while the atrium is filling with blood. When this is completed, the tricuspid valve opens and a wave of contractions starts at the top of the atrium and squeezes the chamber inward and downward, forcing blood through the valve into the right ventricle. The valve then closes. Another wave of contractions starts, this time at the bottom of the ventricle. The blood is now squeezed upward through the semilunar valve, which lies at the entrance of the pulmonary artery leading to the lungs. This process is repeated over and over with each beat of the heart. The same sequence of valve action takes place in the left side of the heart.

Activity 5B

Heart Sounds

The sounds produced by the heart are caused mainly by the vibrations produced when the heart valves close and the blood bounces against the walls of the ventricles or blood vessels. The *stethoscope* makes it much easier to hear these sounds. Some microphones can also be used to magnify the sounds for discussion by the class.

- *The first sound* is produced when the valves between the atria and ventricles close and the semilunar valves open. This sound has a lower pitch and makes a "lub" sound. It occurs at the beginning of systole.
- *The second sound* occurs at the end of systole. This sound is produced by the closure of the semilunar valves and the opening of the valves between the chambers. The pressure in the arteries is higher and the pitch is also somewhat higher. The sound is a short "dub" sound.

Procedure

CAUTION!
Swab the ear pieces of the stethoscope with 70 percent alcohol.

1. Look at Figure 5.16 showing the locations where the sounds are best heard. Listen to your partner's heart, using the four areas suggested by the diagram.

Figure 5.16
Areas where the heart sounds can be heard most clearly

1. the semilunar valve into the aorta
2. the pulmonary semilunar valve
3. the tricuspid valve
4. the bicuspid valve

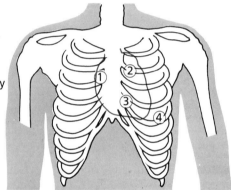

2. *How many distinct sounds can you hear? State the causes of each of the heart sounds that you can hear.*

3. *Which sound persists for the longest time? Did you notice any missed beat? What variations in sound occur between the different areas?*

4. Listen for a pulse in the inner wrist area and neck area. *How do the sounds you hear differ from those of the heart?*

5. Run on the spot for 1 min and then listen to your heart again. *What differences are there in the sounds that you hear? Describe the differences in your notebook.*

Activity 5C

Heart Dissection

The basic structure of the mammalian heart you will dissect is the same as that of a human heart.

Materials

sheep, pig, or cow heart
dissection tools
dissection trays
disposable gloves
goggles

Procedure

CAUTION!
Wear disposable gloves and goggles while performing this dissection.

1. Look carefully at the external features of the heart. Determine which is the front and which is the back of the heart.

2. Look for the stub ends of the blood vessels that have been cut and *identify each*. Note the thickness of the walls in each of the vessels. *Which type of vessel has the thickest wall? Explain why this is so.*

3. Examine the external blood vessels on the surface of the heart. *Explain the function and tell why these vessels are so important.*

4. Using a scalpel, cut the heart down each side from top to bottom, so that each side of the heart is opened and you can see into the atria and ventricles on each side. (See Figure 5.17.)

5. *Identify the chambers of the heart. Note which chambers have thick walls and explain why the differences are present between the atria and ventricles, and the two ventricles.*

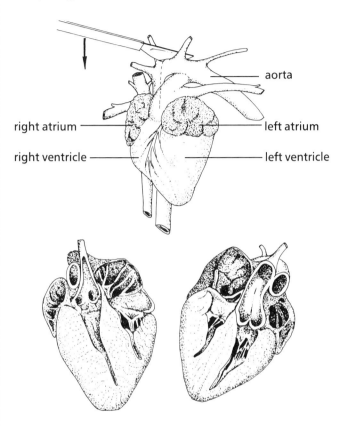

Figure 5.17
Exposing the chambers of the heart

6. Use a blunt probe and push it through each of the vessels entering or leaving the heart. *Locate and identify the valves present.*

7. Locate the valves that are present between the atria and ventricles on each side of the heart. *Describe how the cusps or flaps of these valves are attached and how they are opened. How do these valves differ from the valves found in the pulmonary artery and the aorta?*

8. State the function of the valves in the heart. When the right atrium is filling, would the tricuspid valve be open or closed? When the left ventricle is contracting, would the bicuspid valve be opened or closed?

9. What differences can you find between the aorta and the pulmonary vein?

10. When you have completed your dissection, dispose of the heart and any separated tissues as instructed by your teacher. Clean and dry the dissection instruments, trays, and bench surfaces as directed by your teacher. Wash your hands and lower arms thoroughly with soap and water.

A Round Trip

Consider a single round trip through your circulatory system. During the beat of your heart, your right atrium receives blood collected through your body and delivers it to the right ventricle. From the right ventricle, blood goes to the lungs to pick up oxygen and release carbon dioxide. At the same time, blood rich in oxygen is being received from the lungs by your left atrium. The left atrium sends this blood into your left ventricle. From here, it is returned to the circulatory system for transport to the rest of your body.

Figure 5.18 shows the complete picture of blood flow in the body. Reading about a round trip through your circulatory system takes far longer than the real thing. The average time for one complete trip, from left atrium to right ventricle, is a mere 20 s.

CHECKPOINT

1. *Without looking at your text, draw a quick outline of the four chambers of the heart and label these chambers.*

2. *Add to your sketch the main vessels leading in and out of the heart and label them.*

3. *Draw two lines through the vessels and chambers to show the route that blood takes through each side of the heart.*

4. *What are the functions of the heart valves?*

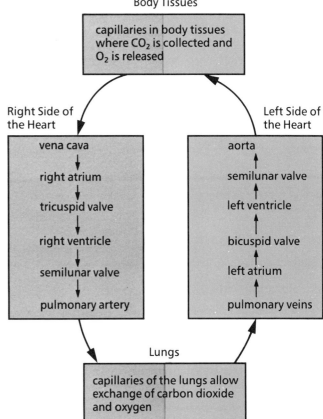

Figure 5.18
The sequence of vessels and valves through which the blood flows. Although only the exchange of gases is explained in this illustration, it is important to realize that nutrients and wastes are also exchanged in the body tissues.

5.5 The Regulation of Heart Rate

As we have seen, the heart has two unique characteristics – it beats automatically, and it beats rhythmically in a continuous manner. The average rate at which a person's heart beats will vary depending upon her or his sex, age, and health. (See Figure 5.19.)

Figure 5.19
The relationship between resting heart rate and age. The upper and lower lines represent the limits within which the normal heart rate may be expected to fall. There are considerable differences between individuals; the figures are approximate only. Note the decrease in heart rate that occurs at the time of birth (indicated at A).

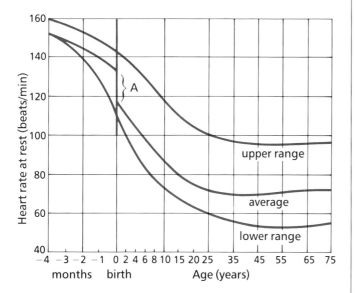

Figure 5.20
The nodes and nerve-conducting pathways of the heart

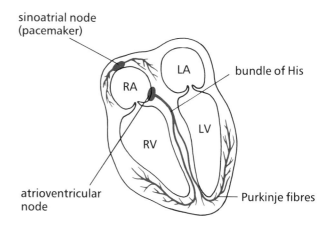

Nervous Control of the Heart

The parts of the heart normally beat in an orderly sequence. First, both atria contract, then both ventricles, then the entire heart relaxes and fills with blood. The period of contraction is called **systole**. The period of relaxation is called **diastole**. You heard this sequence in Activity 5B by listening to the heart through a stethoscope. At the beginning of systole, a low-pitched "lub" can be heard, caused by the turbulence as blood rushes into the ventricle. The softer "dub" marks the end of systole and the beginning of diastole. The ventricles are emptied and blood rushes into the major arteries.

This sequence is controlled by a specialized system of muscle tissue in the heart. (See Figure 5.20.) The **sinoatrial node** (or SA node) is located in the wall of the right atrium, near where the superior vena cava enters with its blood from the body. It is in this node that stimuli arise to begin each beat. For this reason, the SA node is also called the **pacemaker**.

When the pacemaker tissue initiates the impulse, it spreads through both atria and they, too, contract. At the same time, the impulse travels through a special pathway to the **atrioventricular node** (or AV node). When the AV node receives the impulse, it triggers the contraction of the ventricles.

Factors Affecting Heart Rate

If nothing occurred to influence the pacemaker, it would produce a constant steady beat, regardless of the body requirements. There are, however, a number of factors that change this pace, according to the body demands. The stimuli for these changes come from both the parasympathetic and the sympathetic divisions of the autonomic nervous system, centered in the medulla. Sympathetic nerves act to quicken the heartbeat, while parasympathetic stimulation slows the rate. There are many chemical receptors in the blood vessels. Some of these are sensitive to oxygen and others are sensitive to carbon dioxide.

These receptors keep the autonomic nervous system informed of changes in the oxygen content of the blood.

The rate at which the heart pumps depends, for example, on the amount of activity or muscle effort in which the individual is engaged. If you are resting and relaxed, your heart rate will be quite slow; if you run or exercise, your heart will automatically speed up in response to the increased activity. In addition, emotions, such as fear, excitement, shock, and tension, will affect the heart rate.

Certain chemicals can also change the heart rate. Carbon monoxide will slow the heart and excess oxygen will speed it up. Drugs, such as nicotine, caffeine, and alcohol, as well as body hormones (e.g., adrenalin) secreted by the body, will also affect the heart rate. Adrenalin increases both the force of the beat and the rate at which the heart beats. Thyroxin, a hormone from the thyroid gland, increases overall body metabolism, including that of the heart.

Activity 5D

Pulse Rate

The pulse can be used to measure heart rate. It varies widely with amount of exercise, emotional state, and general health. It also varies between the sexes.

Procedure

CAUTION!
Students who do not participate in physical education for medical reasons should not perform the exercise portion of this activity themselves.

1. Take your pulse either at the radial artery or at the carotid artery in the neck. If the radial artery is used, place the fingers firmly over the radial artery along the inside of the wrist, below the line of the thumb. This will press the artery against the bone and the pulse can readily be felt. (See Figure 5.21.)

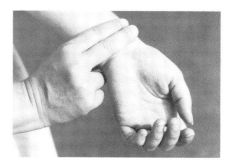

Figure 5.21
Taking the radial pulse

2. Having found your pulse, count the number of beats in 1 min while at rest (lying down), sitting, and standing. Now run in place for 1 min. Make sure that you lift your knees and work hard. Take your pulse *immediately* following the exercise. Count the number of beats in 15 s and multiply your answer by 4 to get the beats per minute. After resting for exactly 2 min, take your pulse again and *record all your results in a table in your notebook*. Use the following table as a guide.

	At rest	Sitting	Standing	1 min Exercise	Recovery rate
Males Class average Females Class average					

A chart of all the results on the chalkboard may enable you to draw some other conclusions. *Compare the results for females and males. Compare the results of students who exercise regularly with those who do not. Compare the results of non-smokers and smokers. Are there major differences in their results? Look especially at the recovery rates. What other factors may affect these results?* (Note: Your heart changes its rate rapidly, so take the pulse immediately after the exercise.)

CHECKPOINT

1. *What is the role of the pacemaker of the heart?*
2. *What causes the two distinct sounds that are heard through a stethoscope when listening to the heart?*
3. *Name and briefly describe three factors affecting heart rate.*
4. (a) *Why can we find the heart rate by holding the wrist?*
   (b) *What characteristic of arteries makes this possible?*

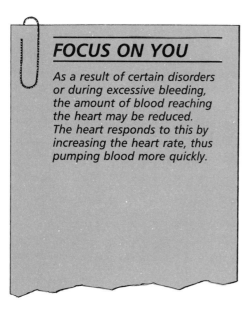

FOCUS ON YOU

As a result of certain disorders or during excessive bleeding, the amount of blood reaching the heart may be reduced. The heart responds to this by increasing the heart rate, thus pumping blood more quickly.

5.6 To Your Good Health

Blood Pressure

If the blood is to reach the hands and feet, the brain, and every part of the body, it must be pumped out of the heart under very considerable pressure. The highest pressure occurs in the aorta, the great vessel that carries oxygenated blood away from the heart. As the blood passes into smaller vessels and the distance from the heart becomes greater, the pressure is greatly reduced.

The pressure in any vessel varies as a result of five major factors:

- **The amount of blood.** If, as a result of some injury, a person has lost a lot of blood, the pressure in the system drops because of the decrease in volume.

- **The heart rate.** The faster the heart pumps blood, the greater the pressure which is built up. The pressure falls as the heart rate decreases, especially during rest or sleep.

- **The size of the arteries.** When the arteries dilate (become larger in diameter), the volume of the vessels increases and the pressure falls. If the arteries constrict, pressure is built up because of the extra resistance to blood flow.

- **Elasticity.** The walls of the arteries must be flexible and elastic. They must be able to expand as a surge of blood is forced out of the heart, and then relax after the surge has passed. If they cannot stretch in this way, they are described as hardened. Hardening of the arteries is a condition common in older people. When the arteries do not expand to accommodate blood flow, the blood pressure in the system is increased. (See Figure 5.22.)

- **The viscosity of the blood.** Viscosity refers to the thickness of the blood. Thick, sticky fluids flow less readily than thin watery liquids. The balance between the number of red blood cells and the amount of plasma present is one factor that controls the viscosity of the blood.

Figure 5.22
Deposits of cholesterol can severely reduce the diameter of blood vessels, causing blood pressure to rise.

Measuring Blood Pressure

The usual blood pressure for young adults ranges between 120/70 and 115/75 mm Hg. (See Figure 5.23.) The numbers refer to the pressure in millimetres of mercury. The numerator of this fraction represents the highest pressure generated when the ventricles contract. It is known as **systolic pressure**. The denominator shows the **diastolic pressure** recorded when the ventricles relax and the elastic walls of the arteries offer the least resistance.

The blood pressure is taken with a **sphygmomanometer** (sfig-moe-man-om-eter). This instrument uses an inflatable cuff to prevent blood passing through the artery of the arm. By using a stethoscope and listening for the sound made by the blood trying to force its way through the constricted artery, the blood pressure can be determined. (See Figure 5.24.) The cuff is inflated to create a pressure above normal in the arm artery (about 160 mm Hg), and it is then gradually lowered. At the point when the blood just manages to squeeze through the con-

Figure 5.23
The relationship between blood pressure and age. The lines show systolic and diastolic pressures that are average for each age group. Variations beyond these limits do not necessarily indicate a disorder.

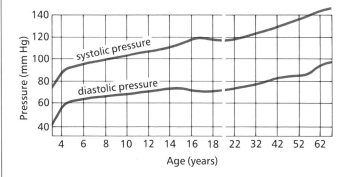

stricted artery, the blood can be heard as it forces the artery open and the systolic pressure is determined. When the sounds disappear, it is a signal that the blood can pass through the artery easily.

Figure 5.24
Taking a blood pressure reading with the sphygmomanometer

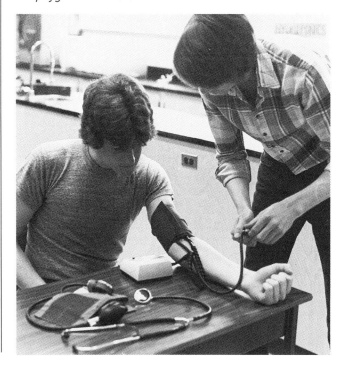

BIOTECH

Heart Attack

It has often been said that we die of the habits we adopt as teenagers. This is true of smoking and death by lung cancer, but also true of lifestyle practices that result in coronary disorders. Heart attacks are the leading killer of both men and women in North America. As you read the list of factors that place a person at risk for heart attack or coronary heart disease, consider your own attitudes and activities which form the elements of your own lifestyle.

Some identified risk factors:

- **Physical activity**. The more sedentary (physically inactive) a person is, the higher the chance of coronary heart disease.

- **Obesity**. Risk of coronary heart disease increases significantly with serious weight problems. The risk is closely associated with high intakes of foods rich in sugar and animal fats which may result in higher levels of cholesterol. Obesity also puts a strain on muscle tissue.

- **Smoking**. Inhaled smoke significantly increases the chances of both heart attack and stroke.

- **Stress**. Prolonged emotional stress and tension aggravate and even precipitate heart attacks.

- **Hypertension**. The higher the blood pressure, the greater the chance of coronary heart disease.

- **Age and sex**. Males are more likely to develop coronary heart disease earlier in life than females. Males also tend to have more serious heart conditions than females.

Location of heart attack pain

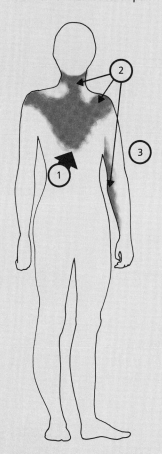

1. Painful squeezing or pressure in centre of chest
2. Pain may spread to shoulder, arm, neck, or jaw
3. Pain often accompanied by sweating, nausea, or shortness of breath

You can significantly reduce your chances of future heart attacks with a sensible lifestyle. Have regular medical checkups throughout your life. Recognize the warning signs, such as increased blood pressure or being very overweight. Be determined to make choices about your life that decrease the risks of heart attack.

The following is a list of things which you should know about heart attacks:

- Delay is dangerous. Minutes count, especially the first few minutes.
- Warning signs
  – Prolonged, heavy pain or unusual discomfort in the centre of the chest. Pain may radiate to the shoulder, arm, neck, or jaw.
  – Sweating may accompany pain or discomfort.
  – Nausea, vomiting, and shortness of breath may also occur.
- Act immediately. Sometimes the signs fade and then return. If one or more of these signs is present, call the doctor and describe the symptoms in detail, or get the person to the hospital emergency room at once. Be prepared to act on the doctor's instructions. Keep a list of numbers – doctor, hospital, ambulance, and other emergency services, such as the police – next to the telephone and in a prominent place in your wallet or purse.
- Take a course in CPR (cardiopulmonary resuscitation). You may save a life.

The artery remains open because the diastolic pressure is sufficient to balance the pressure of the cuff when the heart is relaxed between contractions. At this time, the diastolic pressure is determined.

(The SI unit of measurement for blood pressure is the kilopascal (kPa). Most sphygmomanometers are calibrated using mm Hg, the older unit of measurement. For this reason, the unit mm Hg is used in this text. To convert from mm Hg to kPa, consider 100 mm Hg as equivalent to 13.3 kPa.)

Activity 5E

Blood Pressure

The blood pressure cuff or sphygmomanometer is a valuable diagnostic aid.

Materials

stethoscope
sphygmomanometer

Procedure

1. Wrap the cuff around your partner's arm snugly, just above the elbow. If there is a white ring or dot on the cuff, make sure that it is just over the brachial artery.

    CAUTION! Swab the ear pieces of the stethoscope with 70 percent alcohol.

2. Place the stethoscope just under the rim of the cuff over the brachial artery. Inflate the cuff, first closing the valve on the rubber bulb. Inflate to a reading of about 160 mm Hg, by squeezing the rubber bulb.

3. Slowly release the air by opening the valve just a little. Listen carefully for the first sounds of a pulse.

4. As soon as you hear the first quiet sounds of a pulse, *record the systolic pressure*. Continue to allow the air to escape and to listen to the sounds of the pulse. At the point when you can no longer hear the sounds of a pulse, *record the diastolic blood pressure*.

5. *Record the blood pressures in the following positions or after the activity suggested.* Use a table similar to the one shown.

At Rest, Lying Down	At Rest, Sitting Down	At Rest, Standing	After Exercise, Still Standing

6. Keep the results for females and males separate. *Average the results for each sex. Compare the results of those who exercise regularly with those who do not. Compare results of non-smokers with smokers. What generalities, if any, can you discover? Consider any other factors that may influence these results.*

Questions

1. Refer back to your results on heart rate. What similar patterns, if any, can you describe?

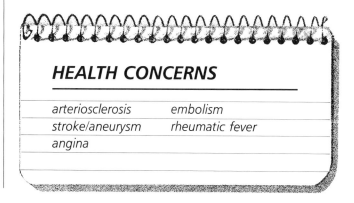

HEALTH CONCERNS

arteriosclerosis   embolism
stroke/aneurysm   rheumatic fever
angina

Hypertension

Hypertension occurs when a person's arterial blood pressure is significantly above average most of the time. This usually involves sustained systolic pressure above 140 mm Hg or a sustained diastolic pressure above 90 mm Hg.

The greatest dangers for people with high blood pressure are cerebral hemorrhage or heart failure. A cerebral hemorrhage occurs when the pressure becomes too great for some small brain arteries to withstand. If an artery feeding a part of the brain has a weak wall, it may burst, thereby preventing that part of the brain from receiving its supply of oxygenated blood. These brain cells may die causing the patient to lose control of some related part of the body.

Monitoring blood pressure is a way of catching medical problems before damage occurs. People with high blood pressure can be helped by certain medications as well as counselling about their lifestyle.

CHECKPOINT

1. *What is the difference between systolic and diastolic blood pressure?*
2. *What factors affect blood pressure?*
3. *What is hypertension and why is it so dangerous?*

BIOLOGY AT WORK
AMBULANCE ATTENDANT
RED CROSS WORKERS
HEMATOLOGY TECHNICIAN
SURGICAL NURSE
MEDICAL TOOLMAKER

BIOTECH

EXERCISE PROGRAMS

Exercise can be anything that provides plenty of physical movement, makes you breathe deeply, and produces a sweat. A vigorous walk up several flights of stairs is exercise, just as much as hockey or basketball. Four factors need to be considered in any healthy exercise program – intensity, frequency, duration, and type of exercise.

Intensity
Heart rate is an excellent indicator of intensity of exercise. Proper exercise should raise your heart rate to a level that produces a *training effect*. For the average person, the training effect begins when your pulse rate, in beats per minute, is at a level of 170 minus your age. If you are 16, your training effect would therefore start at 154 beats per minute (170 less 16). There is a safe upper limit for your exercise program, measured by the difference between 200 and your age. If you are 16, this rate would be 184 beats per minute (200 less 16). There is no benefit in going over this limit. Doing so could overstress the heart. If you adopt a program of workouts, you will find that your fitness level improves and you can gradually work harder within safe limits.

Frequency
Intensive exercising every other day is about right for the maintenance of good cardiorespiratory fitness. The days when you do not exercise as hard allows time for your muscles and connective tissue to recover.

Duration
You should plan on at least 15 min just to "warm up" to the level of activity that gives you the training effect. You should then exercise for a period that is long enough to maintain your pulse rate within the training effect (at least another 15 min). A period of "cool down" after each session of vigorous exercise, such as walking around and gradually relaxing, allows the blood to be redistributed from the muscles to the other organs of the body.

Type of Exercise
Aerobic exercise is the best type of activity for building cardiorespiratory fitness. It provides the body with rich supplies of oxygen to fulfill all its metabolic needs. Aerobic exercise includes walking, jogging, skating, cross-country skiing, cycling, and swimming. If you can talk to a partner as you work out, you are working at the correct level of intensity.

How Fit Are You?
The following test can give you an idea of the approximate efficiency of your heart, lungs, and muscles in response to exercise.

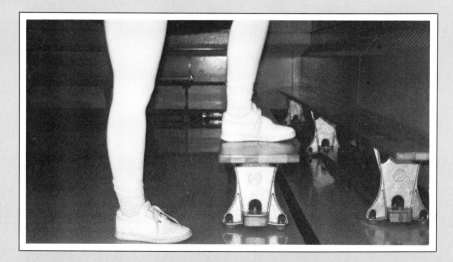

CAUTION!
If you are exempt from physical education for any medical reason, you should not attempt this test.

The Step Test
Choose a bottom stair or a platform step about 20 cm high. Commence the step pattern as follows. Step up with the right foot and bring up the left. Step down with the right foot and bring down the left.

Practice this right-left, right-left pattern at a rate of 24 times a minute. Once you are comfortable with the rhythm, rest for a moment and then commence the test, keeping a steady pace for 3 min.

CAUTION!
Do not continue the exercise if you feel unpleasantly out of breath or dizzy.

Stop after 3 min and wait for exactly 1 min. Then count your heartbeats by counting your pulse over the next 15 s. Refer to the fitness table below and determine your rating.

	Beats per 15 s		Fitness Rating
	Males	Females	
Pulse counted in 15 s	Less than 18 18 to 20 21 to 25 More than 25	Less than 20 20 to 22 23 to 28 More than 29	Excellent Good Average Poor

Chapter 5 / Blood and Circulation

CHAPTER FOCUS

Now that you have completed this chapter, you should be able to do the following:

1. State the components of blood and give the functions of each.
2. Explain how the clotting of blood takes place.
3. Name the four blood types and explain why the mixing of certain blood types causes clumping of the red blood cells.
4. Explain why the Rh factor is important.
5. Differentiate between arteries, veins, and capillaries in both structure and function.
6. Explain how blood is returned from the body tissues to the heart.
7. Draw a simple diagram of the heart and label the parts.
8. Describe the routing of blood through the heart.
9. Explain the function and need for valves in the heart.
10. Explain how heart rate is controlled.
11. List the factors that affect heart rate.
12. Explain the difference between systolic and diastolic blood pressure.

SOME QUESTIONS TO ANSWER

1. In which part(s) of the body are red blood cells and white blood cells produced?
2. (a) What is anemia?
   (b) How can a person's diet determine whether he or she will get some kinds of anemia?
3. If a sample of type A blood is mixed with type B blood, the cells clump together. Explain what causes this reaction.
4. Why are persons with type O blood known as universal donors?
5. What would happen if type A blood was transfused into a person with type AB blood? Explain your answer.
6. What is the advantage of keeping a key ingredient of the clotting reaction inside the platelets? In your answer, include the sequence of events that occurs when blood clots.
7. (a) Compare arteries, veins, and capillaries. Note the differences in both structure and function.
   (b) Consider answer (a) and explain why veins rather than arteries are used for injections, transfusions, and withdrawing blood samples.
8. Draw and label a simple diagram of the heart. Use arrows to indicate the direction in which blood flows through each of the chambers and vessels.
9. (a) List four situations that you have experienced in the last 12 h, which caused your heart to speed up. Give three situations in which your heart rate slowed down.
   (b) What was the importance of your heart's reaction to these situations?
10. Explain the need for valves in the heart.
11. (a) What is the normal heart rate for a person of your age and sex?
    (b) What is the significance of recovery rate of the pulse after exercise?
12. A doctor received the blood test results from two of his patients. For person A, the number of red blood cells was lower than normal for her age and sex; for patient B, the white blood cell count was higher than normal. For each patient, discuss what condition, or disease, could be causing these symptoms. Explain the reasons for the answers you give.
13. How does blood get back to the heart when the pressure in the veins is so low?
14. Why, when an accident victim suffers blood loss, might he or she be transfused with plasma rather than whole blood?

SOME WORDS TO KNOW

Match each of the descriptions given in the left-hand column with a word shown in the right-hand column. DO NOT WRITE IN THIS BOOK.

THE BLOOD

1. A substance required for the clotting of blood
2. The liquid part of blood
3. Present in bright red arterial blood
4. Contains prothrombin
5. Produced by glands and carried to other organs in the blood
6. Pigment containing iron and found in the red blood cells
7. A condition in which only a limited number of red blood cells is present
8. Example of a phagocyte
9. A cancerous disease of the blood-forming organs
10. Blood cells with no nuclei

A hormones
B red blood cells
C white blood cells
D oxygen
E mononucleosis
F plasma
G platelets
H hemoglobin
I anemia
J leukocyte
K lymphocyte
L leukemia
M calcium
N fibrinogen
O antibody
P antigen

THE HEART

1. Chamber from which blood flows into the aorta
2. Chamber which pumps blood to the lungs
3. Pressure created when the ventricles contract
4. The pacemaker
5. Vessels which carry blood away from the heart
6. Blood vessels where the exchange of materials takes place
7. Valve found between the right atrium and right ventricle
8. Pressure created when the ventricles are filling with blood
9. Found around cells in the tissues
10. Separates and disposes of old red blood cells

Select any two of the unmatched words and, for each, give a proper definition or explanation in your own words.

A right ventricle
B left ventricle
C right atrium
D left atrium
E SA node
F AV node
G spleen
H aorta
I vena cava
J arteries
K veins
L capillaries
M tricuspid valve
N bicuspid valve
O systole
P diastole
Q interstitial fluid
R lymphatic system

SOME THINGS TO FIND OUT

1. Write a short report on any one of the following:
   gamma globulin
   first aid and the pressure points for treating bleeding
   Red Cross blood donor clinics
   William Harvey – the discovery of circulation
   coronary by-pass
   the CPR program in first aid
   the intensive care unit
   heart transplants
   tissue rejection.

2. Make a visit to a hematology laboratory and find out about the many tests that are performed on blood.

3. How is blood stored, transported, and prevented from clotting after it has been taken from a donor?

4. What are the latest findings on cholesterol and its formation in the blood vessels?

5. Investigate the many innovations and artificial parts that can be used in the heart, such as pacemakers, valves, etc.

RESPIRATION

He's stopped breathing! This electrifying cry means that there is no time to waste if a life is to be saved. Permanent brain damage will result in less than 5 min if the supply of oxygen to the body is cut off.

In any emergency involving a person's health, breathing is the first concern. People in health or rescue work are trained in breathing assistance techniques. Emergency response vehicles such as ambulances and fire trucks are equipped with respirators.

You must breathe to survive. In this chapter, you will learn why oxygen is so vital. You will also learn how your respiratory system works – and how to take better care of it.

- *Respiration supplies the body's cells with oxygen and removes waste carbon dioxide.*
- *The body uses oxygen to release energy from food.*
- *Air entering the body is warmed, cleaned, and moistened.*
- *The lungs provide a large surface area for the rapid and efficient exchange of gases.*
- *Air pollution, including tobacco smoke, is a serious hazard to health.*

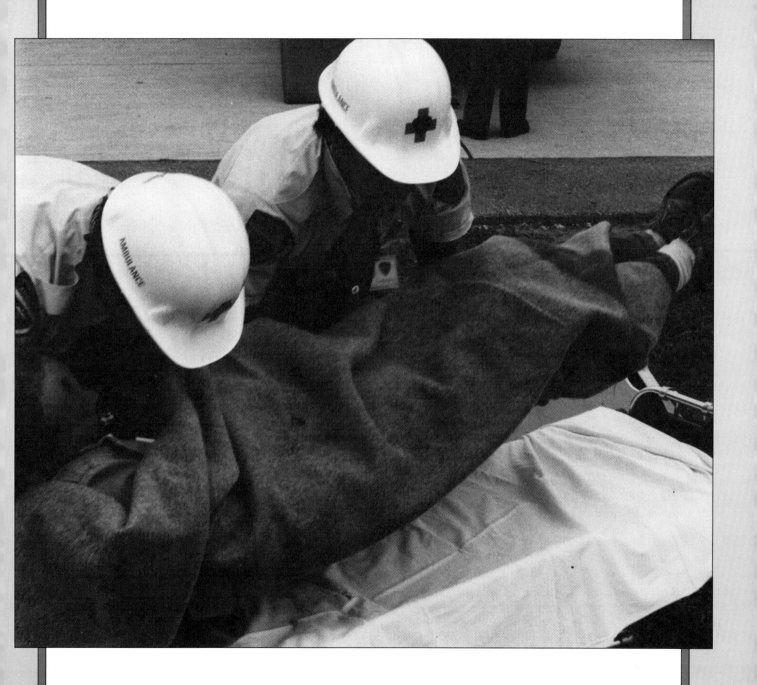

Chapter 6 / Respiration

6.1 The Need for Oxygen

If a machine is to operate, it must be supplied with energy. For example, cars need gasoline. In a similar way, humans operate on chemical energy, which we obtain from food, especially the sugar glucose.

The energy in gasoline is released by combustion – burning in the presence of oxygen. A spark is required to begin the reaction. This type of burning would, of course, not be possible within the body. Instead, the energy in a glucose molecule is released by a slower, more controlled process called **cellular respiration**.

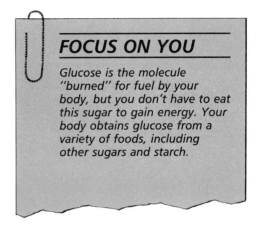

FOCUS ON YOU

Glucose is the molecule "burned" for fuel by your body, but you don't have to eat this sugar to gain energy. Your body obtains glucose from a variety of foods, including other sugars and starch.

Internal Combustion Engine

gasoline + oxygen + spark $\longrightarrow$ carbon dioxide (and other waste products) + water + energy

Cellular Respiration

glucose (sugar) + oxygen $\xrightarrow{\text{enzymes}}$ carbon dioxide + water + energy

As you can see from these word equations, cellular respiration also requires oxygen in order to release the energy from its fuel, glucose. It is the function of the respiratory system to bring this oxygen into the body and to provide a site where oxygen can be transferred into the blood. The circulatory system will then transport the oxygen to all the cells of the body where energy reactions take place. The respiratory system also provides the means by which the waste gas of cellular respiration, carbon dioxide, can be eliminated from the body.

6.2 The Air-Conducting Structures

The structures involved in respiration are the nasal cavity and sinuses, pharynx, larynx, trachea, and the lungs, consisting of the bronchi, bronchioles, and alveoli.

The Nasal Cavity

As air passes through the **nasal cavity**, it is conditioned in three important ways:

- **The air is *warmed*.** The nasal chamber is lined with tiny capillaries filled with warm blood. As the air passes over these surfaces, the heat is transferred to the incoming air. Extending into the nasal chamber, like sagging shelves, are the **turbinate bones**. (See Figure 6.1.) These bones help to increase the amount of surface area. The greater the surface area in contact with the air, the more efficiently the warming is accomplished. On a cold day, this process can

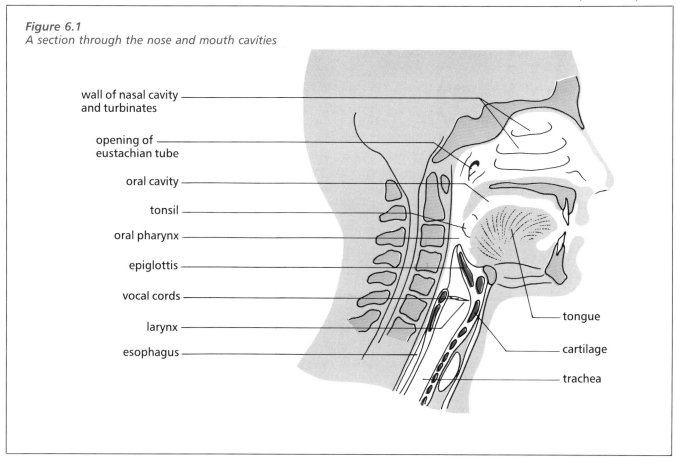

Figure 6.1
A section through the nose and mouth cavities

raise the temperature of air from freezing to almost body temperature by the time it enters the lungs.

- **The air is *moistened*.** The airways are lined with mucous membranes. As the air passes over their surfaces, it picks up moisture from the fluid mucus. The moisture that has been added to the air can easily be seen when we exhale on a cold day and our breath condenses in the cold air.

- **The air is *cleaned*.** The dust particles contained in the incoming air are trapped in the hairs, cilia, and mucus that line all the surfaces of the nasal cavity and the initial airways to the lungs. (See Figure 6.2.)

Figure 6.2
Microscopic view of the cilia (C) lining the nasal passages. Note the particle (P) of dust caught by the cilia.

Activity 6A

The Difference Between Inspired and Expired Air

Problem
How does air entering the nose differ from air leaving the nose?

Materials
thermometer
glass plate
small beaker
limewater
a straw
alcohol for sterilizing the thermometers

Procedure

CAUTION!
The thermometer must be cleaned thoroughly with alcohol before being placed in the mouth.

1. Using the thermometer, first find the air temperature of the room. *Record this result.*

2. Place the thermometer between your teeth or lips about 2 to 3 cm into your mouth. Do not allow the thermometer to touch the tongue. Breathe out rapidly so that the air passes over the bulb of the thermometer, and determine the temperature of the exhaled air. *Record this result, then, by subtraction, find out the increase in temperature caused by the air being warmed in the lungs.* (Note: You must read the thermometer immediately after it is removed from the mouth.)

3. Breathe out onto a glass plate. *Note what collects on the surface of the glass and explain what you see.*

4. Fill the beaker half full of limewater. *Gently blow through the straw into the limewater. The contents of the beaker will turn "milky" as a white precipitate of calcium carbonate forms. What gas is tested for with this experiment?*

5. *Summarize your results.*

Questions

1. What three changes between the composition of exhaled air and inhaled air have these tests demonstrated?

2. What structures in the respiratory system are responsible for warming and adding moisture to the air entering the lungs?

3. Where does the exchange of carbon dioxide and oxygen take place?

The Pharynx

The **pharynx** (fair-inks) forms a tube common to both the respiratory and digestive systems. It starts at the back of the nasal cavity and extends down to the larynx (voice box). The upper part of the pharynx is covered with ciliated cells that trap the fine particles in the air. It also contains the **tonsils** and **adenoids**, each consisting of a mass of lymphoid tissue which helps to prevent the entry of infectious pathogens through this opening into the body.

The lower portion of the pharynx passes behind the mouth cavity and forms a passageway for both food and air. Its walls are lined with epithelial cells which can stand up to the rough wear and tear of foods passing through. This tough lining extends to the end of the pharynx, which divides into two tubes, one carrying food (the esophagus), the other air (the trachea).

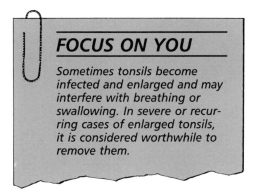

FOCUS ON YOU

Sometimes tonsils become infected and enlarged and may interfere with breathing or swallowing. In severe or recurring cases of enlarged tonsils, it is considered worthwhile to remove them.

The Larynx and Epiglottis

The **larynx** (lair-inks) is a boxlike structure located at the opening to the respiratory passageway. It is formed by several groups of cartilage. The largest of these, the **thyroid cartilage**, forms the framework of the Adam's apple. Above this is a leaf-shaped flap that forms the **epiglottis**. The epiglottis seals the opening into the respiratory tract during swallowing, thereby preventing the passage of food into the lungs. (See Table 6.1.)

The larynx contains two stretched bands of cartilage controlled by muscles, the **vocal cords**. (See Figure 6.3.) When we speak, air passes out of the lungs and through the larynx. It causes these cords to vibrate and produce sounds, in much the same way as we can make a blade of grass produce sounds when we hold it tightly between our thumbs and blow on its edge. The pitch of sounds made by the vocal cords can be changed by tightening or loosening the muscles that hold the cords in place. The thickness and length of the cord also affect the quality and pitch of the sounds produced. As the sounds pass through the mouth, they are moulded by the shape of the mouth and the position of the tongue into words that we can recognize.

The Trachea and Bronchi

The **trachea** (trae-kee-ah) is a tube, about 12 cm in length, which extends from the larynx into the chest cavity, where it divides into the right and left **bronchi** (brong-ky). The trachea is constructed of smooth muscle in which C-shaped rings of cartilage are embedded. (See Figure 6.3.) The rings serve primarily to prevent the trachea from collapsing if there are changes of pressure in the tube.

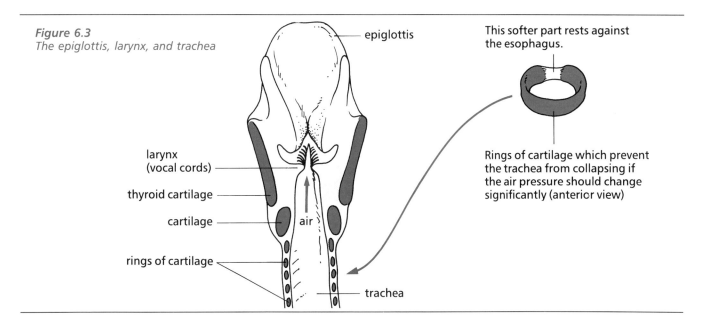

Figure 6.3
The epiglottis, larynx, and trachea

Chapter 6 / Respiration

Table 6.1 Helping a Choking Victim

CAUTION! Choking is an immediate problem. If the person is not getting air, there is very little time to get help and your swift action could save a life. **You should seek a proper demonstration of these techniques from a qualified instructor before trying them on your own. They are techniques That everyone should know.**	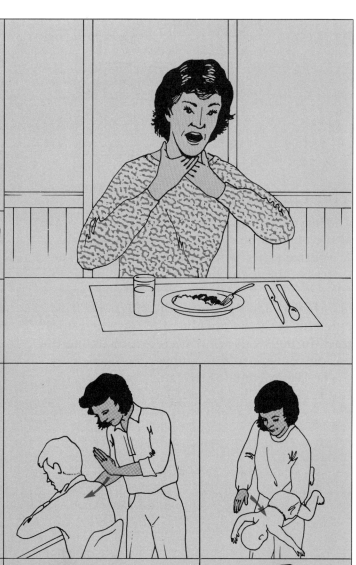
The usual sign that a person is choking is a characteristic holding of the neck with both hands and the inability to speak or draw a breath.	
The problem is usually caused by food or some object that is lodged in the airway. Trying to pull out the obstruction with your fingers is likely to push the object further down.	
Strike the victim sharply between the shoulder blades. The person should be bent forward so that the dislodged material is shot out of the mouth. Support the person so that your blows do not cause her or him to fall forward. You can also lay the person down with the head over the edge of a bed or table while you try to dislodge the blockage. If the victim is an infant, she or he may be held over the knee while you strike the back. If the infant is very small, it is also necessary to support the head.	
An alternative method is known as the **Heimlich manoeuvre**. Stand behind the victim and wrap your arms around the upper abdomen. Grasp your hands together so that they form a hard knot and pull in with an upward thrust, just below the point of the sternum. Use only one hand when helping small children in this manner. The intent is to cause a sharp expulsion of air that will force the obstruction out.	 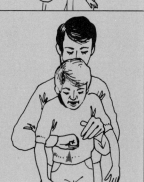

The right and left bronchi are also supported by small rings of cartilage. The bronchi, in turn, branch into smaller tubes, forming what is often called the bronchial tree. The smallest branches are called the **bronchioles**, which lack cartilage. They become smaller and smaller in diameter and more numerous, spreading throughout the entire lung tissue. (See Figure 6.4.) Finally, they lead into tiny chambers where gas exchange takes place. These chambers, or air sacs, are called the **alveoli** (al-vee-oh-li). (See Figure 6.5.) (See Table 6.2.)

Figure 6.4
The trachea and bronchial "tree". The inset shows the capillaries surrounding the alveoli where oxygen and carbon dioxide are exchanged.

The Trachea and Bronchial Tree

- pulmonary artery brings deoxygenated blood from heart, delivers blood to capillaries
- trachea brings air into lungs
- pulmonary veins take oxygenated blood back to the heart
- the pleura (two membranes with fluid between them)
- bronchioles divide into smaller and smaller branches
- alveoli (see inset)
- lining of alveoli bathed in moisture
- air
- bronchiole
- carbon dioxide-rich blood
- oxygen-rich blood
- alveoli
- capillaries

The Alveoli and Capillaries

Figure 6.5
A section of a normal lung showing some alveoli. Note the thin walls of the alveoli and the small blood vessels in those walls. How do these features help the process of gas exchange in the lungs?

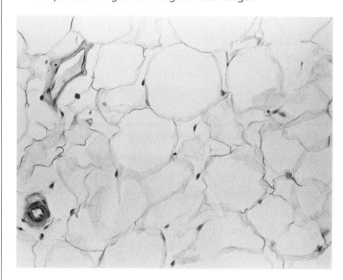

Table 6.2 Comparison of the Numbers and Size of Airway Tubes in the Respiratory System. The branching structure of the bronchial tree makes it very effective. As the number of tubes increases, although the diameter of each tube decreases, the total cross-sectional area increases.

Organ	Number of Tubes	Diameter (mm)	Total cross-sectional area (cm²)
trachea	1	18 to 25	2.5
bronchi	2	12	2.3
small bronchi	1020	1.3	13.4
bronchioles	262 000	0.5	534
alveolar ducts	4 200 000	0.4	5580
alveoli	300 000 000	0.2	50 to 70 m²

The Lungs

The lungs are two cone-shaped organs moulded into the form provided by the thoracic cavity. They are well-protected by the surrounding ribs, sternum, and spine. The base of each lung lies in contact with the diaphragm and the top of each lung reaches just above the clavicles.

The Pleura

The lungs are contained within the **pleura**, two membranous sacs which surround the lungs. The outer membrane lines the inner surface of the chest wall and covers the upper surface of the diaphragm. The inner membrane adheres to the surface of the lungs. These two membranes are so close together that only a very thin film of fluid separates them.

The pleura help to isolate each lung. The film of fluid has a lubricating function. It reduces friction produced when the lungs move against the walls of the thoracic cavity. Because two smooth surfaces adhere closely together when there is a film of moisture between them, when the rib cage expands it pulls the lung wall with it. If you take two glass slides, wet them, and place the moistened surfaces together, you will find that it is quite difficult to separate them. This same action "glues" the lungs to the walls of the rib cage.

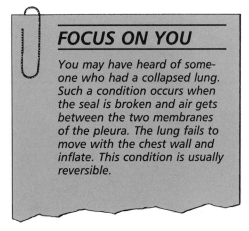

FOCUS ON YOU

You may have heard of someone who had a collapsed lung. Such a condition occurs when the seal is broken and air gets between the two membranes of the pleura. The lung fails to move with the chest wall and inflate. This condition is usually reversible.

Inflammation of the pleura is known as **pleurisy** and may be caused by pneumonia, tuberculosis, or influenza. When the membranes become inflamed, breathing becomes difficult and painful. Coughing, fever, and rapid, shallow breathing are common symptoms.

The right lung is divided into three lobes and fills most of the right side of the thorax. The left lung has only two lobes and shares some of its space with the heart.

CHECKPOINT

1. Why does the body need oxygen?

2. (a) What three changes occur in air as it passes through the nasal chamber?
   (b) What structures are responsible for these changes?

3. (a) Where are the vocal cords found?
   (b) How do they produce sound?

4. What two tubes form a junction with the pharynx?

5. What prevents the trachea and bronchi from collapsing when air pressure drops?

6.3 The Mechanism of Breathing

Moving Air In and Out of the Lungs

For air to enter the lungs, two basic actions must occur, both of which have the effect of increasing the volume of the thoracic cavity. One involves the diaphragm. The **diaphragm** (dy-ah-fram) is a thin, dome-shaped sheet of muscle, approximately level with the bottom of the ribs, which is stretched across the bottom of the thoracic cavity separating it from the abdominal cavity. This sheet of muscle is curved upward in the middle, like an upside-down saucer. As we breathe in, the sheet is pulled downward, which tends to flatten it out, thereby making the cavity above it larger in volume. The second action causes the rib cage to move upward and outward. This results from contraction of the **intercostal muscles** which lie between the ribs. (See Figure 6.6.)

Figure 6.6
The mechanics of breathing

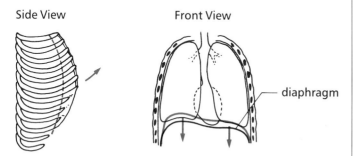

Inspiration. The ribs move upward and outward; the diaphragm moves downward and flattens. Both of these actions increase the volume of the chest cavity and decrease the pressure. As a result, air rushes in to equalize the pressure.

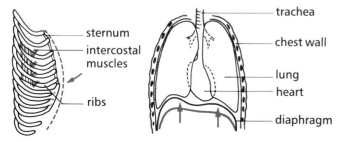

Expiration. The ribs and diaphragm return to their former relaxed positions. This decreases the volume of the chest cavity and increases the pressure. Air is forced out.

Inspiration (Breathing In)
The volume of the lungs increases as the chest wall moves upward and outward, and the diaphragm moves downward. As the volume increases, the pressure decreases. As the pressure decreases below outside pressure, air rushes in to equalize the pressure inside the lungs.

At the start of an intake of breath, air moves rapidly into the lungs. Then, as the two pressures become equalized, the air moves more slowly. The process of inspiration requires that muscles actively contract.

Expiration (Breathing Out)
As the diaphragm relaxes, it pushes up to regain its shape. The intercostal muscles in the chest wall relax and the ribs move down and inward. These movements decrease the volume of the lungs. As the volume decreases, the pressure inside increases. The increase in pressure pushes air out of the lungs until the internal and external pressures are equal once more.

Breathing out requires no muscle contraction; it is the result of muscle relaxation.

Activity 6B

Listening to Breathing Sounds

Two kinds of breathing sounds can be heard through a stethoscope.

- **Vesicular breathing.** This soft rustling sound is caused by air passing into the alveoli and small vessels in the lungs. The inspirational sounds last about twice as long as the expirational sounds.

- **Bronchial breathing.** This second type of sound is a loud harsh sound like a sharp whispered "hah". This is heard when air passes through the larger airways.

Materials

stethoscope
alcohol and cotton swabs for cleaning ear pieces

Chapter 6 / Respiration

Procedure

CAUTION! Swab the ear pieces of the stethoscope with 70 percent alcohol.

1. Tilt the head back and place the stethoscope gently over the trachea below the Adam's apple (larynx). Move the stethoscope up and down to find the best location and describe the sounds that you hear. Try taking deep breaths as well as normal breaths and *note any differences in what you hear.*

2. Place the stethoscope on your partner's back between, and at the base of, the shoulder blades. This is where the bronchi divide to each lung. Listen for the sounds that can be heard here.

3. Place the stethoscope on the side of the chest wall under the arms and listen. Then try other areas on the back below the shoulder blades and on the upper chest.

CAUTION! Students who are currently excused from physical education for medical reasons should not attempt this portion of the experiment.

4. Run in place for 1 to 2 min until you are breathing heavily, then repeat the exercise.

Observations

1. *Draw a sketch outline of the chest and neck. Using a "B" for bronchial and a "V" for vesicular, mark the locations in which you heard each kind of sound.*

2. *Describe in your own words the characteristics of the two kinds of sounds you heard.*

Questions

1. Why can we hear vesicular breathing more clearly under the arms while bronchial breathing is heard more clearly between the shoulder blades?

2. What differences did you detect between the sounds heard during quiet breathing and breathing after more strenuous activity?

Lung Capacity

The average number of breaths per minute varies between 14 and 20 for a healthy adult. The amount of air moved by a normal individual breathing while at rest is called the **tidal volume**. This is only a portion of the potential lung capacity. If you breathe in and out normally and then, at the end of a normal exhalation, forcibly push out as much extra air as you can, the air you remove is called the **expiratory reserve volume**. Similarly, the amount of extra air you can forcibly pull in at the end of a normal inhalation fills the **inspiratory reserve volume**. These three volumes together make up the **vital capacity** of the lungs. (See Table 6.3.) No matter how hard you try to push air out of the lungs, there will always be a small amount left in the spaces and tubes. This is called the **residual air capacity**. (See Figure 6.7.)

Table 6.3 The Respiratory Capacities of an Average Adult

Expiratory reserve volume	1500 cm^3
Tidal volume	500 cm^3
Inspiratory reserve volume	2000 cm^3
Vital capacity	4000 cm^3

The average residual air capacity is about 1500 cm^3, making the total capacity of the lungs about 5500 cm^3. The vital capacity of each individual will vary according to an individual's size, build, sex, physical condition, and age. The vital capacity can vary among adults from 1500 cm^3 to 7000 cm^3. For adults in good physical condition, the average vital capacity for males is about 4500 cm^3 and for females about 3100 cm^3. (See Figure 6.8.)

Chapter 6 / Respiration

Figure 6.7
The action of the lungs is rather like a bellows. When we make small movements with the handles of a bellows, only small amounts of air are drawn in and pumped out. Similarly, when we are resting, only a small portion of each lung is used. When we exercise vigorously, we require more air, and larger portions of the lungs are used. There is always a small amount of air left in the lungs even after a forced expiration.

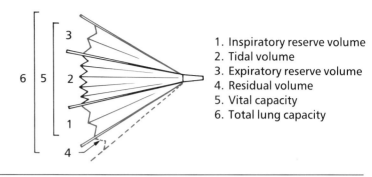

1. Inspiratory reserve volume
2. Tidal volume
3. Expiratory reserve volume
4. Residual volume
5. Vital capacity
6. Total lung capacity

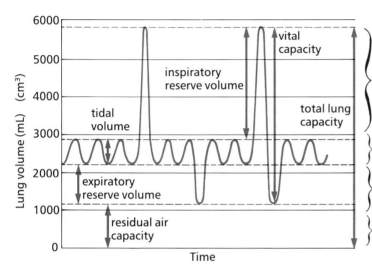

Figure 6.8
Lung volumes and capacities

Extra air available during exercise

Air used when body is not active

Extra air also used during very deep breathing

This volume is always present. It represents the air left in the tubes and alveoli even after the very deepest expiration.

Activity 6C

Measuring the Capacity of the Lungs

Problem

What are the tidal volume, expiratory and residual volumes, and the vital capacity of your lungs?

Materials

spirometer

Procedure

- **Tidal volume** is the amount of air inhaled or exhaled during normal, quiet breathing.

1. Sit by the spirometer and breathe normally for about 1 min. (See Figure 6.9.) After taking a normal breath, place the mouthpiece between your lips and exhale into the tube. Try to be natural and not force the air out in any way. Repeat this three times and average the results. *Record the number of litres of air in your notebook.* (Tidal volume = TV.)

Figure 6.9
A classroom spirometer used to determine lung capacity

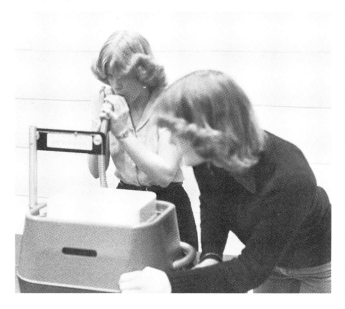

- **Expiratory reserve volume** is the amount of air that can be forcibly exhaled after a normal exhalation.

2. Stand up and breathe normally for 1 or 2 min. After a normal exhalation, place the tube in your mouth and forcibly remove all the extra air that you can. Repeat this three times and average your results. *Record the number of litres of air in your notebook.* (Expiratory reserve volume = ERV.)

- **Inspiratory reserve volume** is the amount of air that can be forcibly inhaled after you have already taken in your normal breath of air.

3. Stand up and breathe normally for 1 or 2 min. Now breathe in as deeply as you can. Breathe out into the mouthpiece slowly as you would normally, but do not forcibly remove any of the extra air. You can find the inspiratory reserve volume by subtracting your tidal volume from this result. *Record the number of litres of air in your notebook.* (Inspiratory reserve volume = IRV.)

- **Vital capacity** is the amount of air that can be forcibly removed from the lungs after the largest possible inspired breath.

4. Stand up and breathe normally as before. Now breathe in as deeply as possible. Place the tube to your mouth and exhale as hard and long as you can to try to empty all the air from your lungs. Repeat three times and average your results. *Record the number of litres of air in your notebook.* (Vital capacity = VC.)

A cross-check can be made by adding:
IRV + ERV + TV = VC

Observations

You should now have a table showing the different volumes. *Compare your scores with those of other students. If you have completed any of the other experiments such as respiration rates, thoracic index, etc., you might try to make a large chart on the chalkboard and see if there is any correlation in the results. Try to compare students by sex and size.* (Mass, body size, posture, and the types of activity in which the student is involved also affect the results.)

Questions

1. Check to see if there is any correlation in the results you obtain between smoking and non-smoking students and their vital capacity, or between students involved in regular exercise programs and nonactive students and their vital capacity.

2. If some students have very low scores, inquire if they have any temporary respiratory infections or suffer from such problems as asthma.

CHECKPOINT

1. (a) What are the small air sacs in the lungs called?
   (b) What important process takes place within them?

2. Why are there so many small air sacs in the lung instead of one large hollow chamber?

3. What is the function of the pleura?

4. List three simple steps that take place as air is passed into the lungs.

5. Define the following terms:
   (a) tidal volume
   (b) inspiratory reserve volume
   (c) expiratory reserve volume
   (d) vital capacity.

6.4 The Exchange of Gases

The bronchioles end, as we have already discussed, in air sacs called alveoli. Each tiny alveolus is surrounded by a network of capillaries. It is here that the exchange of gases takes place. Oxygen and carbon dioxide pass through the walls of the alveoli to enter, or leave, the capillaries of the circulatory system. This exchange can take place because the walls of the air sacs are extremely thin and moist.

The Exchange of Carbon Dioxide and Oxygen in the Lungs

At this point, the partnership between the respiratory system and the circulatory system comes into effect. The respiratory system has brought oxygen molecules to the exchange site in the alveoli. (See Figure 6.10.) Now oxygen will cross the membranes, enter the bloodstream, and be transported to the cells which require oxygen for their activities.

You will remember that diffusion is defined as a movement of molecules from an area of high concentration to an area of low concentration. The air we breathe into the alveoli is made up of nearly 80 percent nitrogen, which the body

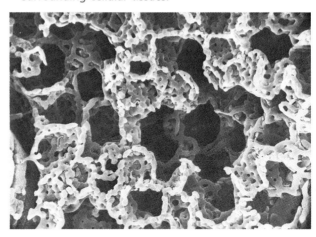

Figure 6.10
The extensive network of capillaries surrounding the alveoli has been exposed by removing all the surrounding cellular tissues.

cannot use and, therefore, it is exhaled. About 18 to 20 percent of the air is oxygen, which is a much higher concentration of oxygen than that found in the blood. (See Figure 6.11.) Due to these differences in concentration, the oxygen diffuses from an area of high concentration – in the alveoli – to an area of lower concentration – in the blood of the capillaries. (The blood arriving at this point will be high in carbon dioxide, but has very little oxygen due to the process of cellular respiration.)

The oxygen must be dissolved to cross the membrane from the alveoli to the capillaries. The walls of the alveoli are bathed in a film of moisture in which the oxygen gas becomes dissolved. As the dissolved oxygen molecules cross into the capillaries, they pass into, and mix with, the blood plasma, the watery part of the blood. The plasma could not possibly, however, transport all the oxygen that the body needs. Therefore, the oxygen in blood plasma is quickly picked up by hemoglobin molecules contained in red blood cells. Each hemoglobin molecule has four sites to which the oxygen molecules can be attached. Usually, when the hemoglobin molecules leave the lung area, 99 percent of these carrying sites are filled with oxygen.

LIFE · SIGNS

DROWNING

Exams were over and a party was planned at a cottage on Sandy Lake. Everyone was in high spirits and excited by the prospect of summer fun ahead.

The water was warm and inviting, and everyone was anxious to swim, water ski, and sail. All the friends could swim, although not all were excellent swimmers. As the afternoon progressed, the wind dropped and the lake became flat and calm. Two of the girls had been sailing and their boat was drifting, not far from shore. David and Riad decided to swim out to talk to them.

Distances over water can be very hard to judge, and David made a poor assessment of how far the boat was from shore. When they were about halfway to the sailboat, David realized that he was in trouble. He looked behind and the shore seemed a long way off, so he struggled on. His breath was coming in short gasps and his tired arms were flailing. Panic rose in his throat and he called to Riad, who was by now well ahead and failed to hear his cries. The girls had seen him, however, and were shouting to Riad.

David started to scream and stopped swimming. He sank below the lake surface, gulping in mouthfuls of water as he tried to shout. He came spluttering to the surface, gasping for air, then started to sink before he could scream again. David didn't recall much more until he found himself in the boat with the worried faces of his friends bending over him.

Riad was a fast swimmer and had taken lifesaving instruction from the Red Cross. He had searched and found David beneath the water and brought him to the surface. He then gave David mouth-to-mouth resuscitation while holding his unconscious body until help arrived. The two girls had paddled the sailboat over to the pair in the water as quickly as they could and then hauled them both into the boat.

All three friends took turns giving David artificial respiration until he started to breathe normally again and regained consciousness. David was in shock as they paddled the boat to shore. On shore the parents who owned the cottage took over, first treating David for shock and then taking him to the hospital for a medical checkup.

There is no doubt that Riad's knowledge of life-saving and his swimming ability saved David's life.

Rescue Breathing (also called mouth-to-mouth resuscitation)
Learn this technique from a qualified instructor before you have to use it!

Use rescue breathing when persons have stopped breathing as a result of:
- drowning
- choking
- suffocation
- excessive drugs
- electric shock
- heart attack
- gas poisoning
- smoke inhalation.

Quickly remove the victim from the cause or remove the cause from the victim.

Start immediately. The sooner you start, the greater the chance of success. Apply rescue breathing anywhere:
- on a dock
- on the beach
- in a boat
- from a boat
- standing in water
- kneeling in water
- on the ground
- in a car
- on a hydro pole
- in a chair
- on a bed
- on the street.

Send someone for medical aid.

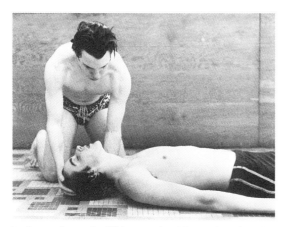

1. Open airway by lifting neck with one hand and tilting the head back with the other hand.

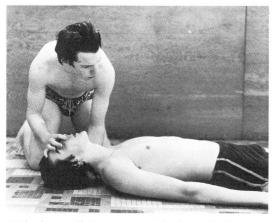

2. Pinch nostrils to prevent air leakage. Maintain open airway by keeping the neck elevated.

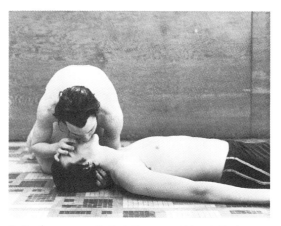

3. Seal your mouth tightly around the victim's mouth and blow in. The victim's chest should rise.

4. Remove mouth. Release nostrils. Listen for air escaping from lungs. Watch for chest to fall.

Repeat last three steps 12 to 15 times per minute. Continue until medical help arrives or breathing is restored.

Time Breathing Stopped (min)	Chances of Recovery
1	98 out of 100
5	25 out of 100
10	1 out of 100
11	1 out of 1000
12	1 out of 100 000

Figure 6.11
The composition of inspired and expired air

Inspired air

500 mL atmospheric air
oxygen 21 %
nitrogen 78 %
carbon dioxide 0.04%
Other gases in small amounts

150 mL of air fills the spaces in the tubes of the respiratory system. The composition does not change as it is not in contact with respiratory surfaces.

350 mL of air reaches the alveoli. Here it diffuses into the moisture on the respiratory surfaces and the O_2 passes into the blood, while CO_2 is taken up from the blood.

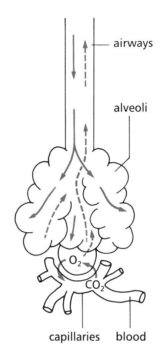

Expired Air

oxygen 16%
nitrogen 78%
carbon dioxide 5%
Other gases in small amounts

Some of the "dead air" from the tubes will be mixed with the air from the alveoli.

Alveolar air. This has been in contact with the alveoli. In the alveoli the air has a composition of:

Oxygen 14.5%
Nitrogen 80 %
Carbon dioxide 5.5%

The Transport of Gases in the Blood

The blood, now rich in oxygen, is rapidly carried back to the left side of the heart and then forcefully pumped into the large arteries and arterioles. These vessels are all thick-walled, so until the blood reaches the thin-walled capillaries in the tissues of the muscles, brain, or other body parts, the oxygen cannot diffuse into the cells where it is needed. In capillaries, the blood is moving slowly enough, and the vessel walls are thin enough to permit such diffusion.

The Exchange of Gases in the Cells and Tissues

The concentration of oxygen in the capillaries is high. Outside, in the tissues, oxygen is in low concentration, because it is being used by the activities of the surrounding cells. Therefore, there is movement from an area of high concentration to one of low concentration as the oxygen molecules diffuse out of the capillaries into the tissues around them. These same tissues, although they are low in oxygen, contain high concentrations of carbon dioxide, which is the waste gas of cellular metabolism. Carbon dioxide, therefore, diffuses into the bloodstream where the concentration of this gas is low. It is then carried back to the lungs, where it diffuses into the alveoli and is finally exhaled. (See Figure 6.12.)

Although a small amount of carbon dioxide is carried by the hemoglobin, most is carried in the plasma of the blood as carbonic acid and by ions in the plasma that form certain salts called bicarbonates. Carbonic acid is produced by combining carbon dioxide and water ($CO_2 + H_2O \rightarrow H_2CO_3$).

Figure 6.12
The exchange of gases between the capillaries and surrounding cells

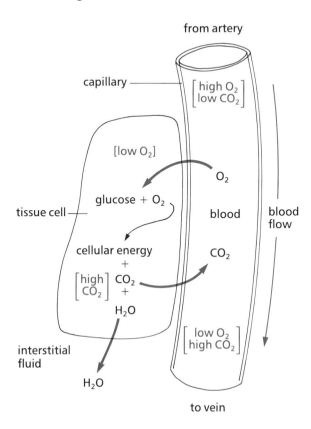

There is a high oxygen concentration inside the capillaries, compared to a low pressure inside the tissue cells. The CO_2 level inside the cells of the tissues is high, while the CO_2 level is low inside the capillary. CO_2 diffuses out and the oxygen diffuses into the cells.

When free oxygen has been delivered to the body tissues and it has entered the cells, it quickly combines with the many cellular molecules present. These molecules cling tightly to the oxygen, preventing it from diffusing back out of the cell. Carbon dioxide does not combine nearly as readily with the other molecules, so it is free to leave the cell as its concentration increases.

Nervous Control of Breathing

Whenever we increase our activity rate, we breathe faster. What causes this change? We don't consciously think, "I'm going to run. I must now start breathing faster." The reaction is automatic.

As we engage in some strenuous activity, the rate of cell metabolism increases and the waste products of metabolism, including carbon dioxide, begin to accumulate; thus, the cells, and consequently the blood, show an increased carbon dioxide content. Although a low concentration of oxygen and a high concentration of carbon dioxide both act to stimulate breathing, the carbon dioxide level has the more important effect. Carbon dioxide levels in the blood passing through the brain are monitored by the respiratory control centre in the central nervous system. This respiratory centre is located in the medulla oblongata and the pons. It normally operates with a regular rhythm, but is greatly influenced by information passed to it from various parts of the body. A constant stream of impulses regarding the oxygen needs of the body and the carbon dioxide concentrations present in the tissues are processed by the central nervous system. The respiratory centre responds to these signals by increasing or decreasing the breathing rate to meet the body's demands.

When high carbon dioxide levels are recognized, the cells in the medulla send out nerve impulses to the diaphragm and intercostal muscles to speed up their action. This causes a more rapid exchange of carbon dioxide and oxygen, and gradually returns the concentration levels of each to normal.

CHECKPOINT

1. Give the approximate proportions of nitrogen, oxygen and carbon dioxide in
   (a) inspired air
   (b) expired air.

2. How are oxygen and carbon dioxide carried in the blood?
3. How are oxygen and carbon dioxide exchanged
   (a) between the alveoli and the capillaries?
   (b) between the capillaries and the cells and tissues of the body?
4. What gas has the major effect on increasing the rate of respiration?
5. Where is the respiratory centre that controls the rate of respiration?

6.5 Air Pollution and Healthy Lungs

You rarely breathe in clean, pure air. You probably would not want to, for airborne substances give us information and enjoyment. Consider the scent of flowers, cooking food, or spring earth. Unfortunately, our atmosphere also contains harmful chemical substances. When jogging along the street, you constantly inhale traffic exhaust fumes; in the country, you may inhale pollen and other allergens; on a work site, you may be breathing in several kinds of dust particles. We cannot stop breathing, and our senses are of little help in detecting many of the chemical pollutants in the air. Many of these chemical substances are toxic or potential health hazards. We can filter out only some of the dust and try to avoid areas with a foul smell.

Some Adverse Effects of Airborne Pollutants

- The hairs and cilia lining the airways may become clogged with dust particles. This may become severe enough to cause the cilia to harden and be unable to dislodge and remove particles of dirt.
- Mucous secretions increase with irritation in order to protect the lining of the lungs and airways. This may clog the cilia and hairs. Eventually, the mucus takes up space within the airways, reducing the passage of air and providing sites for bacteria to multiply.
- Coughing increases as the body tries to free the airways of the irritating materials.
- Normal lung tissue is spongy and flexible, but as particulate matter increases, the lung tissues become stiff and less flexible. This makes breathing more difficult. To compensate, the heart works harder to move the blood to oxygen-hungry tissues. The extra effort can cause the heart to enlarge and possibly fail.
- Pollutants can irritate the bronchial tubes causing them to constrict and obstruct air flow. This may cause the delicate lung tissues to expand under pressure and rupture the tiny alveoli, decreasing the surface area available for oxygen absorption, and causing emphysema.
- Airborne pollutants which are carcinogenic or toxic can be absorbed across the lung surface.

Activity 6D

Finding Out About Air Pollution

Here are some common substances that may be found in the air we breathe.

carbon monoxide   nitrogen oxide
sulphur dioxide   dust
pollen   carbon particles
asbestos   lead

Make a table like the one at the top of page 203 and, beside each pollutant, state what process generates it and indicate the effect which that pollutant has on the body. You will find information in textbooks and the library.

Substance	Source	Effect on the Body
Carbon monoxide	Automobile exhaust	Attaches to hemoglobin, causes tiredness, drowsiness, loss of consciousness, death

SAMPLE ONLY

Activity 6E

How Clean is the Air You Breathe?

Make a list of eight substances that you know, or think, may be present in the air you breathe on a regular basis at home, school, work, or in other locations. State whether that substance is harmful or safe, how it is generated, whether the substance can easily be detected by smell or sight, or whether it is odourless or invisible and hard to detect.

If you need some ideas, consider chalk dust, chlorine gas at the pool, tobacco smoke, paint fumes, insulation fibres, gases from vinyl or plastic coverings on new furniture, etc. (Note: Many substances are considered safe for limited exposure, but extended exposure, or even a lifetime of working in special conditions, can be a severe health hazard.)

Substance	Source	Harmful or Safe	Hidden or Easily Detected

SAMPLE ONLY

A Specific Pollutant: Smoking and Your Health

Tobacco is big business. Anyone who smokes a package of cigarettes a day will contribute about $15 000 during their lifetime to the treasury of the tobacco industry which has a total income exceeding $8 billion a year. Tobacco companies spend more than $100 million annually on promotional campaigns. By contrast, Terry Fox raised just over $22 million dollars to fight cancer, a fraction of the advertising budget spent to promote a proven cancer-inducing habit!

Smoking also affects non-smokers because a smoke-filled environment has a measurable effect on everyone in the room. Non-smokers must breathe in smoke-laden air and this affects their heart rate and blood pressure, as well as the amount of carbon monoxide and carbon dioxide in their blood. Smoke also causes eye irritation, headaches, sore throats, coughs, and allergic reactions in some people.

The smoke from a cigarette left in an ashtray will give off approximately twice the tar and nicotine that a smoker takes into his or her lungs while smoking a whole cigarette. The children of smokers are ill more often than the children of non-smokers, particularly with diseases that affect the respiratory organs.

The Effects of Tobacco Smoke

More than 1000 known chemical compounds have been identified in tobacco, including more than 200 toxins. Some of these compounds are found in the smoke, some remain in the ashes, and still others are formed during combustion. It is the composition of the smoke that enters the lungs that is the chief cause of concern.

The major compounds found in cigarette smoke are tars, nicotine, phenols, carbon monoxide, hydrocarbons, arsenic, and more than 15 other agents known to cause cancer (carcinogens). The effect of these compounds will vary according to whether the smoke is inhaled or not. Usually pipe and cigar smoke is too strong and acrid to inhale, but cigarette smoke is quite commonly inhaled.

As smoke passes through the bronchi, many chemical substances are deposited on the bronchial walls and it is in these locations that cancer frequently develops. As we have already seen, the walls of the respiratory passageways are lined with cilia. Several substances contained in smoke irritate the ciliated cells. Nicotine can actually paralyze cilia, preventing them from cleaning air as it enters the lungs.

Lung Disease and Smoking

When cells are exposed to constant irritation over a long period of time, carcinogens produce changes in the cells which cause them to multiply at an unusually rapid rate. These cells are abnormal and obstruct the work of the normal cells around them. (See Figure 6.13.)

Figure 6.13
A comparison between a normal lung (left) and a cancerous lung (right). The tumour is seen as a smooth, grey body. The dark patches are greatly enlarged air spaces caused by emphysema – the breakdown of alveolar walls.

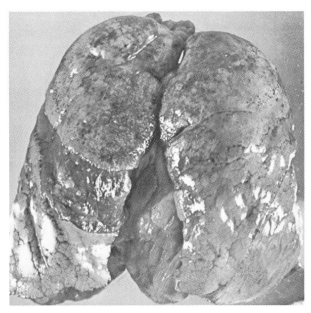
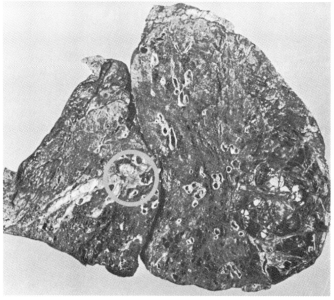

An increased incidence of cancer deaths among smokers has been demonstrated repeatedly by many studies. Each package of cigarettes must, by law, carry a warning that smoking is injurious to health.

Cancer is not the only disease caused, or aggravated, by smoking. Bronchitis, emphysema, ulcers, cirrhosis of the liver, and heart disease are all examples of disorders that have been attributed either directly or indirectly to smoking.

Smoking During Pregnancy
A woman who smokes while pregnant risks affecting both her own life and that of the unborn baby. More stillbirths, premature deliveries, and spontaneous abortions occur when pregnant women smoke than when they don't. In addition, the babies of smoking mothers are, on average, of lower weight at birth than those born to women who do not smoke. Such babies also have a higher incidence of neurological damage.

CHECKPOINT

1. List four compounds found in tobacco smoke and describe the adverse effects that each of them has on the body.

2. List four respiratory disorders that have been attributed to smoking.

3. List three air pollutants and describe the effects they have on the health of people.

FOCUS ON YOU

Carbon monoxide, a colourless, odourless, and poisonous gas is produced by the burning of gasoline, cigarettes, and other materials. It clings 200 times more strongly than oxygen to hemoglobin molecules and will not easily let go — preventing oxygen from being transported.

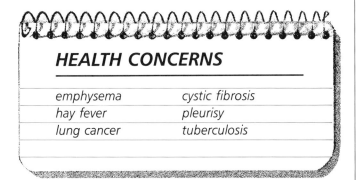

HEALTH CONCERNS

emphysema	cystic fibrosis
hay fever	pleurisy
lung cancer	tuberculosis

BIOLOGY AT WORK

- RESPIRATORY UNIT TECHNICIAN
- SMOKING ADDICTION THERAPIST
- INDUSTRIAL SAFETY TECHNOLOGIST
- OCCUPATIONAL HEALTH TECHNOLOGIST
- ANESTHESIOLOGIST

FOCUS ON YOU

Most businesses and workplaces now ban smoking on their premises for two reasons — the effect of "second-hand" smoke on the health of non-smokers, and the damage smoke does to delicate equipment such as computers. The message is clear — if you want to have your choice of good jobs, don't become a smoker.

Chapter 6 / Respiration

CHAPTER FOCUS

Now that you have completed this chapter, you should be able to do the following:

1. Write the word equation for cellular respiration.
2. Describe the location, structure, and function of the parts of the respiratory system.
3. State the differences in composition of inspired and expired air.
4. Explain the mechanical processes of inspiration and expiration.
5. Given data, calculate total lung capacity, expiratory reserve volume, inspiratory reserve volume, and vital capacity.
6. Explain how oxygen and carbon dioxide are exchanged in the lungs and in the tissues.
7. Identify the components of air and name several common air pollutants and the problems they may cause.
8. Recognize the characteristics of a healthy respiratory system and explain how smoking adversely affects the efficiency and health of this system.

SOME QUESTIONS TO ANSWER

1. (a) Why does the body need oxygen?
   (b) Where and how is it utilized in the body?
2. List the structures, in the correct sequence, through which a molecule of oxygen would pass from the point of entry into the respiratory system until it reaches the alveoli.
3. In point form, explain the mechanical process of breathing in air.
4. Explain the following terms:
   (a) inspiratory reserve volume
   (b) expiratory reserve volume
   (c) tidal volume
   (d) vital capacity.
5. Sketch two or three alveoli and their associated blood vessels.
6. What is the functional value of the pleura?
7. What processes take place in the nasal chamber?
8. Briefly describe the structure and function of the larynx.
9. List the gases in normal air and the approximate proportions in which they are found.
10. (a) Give two examples of air pollutants that affect the respiratory system.
    (b) State what processes produce these pollutants and how they affect our bodies.
11. How does inspired air differ from expired air? Why is it possible to use our expired air to revive a person by mouth-to-mouth resuscitation?
12. Describe the characteristics of a healthy respiratory system.
13. (a) Explain how O_2 and CO_2 are exchanged between the alveoli and the capillaries.
    (b) Why does the reverse exchange take place between the capillaries and the working cells and tissues of the body?
14. How do cilia, mucus, and coughing help in keeping the respiratory system healthy?
15. List four respiratory disorders and give the cause and effects of each.

Chapter 6 / Respiration

SOME WORDS TO KNOW

Match each of the descriptions given in the left-hand column with a word shown in the right-hand column. DO NOT WRITE IN THIS BOOK.

1. Tiny sacs where gases are exchanged
2. Muscles that raise the ribs
3. Tubes that branch off from the trachea into the lungs
4. A membrane which surrounds each of the lungs
5. The amount of air filling the lungs (not including the residual air)
6. The part of the brain which controls the rate of respiration
7. Tube which connects the nasal and oral cavities with the trachea
8. A sheet of muscle which separates the thoracic and abdominal cavities
9. The part where the vocal cords are found
10. The gas which has the major influence on controlling the rate of respiration

Select any two of the unmatched words and, for each, give a proper definition or explanation in your own words.

A diaphragm
B lungs
C trachea
D bronchioles
E bronchi
F alveoli
G pharynx
H intercostals
I sinuses
J pleura
K vital capacity
L reserve volume
M oxygen
N carbon dioxide
O medulla
P cerebrum
Q larynx

SOME THINGS TO FIND OUT

1. Select one of the following and prepare a report on it:
   The Canadian Lung Association
   the rights of a non-smoker
   air pollution and respiratory disorders
   cancer of the lung, lips, tongue, or larynx.

2. There are a number of reflexes associated with the respiratory system. Research what happens when we yawn, sneeze, cough, sigh, or hiccup.

3. Make a list of common carcinogenic materials found in the environment and find out how they affect the human body.

YOU ARE WHAT YOU EAT: DIET AND NUTRITION

Any factory producing consumer goods for market needs a constant supply of raw materials – but not just any raw materials. The needs of a particular company are very specific, both in amount and type. The shopping list of a video tape manufacturer, for example, would be vastly different from that of a furniture manufacturer.

Humans also need a constant supply of raw materials, called nutrients, both for energy and for the construction of body tissues. The substances required are very specific, and failure to obtain even one could mean some bodily process would fail. The function of our diet – the food we consume – is to supply the needs of our body.

KEY IDEAS

- *A variety of foods is required to supply the nutrients needed by the body.*
- *Complex food compounds are made up of simpler molecules.*
- *Wise shoppers read and understand food labels.*
- *Body mass can usually be controlled by sensible modification of dietary intake and energy expenditure.*

Chapter 7 / Diet and Nutrition

7.1 What Determines a Diet

We often hear people talk about going on a diet. In fact, we are all on a diet, all the time. The amount and type of food that we each eat from day to day defines our **diet**.

There are many factors that influence the choice of food we buy and eat. For instance, the food purchased generally reflects the likes and dislikes of the family. Food costs are another important consideration. As one item becomes expensive, it may be replaced by a less expensive substitute. Advertising and display merchandising also influence the purchase of foods.

Fortunately, there has recently been an increasing trend to the selection of foods for their nutritional value. To do this well, the basic needs of the human body and simple principles of nutrition should be understood. The body requires certain elements, in various combinations, to provide energy for the growth and repair of tissues and the general maintenance of the body systems. The food substances that provide these are of six different types:

- carbohydrates
- fats
- proteins
- vitamins
- water
- minerals

7.2 Carbohydrates for Energy

Carbohydrates are the body's major source of energy. As their name suggests, **carbohydrates** are formed from carbon and hydrates. The hydrate portion is composed of hydrogen and oxygen in the same proportions as they are found in water (two parts hydrogen to one part oxygen).

The simplest carbohydrates are called **monosaccharides** (mon-oe-sac-ah-ride) (*mono* means one, *saccharide* means sweet), or single sugars. (See Table 7.1.) Glucose is one of the most important monosaccharides. Others occur naturally in many fruits and vegetables. More commonly, however, they are found bound together in pairs called **disaccharides**, or double sugars. For example, table sugar (sucrose) is made up of two monosaccharides, glucose and fructose, joined together.

Table 7.1 Sugars – The Simplest Carbohydrates

Monosaccharides are single molecules of sugar.
glucose   fructose   galactose
Disaccharides are composed of two monosaccharides linked together.
glucose + glucose → maltose
glucose + fructose → sucrose
Polysaccharides are composed of many monosaccharide units joined together into long chains.
glucose + glucose + glucose + glucose → starch
Many of these chains contain hundreds of monosaccharide units. Some chains are branched or have units in alternating positions.
Other examples of polysaccharides are glycogen, cellulose, and pectin. The two latter examples are not digestible by enzymes in the human body.

When many monosaccharide or disaccharide units are joined together into long chains, they are called **polysaccharides** (*poly* means many). The starch and cellulose found in plants are examples of these complex molecules. Since a large part of our diet is plant material, these substances form a significant portion of our daily food intake.

Starch is easily broken down into simple sugars by the action of enzymes during digestion. Cellulose cannot be digested by the human body and forms part of the fibre that passes, undigested, out of the body.

Another important polysaccharide is **glycogen**. Although this substance is rarely found in the food we eat, it is the major form in which carbohydrates are stored in the body. Glycogen is composed of chains of glucose molecules. These chains can easily be broken down, releasing glucose as demands for energy are made by various parts of the body.

Glucose and Blood Sugar

The glucose molecule is of great importance to the human body. A special body system monitors its presence in the blood at all times. The amount present is referred to as the **blood sugar level**. After a heavy meal, large quantities of this sugar may be present in the small intestine. These molecules are absorbed and carried through the bloodstream to the liver. In the liver, extra glucose, which is not immediately needed by the body, is converted to glycogen and stored until it is required. The amount of glucose in the blood is influenced by **insulin**, a hormone produced by the pancreas.

Muscles are often called upon to react quickly in an emergency, without special warning. For this reason, glycogen is also stored in the muscle tissues for instant conversion and use.

Sources of Carbohydrates

Foods made from grains, along with starchy vegetables and fruits, are rich sources of carbohydrates. (See Figure 7.1.) Carbohydrates are often condemned as being the major contributors of body fat in overweight people. While excess carbohydrates are converted to fat and stored, any excess food that is eaten, whether in the form of carbohydrates, fats, or protein, may also be converted to fat for storage.

Figure 7.1
Common dietary sources of carbohydrates

LIFE · SIGNS

DIABETES

Wandering through the house, Latisha found her mom sitting on the floor in front of the bookcase, flipping through the pages of a photograph album. "Looks more interesting than homework," she joked.

Her mom glanced up. "Hard assignment?"

Latisha sat on the floor beside her. "Just complicated. I need to interview someone whose life has been affected by scientific discoveries. It's Miss Hanson's latest idea."

Rather than answering, her mom turned the pages of the album in her lap to the front. "There I am at 14, Latisha."

Latisha hadn't seen these pictures for years. "Whew. You were so thin. Were you on a diet?"

"Far from it. I was eating huge amounts of food and drinking constantly. But I couldn't use the food I was eating because my pancreas had stopped making insulin of its own."

Latisha nodded. "That's when you found out you had diabetes. But what happened then?"

"My doctor admitted me to the hospital. I was given injections of insulin and in just two weeks I gained 9 kg."

"So you were back to normal," Latisha said quickly.

Her mom smiled. "Well, I could do all the things I wanted to do again, but normal isn't exactly the word I would choose. Before I could safely leave the hospital, I had to learn how to control my body's blood sugar (glucose) by myself. You see, insulin controls how food energy is used in a person's body by controlling the amount of glucose in the blood. Insulin acts to take glucose out of the blood by increasing the rate at which it goes into muscle, fat, and liver cells. If glucose were to stay in the blood, the kidneys would remove it very quickly by adding it to urine, and body cells would starve. My doctor discovered my diabetes by finding large amounts of glucose in my urine.

"To copy this system using injections of insulin meant that I had to learn to test my own blood for its glucose level. It also meant that a careful balance had to be worked out between the amount of food I ate, the amount of exercise I did (because exercise uses glucose more rapidly), and the amount of insulin I needed to keep my blood glucose at normal levels at all times of day."

"That's why you never have seconds on birthday cake. And you had to learn to give yourself needles every morning." Latisha winced sympathetically. "I couldn't do that."

"You could if your life depended on it, believe me," her mom countered practically. "And in a way, yours did, didn't it?"

Carbohydrates are an essential part of the diet. When the body does not have adequate supplies of carbohydrates, fats and, in extreme cases, structural proteins, will be broken down to fuel the body's energy needs. You may have heard that athletes anticipating a large demand for energy, such as a marathon swimmer before a swim, may deliberately "load" their bodies with carbohydrate-rich foods in an effort to prevent the loss of important fat and protein from their muscles. The effectiveness of this practice has not been proven.

Not all carbohydrate sources are equally valuable. Each North American consumes, on average, more than 45 kg of sugar per year. Much of this sugar is eaten in the form of sweetened soft drinks, cakes, candies, syrups, and confections. It is important to realize that these foods have little or nothing else to offer in the way of nutrients besides quick energy. If eaten to satisfy hunger, they could be taking the place of better, and sometimes less expensive foods which also provide essential vitamins, minerals, and proteins.

began to bleed, or hemorrhage. It was like looking through a fog – a fog with dark spots in it. Fortunately, the blood vessels were sealed using a laser and my loss of vision was only temporary. I was lucky. There are a lot of other possible complications that aren't so easily fixed – including kidney failure and heart disease."

"Is that why you use your monitor so often?"

"Yes. It's been a great help. Just by testing a drop of blood now and then, I can find my blood sugar level and correct it by eating a little more or less, or by contacting my doctor about adjusting my insulin dose." Latisha's mom paused. "Now what was that about an interview?"

"Thanks, Mom." If Latisha's quick hug was a little tighter than usual, her mom was too tactful to mention it.

In class on Friday, Latisha gave her presentation, including some additional information from the local Diabetes Foundation. "Most diabetics can take their insulin in a tablet. Some don't need insulin at all, but can adjust their diet to compensate. All types benefit from at-home monitors to measure their blood sugar.

"The main thing is to realize that a person with diabetes has to be very careful," Latisha concluded. "It's not fair to tempt them with foods they know they shouldn't eat, or to make it difficult for them to keep their physical activity within the limits set by their doctor. Sometimes good friends are part of a diabetic's balance, too."

Latisha thought for a moment. "I remember Dad saying you had a difficult time when you were pregnant with me. Why was that?"

"The balance between the glucose I need and the insulin I take is a delicate one. It wasn't as easy to keep that balance during my pregnancy. One consequence was that the blood vessels behind my eyes

Activity 7A

Sources of Carbohydrates

Problem

Which foods contain the carbohydrate starch?

Materials

spot plates
iodine solution
starch solution
potato
apple
carrot
sucrose (table sugar) solution
additional food samples selected from lunches

Procedure

1. On a spot plate, place the following substances in a marked sequence: water, starch solution, potato, apple, carrot, sucrose solutions, and any other food sample approved for this activity. Place a piece of white paper under the spot plate so that the results may be seen more easily. (Note: Potato or carrot may be cut into convenient portions by using a cork borer to make a cylinder of the sample, then cutting it into small slices.)

2. Add a few drops of iodine solution to each sample. Wait a few minutes and then examine the samples for any colour changes and the presence of granules.

3. *Record your results in a suitable table in your notebook.*

Substance	Colour change	Is starch present?

Questions

1. State the recognized test for starch.

2. (a) Is starch soluble?
   (b) What evidence is there for your answer?

3. (a) Name eight foods that contain starch.
   (b) Are they all solids?

4. (a) Based upon the class results for foods from lunches, what are the most common sources of carbohydrates eaten by you and your classmates?
   (b) Would you classify these sources as very valuable or not? Explain your answer.

Activity 7B

Testing for Sugars

Sugars form a subgroup of carbohydrates. Monosaccharides are the basic units of which all carbohydrates are composed.

Materials

Benedict's solution
25-mL test tubes
solutions of glucose, corn syrup, brown sugar, and starch
hot plate
water bath
dilute hydrochloric acid
goggles

Procedure

1. Place about 3 mL of each of the solutions (water, glucose, corn syrup, brown sugar, and starch) in different test tubes. Mark the test tubes for identification.

2. Add about 2 mL of Benedict's solution to each test tube.

 CAUTION! Wear eye protection when heating liquids.

3. Place the test tubes in a hot water bath and heat gently. *Observe and record any colour changes that take place and the sequence in which they occur. Use a table similar to the one shown.*

Sample	Original colour	Changes	Final colour

BIOTECH

Food Energy

The unit by which we measure food and body energy is the **kilojoule**. Energy is usually defined as the ability to do work or produce heat. The joule is the unit of energy, but, in order to avoid very large numbers, the kilojoule, which is equal to 1000 J, is used.

Until recently, the generally accepted unit for energy was the calorie. One calorie is equal to approximately 4.2 kJ.

The following examples will help you "think" in joules. It requires about 400 kJ of energy to meet the needs of an average adult for 1 h.

Average Kilojoule Requirements Per Day

Sex	Age (years)	Weight (kg)	Height (cm)	Energy (kJ)	Protein (g)
Male	15 to 18	60	175	13 400	54
	19 to 22	66	175	13 400	54
	23 to 50	70	175	11 300	56
Female	15 to 18	54	165	8800	48
	19 to 22	58	165	8800	46
	23 to 50	58	165	7900	46

(Adapted from Food and Nutrition Board. RDA revised 1979. (NRC))

- 100 g of carbohydrate provide about 1680 kJ.
- 100 g of protein provide about 1680 kJ.
- 100 g of fat provide about 3780 kJ.
- A chocolate milkshake (350 mL) contains about 2200 kJ.
- A piece of fudge (3 cm^3) contains about 480 kJ.
- 20 french fries provide about 1300 kJ.
- One glass of milk (250 mL) contains about 800 kJ.

CAUTION!
Rinse spills of Benedict's solution or hydrochloric acid with water. Avoid getting these substances on your skin.

4. Take another sample of starch solution. Add a few drops of dilute HCl, and boil for a few minutes. Test with Benedict's solution. *Record all your observations and results.*

Questions

1. State the general test for sugars.
2. What significance might there be to the sequence of colour change from the original blue to the last colour change?
3. Devise an experiment that will test for sugars in a quantitative way.
4. Explain the results that you obtained after boiling the starch in HCl.
5. Explain the differences between starches and sugars.

CHECKPOINT

1. List the six major categories of food substances.
2. Explain the differences between monosaccharides, disaccharides, and polysaccharides.
3. What is the main function of carbohydrates in the body?
4. List five good sources of carbohydrates.
5. State a simple test for the presence of starch.

7.3 Fats for Storage and Vitamin Metabolism

When the body turns surplus carbohydrates into fats for storage, these fats are stored in the **adipose** tissues of the body. They exist in a liquid state (oil) inside the cells, since the temperature of the body is considerably higher than that at which most fats liquefy.

Fats are present in the foods we eat. (See Figure 7.2.) In plants, such as corn or cotton, fat in the form of oil is usually stored in the plant seeds. In animals, the body fat represents the animal's surplus energy store.

Fats, like carbohydrates, are used to provide energy for body activities. Fats contain twice the amount of energy per gram as either carbohydrates or proteins. Fats are also necessary to dissolve the fat-soluble vitamins A, D, E, and K, and provide a vital source of linoleic acid, required for proper growth. Because fats take longer to digest, staying in the stomach for about 3.5 h, this gives us a feeling of fullness and satisfaction after eating.

Natural fats and oils are usually composed of two main parts – an alcohol, called glycerol, and three fatty acids. The fatty acid molecules consist of long carbon chains with a special combination of atoms, which makes the molecule an acid, tacked onto one end. When three of these fatty acid molecules bond with a glycerol molecule, they form a **triglyceride** (try-gliss-er-ide). (See Figure 7.3.)

Saturated and Unsaturated Fats

All the available bonding sites on the carbon molecules of some fatty acids are filled by hydrogen atoms. In others, some of the carbon atoms have double bonds between them; these have fewer hydrogen atoms. When all of the sites are filled and no double bonds are present, the fatty acids are called **saturated**. Those with some free spaces, and double bonds, are called **unsaturated** fatty acids. The type of fatty acid involved determines the type of fat formed. Fats are thus labelled accordingly as either saturated or unsaturated. (See Figure 7.3.)

Figure 7.2
Sources of fats. The fat in some foods, such as eggs, may not be visible to the eye.

Figure 7.3
The composition of fats

a saturated fatty acid (stearic acid)

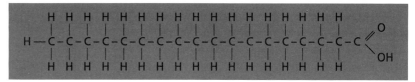

Stearic acid melts at 69.6°C.

an unsaturated fatty acid (linoleic acid)

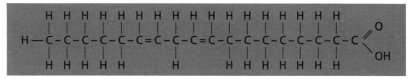

Linoleic acid melts at -5° C.

An unsaturated fatty acid contains some double bonds and a smaller number of hydrogen atoms than a saturated fatty acid. It also has a lower melting point.

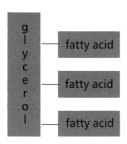

Diagrammatic representation of a triglyceride molecule with three fatty acids.

Saturated fats are solid at room temperatures. Animal fats tend to be made up of a higher proportion of saturated fatty acids, while vegetable fats and oils consist more of unsaturated fatty acids. This is why vegetable fats are usually liquid at room temperature.

Cholesterol
Cholesterol is a white, waxy substance which has several useful functions in the body. Under normal conditions, including a proper diet, the liver produces all the cholesterol needed by the body. The cholesterol is transported through the bloodstream.

Cholesterol has been linked to the occurrence of coronary heart disease. Under certain conditions, cholesterol clings to the walls of large blood vessels, thus reducing their diameter and restricting the flow of blood. The gradual reduction in the diameter of these important blood vessels (*atherosclerosis*) is a leading cause of heart attacks.

All meat, poultry, shellfish, dairy products, and egg yolks contain cholesterol. (It is not present in plant material.) The liver, you recall, naturally produces this substance. What is of concern is the amount of cholesterol which circulates in a person's blood – the *blood cholesterol level*. A high blood cholesterol level can mean a person is at serious risk of developing heart disease. The type of fatty acids in the diet has been shown to affect this level, with polyunsaturated fats tending to reduce the amount of blood cholesterol, and saturated fats significantly increasing it.

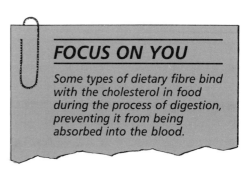

FOCUS ON YOU

Some types of dietary fibre bind with the cholesterol in food during the process of digestion, preventing it from being absorbed into the blood.

Fats in the Body

Fats are both good and bad for a person. Internal fat protects and cushions the body organs. Subcutaneous layers of fat help to insulate the body against heat loss through its surface. If too much fat is deposited, however, an individual may become overweight, thus placing extra strain on the heart, muscles, and bones, as well as causing some emotional distress.

Activity 7C

The Identification of Fats and Oils

Problem

Which foods contain fats or oils?

Materials

25-mL test tubes
Sudan IV indicator
brown paper
bromothymol blue indicator
samples of olive oil, vegetable oil, margarine, cream, etc.
soap solution
food items from lunches (with teacher's approval)
goggles

CAUTION!
Avoid getting the indicating solutions Sudan IV and bromothymol blue on your skin.

Procedure

CAUTION!
Wear eye protection when mixing liquids.

1. Pour about 2 mL of any oil into a test tube and add a few drops of Sudan IV. Shake gently.

2. Pour about 2 mL of water into another test tube and add a few drops of bromothymol blue indicator. Shake the tube gently.

3. Tip the contents of the first tube into the second tube and shake them together. Observe what occurs, then wait a few minutes and note any changes that you see. *Record your observations regarding the mixing of oil and water, and the mixing of the colour indicators with oil or water.*

4. Add a few drops of soap solution or detergent and shake the test tube. *Record your observations.*

5. Look up the word "emulsification". *How does it apply in this experiment?*

6. Rub a drop or two of each sample onto a piece of brown paper. Mark each spot with the name of the sample. Hold the paper up to the light. *Record what you observe and state a simple test for fats. What word is used to describe the condition when light passes through a substance but no image can be seen?*

7. Add small quantities of the samples given to different test tubes and add about 2 mL of water to each. Add two drops of Sudan IV to each tube. Shake gently. Observe the tubes after a few minutes. Any oil drops present will take up the red stain. *Record which of the substances contain fats or oils.*

CHECKPOINT

1. What functions does fat perform in the body?
2. What are the two main sources of fats?
3. What are the differences in structure and function between saturated and unsaturated fats?
4. Give a simple test for the presence of fats.

7.4 Proteins for Building and Repair

Proteins are essential for the building, repair, and maintenance of body tissues. For example, proteins form a major part of the muscles, the internal organs, the brain, nerves, skin, hair, and nails. They play an important role in fluid balance, as they are large molecules that do not readily cross cell membranes. This preserves the balance of fluids and nutrients inside and outside the cells. Proteins are also needed to make enzymes, antibodies, and hormones. They further help in maintaining the pH balance of the body and, if necessary, can supply energy for body activities.

Proteins are giant molecules made up of hundreds, sometimes thousands, of units called **amino acids**. These are composed of nitrogen, carbon, hydrogen, and oxygen atoms. There are only 22 different kinds of amino acids in the human body. The order, or sequence, and number of these amino acids determine the type of protein. (See Figure 7.4.)

When proteins enter the body in the food we eat, they pass into the digestive tract. There, enzymes disassemble the long complex chains into separate amino acid units. These are small enough to pass through the pores of cell membranes. After travelling through the blood, they are then reassembled into new chains inside various body cells.

Essential Amino Acids

The human body can make some of the amino acids it requires, but is unable to make others. These must be supplied in the diet. Amino acids that the body cannot make for itself are called **essential amino acids**. Lack of any one of these essential amino acids blocks the production of proteins that contain this unit. (See Table 7.2.)

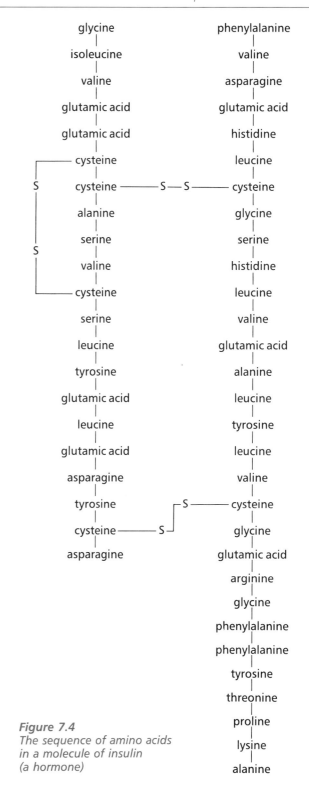

Figure 7.4
The sequence of amino acids in a molecule of insulin (a hormone)

Table 7.2 The Essential Amino Acids

Methionine	Threonine
Tryptophan	Isoleucine
Leucine	Lysine
Valine	Phenylalanine

The quality of a protein, in terms of its food value to the human body, is determined both by the number of essential amino acids it contains and by the current needs of the body for particular types or amounts of amino acids. At any one time, the value of a protein may depend on how important it is for tissue growth and maintenance.

Sources of Protein

The sources of food supplying essential amino acids may be conveniently divided into two groups – animal and plant proteins. (See Figure 7.5.)

Animal proteins, which include those found in meat, fish, eggs, and other animal products, contain rich amounts of the essential amino acids. Plant or vegetable foods, such as flour, rice, cereals, peas, beans, and nuts, also contain protein; however, the proteins are usually incomplete, either entirely lacking or containing only small amounts of essential amino acids.

Fortunately, our desire for variety in our diet helps to ensure that we get our quota of essential amino acids. Sources low in one essential amino acid must be coupled with another source which contains that missing amino acid. Foods must be eaten together or within a short period of time, because the body cannot stockpile some amino acids while waiting for others to arrive.

Body Protein Requirements

During adult life, proteins are required primarily for the purpose of repair and maintenance of body tissues. While the body is growing to maturity, however, its demand for protein is much greater because new cells must be produced as well as old ones repaired. The more rapid the growth rate, the more protein is needed. It is

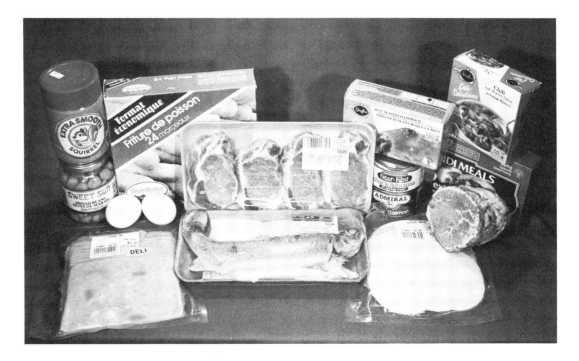

Figure 7.5
Common sources of protein

especially important that the body has sufficient protein during the growth spurts that occur in the first two years of life and again in adolescence. Pregnant women and nursing mothers also have higher protein requirements because of the rapid cell production taking place in the unborn child and because of the need to supply the nutrients necessary in mother's milk.

Protein as an Energy Source
When fats and carbohydrates are eaten in excess of body needs, they may be stored in the tissues for future use. Proteins cannot be stored in the same way. If the body demands amino acids which are not available from the food intake (during an illness, for example), tissue proteins are broken down for use in the emergency. Any surplus of amino acids not used for energy purposes is lost, because amino acids cannot be stored. Should the amount of available amino acids exceed the body's requirements, their molecules are converted first to glucose and then by further reactions to fat for storage. During this process, the nitrogen component of the amino acids is excreted, which means that the body cannot reconvert these fats to proteins. To use proteins for energy is thus an inefficient use of a valuable food component.

Activity 7D

Identification of Proteins

Protein molecules are very large and, to be absorbed, they must be broken down into smaller units called amino acids. Most tests for proteins indicate not the presence of the whole protein, but the presence of a specific amino acid in the protein chain.

Problem
Which foods contain proteins?

Materials
egg white mixed with an equal volume of water
milk
eggshell
gelatin
meat emulsion
Biuret reagent
25-mL test tubes
water bath
goggles
hot plate

Procedure
1. Add 2 mL of each substance listed to a different test tube and mark them so that you can identify what is in them.

   CAUTION!
   Be careful when using Biuret reagent. Wear eye protection. If any reagent splashes on your skin or clothing, wash under flowing water and inform your teacher immediately.

2. Add five drops of Biuret reagent to each tube.

3. Place the tubes in a warm water bath or gently heat for a few minutes.

4. *In your notebook, make a suitable table in which to record your results.*

5. *After a few minutes, record any colour changes or observations in your notebook.*

Questions
1. State the test and colour change for Biuret reagents.

2. What control could have been used for this experiment?

CHECKPOINT

1. *What are the major functions of proteins in the body?*
2. *Why are essential amino acids so important?*
3. *List four good sources of protein in the diet.*
4. *What happens when we have an excess of protein in our diet?*
5. *What happens if the body uses protein as an energy source?*
6. *State a test for the presence of proteins.*

7.5 Vitamins for Health

More than 200 years ago, in the days of wooden sailing vessels, it was not unusual for two-thirds of a ship's crew to die during a voyage as the result of a mysterious disease called scurvy. Scurvy was not a problem, however, if citrus fruits were part of the sailor's diet.

It was a very long time before the chemical substance in citrus fruits, important for the prevention of scurvy, was identified as ascorbic acid, that is, vitamin C. We now know that many other similar substances exist and are vital to health. These compounds are known as **vitamins**.

There are a few vitamins which the body itself can produce. Vitamin D is produced when the skin is exposed to sunlight. It is also found in milk, since cows produce it while feeding in sunny pastures. Ultraviolet light is the important agent in this process (although ultraviolet rays can also have a harmful effect upon the skin). Some of the types of beneficial bacteria living in our large intestine produce vitamin K and some B vitamins. These are then absorbed through the lining of the colon for use by the body. The remaining vitamins needed must be obtained from the foods we eat.

Sources of Vitamins in the Diet

A well-balanced diet should provide all the required vitamins without the need to resort to vitamin supplement pills. There are several reasons, however, why even a balanced diet may be short of some vitamins.

Garden soils are sometimes lacking in essential elements. Vegetables grown in such soils may lack the vitamins that would ordinarily be present. Fruits and vegetables transported long distances, or fruits harvested before they are ripe, can also be low in vitamin content. Freezing and storing foods depletes vitamins, as do cooking and canning processes. It is not uncommon to find that the water in which vegetables are boiled contains more vitamins than the food which is eaten. (See Table 7.3.) Vegetables should be cooked lightly and served quickly. In some vegetables and fruits, the best vitamin supply lies in the skin. This supply is thrown away if these vegetables and fruits are peeled.

Table 7.3 Factors Affecting Vitamin Retention During Food Manufacture

Vitamin	Soluble in	Loses Effective Properties When Exposed to			
		Acids	Bases	Light	Heat	Oxygen
A	Fat					x
D	Fat					
E	Fat		x	x		
K	Fat	x	x	x		x
Thiamin	Water		x		x	x
Riboflavin	Water		x	x		
Niacin	Water					
B_6	Water		x	x		x
B_{12}	Water		x			
C	Water		x		x	x

If you decide to take a vitamin supplement, read the label. High price or claims of "natural" or "organic" ingredients are no indication of value. "Natural" is a much abused term.

Vitamins are needed in very small amounts, but they have very important functions. Although there are many attitudes about nutrition, experts agree that many vitamins are interdependent. If one type of vitamin is in short supply, it affects the efficiency of many of the others. Conversely, an overdose of some vitamins may have serious effects.

Types of Vitamins

As has been indicated, there are two types of vitamins – **water-soluble vitamins** and **fat-soluble vitamins**. Vitamin C and the B complex vitamins, which include riboflavin, niacin, and vitamins B_6 and B_{12}, are all soluble in water. Vitamins A, D, E, and K are soluble in fats. If fats are not present in the digestive tract when the vitamins are taken in, they cannot be absorbed. Vitamin supplements should, therefore, be taken at meals, rather than with a glass of water on an empty stomach. Fat-soluble vitamins present in foods are already in oil solutions, since they are found in fatty foods, such as butter.

Storage of Vitamins

Only a few vitamins, for example A and D, are stored in the body. The others, if not absorbed, are excreted. Water-soluble vitamins, if present in excess, are simply excreted into the urine; fat-soluble vitamins, if present in excess, may cause poisoning.

Table 7.4 on page 224 lists the major vitamins, their value or function, and some foods in which they are found.

Activity 7E

Vitamin C

Problem

What are the relative quantities of vitamin C present in various foods? Vitamin C (ascorbic acid) is present in most citrus fruits, tomatoes, and leafy green vegetables. One of its chemical properties is that it is a reducing agent, that is, it removes oxygen from, or adds oxygen to, other chemicals. Indophenol is a blue dye that is bleached by ascorbic acid because of the reducing properties of vitamin C.

Materials

apple juice
orange juice
drinks with vitamin C added
rosehip extract
0.125 percent ascorbic acid
distilled water
indophenol solution (0.003 mol/L)
test tubes and rack
water bath
hot plate
eyedroppers
goggles

Procedure

CAUTION!
Avoid getting indophenol solution on your skin. Rinse spills with water.

1. Add 20 drops of indophenol to a test tube. Add the ascorbic acid solution drop by drop and gently shake the tube after each drop. Count the number of drops required to turn the indophenol colourless or pale pink. *Record the*

Table 7.4 The Vitamins

Vitamin	Function	Deficiency Symptoms	Daily Requirements (16-18 years of age)	Food Source
A	Needed to maintain healthy skin, hair, eyes, etc., improves resistance to infections, helps break down fats	Rough, dry skin, low resistance to infections, night blindness	Females 4000 IU Males 5000 IU	Whole milk, liver, butter, carrots, eggs, green and yellow vegetables
D	Needed for calcium and phosphorus absorption to produce good bones and teeth	Rickets, softening of bones in adults, and poor teeth	400 IU	Fish liver oils, sardines, salmon, liver, some made by the skin in sunlight
E	Helps in the formation of red blood cells, muscle, and other tissues, prevents abnormal breakdown of fat	Circulation problems, loss of sexual and body vigour, muscular and heart problems	Females 8 mg Males 10 mg	Vegetable oils, whole grain cereals
K	Aids in blood clotting	Rare, generalized bleeding	Not required daily	Green vegetables
Thiamine (B_1)	Needed for oxidation of carbohydrates, insures proper use of sugars	Loss of energy, depression, poor appetite, skin problems	Females 1.1 mg Males 1.5 mg	Whole grain cereals, dry yeast, pork, fish, lean meat
Riboflavin (B_2)	Needed for energy metabolism in cells, helps synthesize fats	Tissue damage, eye strain, fatigue, itching, sensitivity to light	Females 1.4 mg Males 1.8 mg	Liver, milk, cheese, green leafy vegetables, beans
Niacin (B_3)	Involved in energy reactions in cells	Lack of concentration, headaches, insomnia, backache, poor memory	Females 14 mg Males 20 mg	Meat, poultry, fish, whole wheat and enriched grains
Vitamin B_{12}	Builds genetic molecules, essential for proper functioning of the nervous system	Anemia, bowel disorders, poor appetite, and poor growth	1.9 mg	Liver, kidney, fish
C	Helps to maintain normal development of bones, teeth, gums, and cartilage	Scurvy, bleeding gums, easy bruising, low resistance to infections	45 mg	Citrus fruits, green vegetables, potatoes

number of drops required. Make sure that the test tubes are dry and that you count the drops accurately.

2. Repeat, using distilled water. This will be your control. *Record your results.*

3. Add 20 drops of indophenol to a clean, dry test tube and then add, drop by drop, one of the samples of fruit juice to be tested. Count and *record the number of drops required to turn it pale pink or colourless.*

4. Repeat using the other samples. *Record your results.*

CAUTION!
Wear eye protection when heating liquids.

5. Take a fresh sample of one of the fruit juices, boil for about 5 min, and then test it as before. *Record your results.*

6. Leave out on the bench overnight a sample of one of the juices you have tested. Test it after 24 h and *compare the results with the fresh juice.*

Questions

1. If a sample of juice A requires eight drops to bleach the indophenol, and a sample of juice B takes 12 drops, which has the stronger concentration of vitamin C? Why?

2. What effect does exposure to air have on vitamin C? Explain your observations.

3. What is the effect of boiling vitamin C? If vegetables contain vitamin C, what happens when they are cooked? What can be done to minimize this problem?

4. Which of the samples tested by you had the most vitamin C? Did the label indicate the amount present? According to your experimental results, do you think that the labels are accurate indications of the amount present?

5. Was there any relationship between the price of the juice and the amount of vitamin C present? Could you determine which juice was a "good buy" for vitamin C content?

CHECKPOINT

1. *What are the two main types of vitamins? Give two examples of each.*

2. *Why might some fruits or vegetables have fewer vitamins than expected?*

3. *Select any two vitamins, and state their value to the body and two good food sources of supply.*

7.6 Other Dietary Requirements

Water

Water is essential for life. Although it is possible to live without food for several weeks; without water we would die in a few days.

The Need for Water
We need water to dilute and help dispose of the body wastes and toxins in our systems. Each body cell must be bathed in fluid, to enable dissolved nutrients and wastes to pass in and out of the cells. Water is needed to move the nutrients, wastes, and blood cells along in the bloodstream. If water is not available for these functions, we must take it in. Our body signals its need for water by the sensation of thirst.

In hot weather, when we lose water by sweating to keep cool, there is an increase in our desire to drink. Fevers, vomiting, diarrhea, or the use of antihistamines to control allergies, all cause water to be lost from the body and thus increase our thirst.

Water Loss

Loss of water occurs in several ways. The largest amount passes out of the body as urine, heavily charged with dissolved wastes. Some is also lost when solid wastes are excreted. Water is lost by evaporation from the skin to assist in cooling the body, and from the lungs in expired air.

Water is replaced by fluids in the diet. Some foods, such as fruit and vegetables, are composed primarily of water. Watermelon is about 90 percent water, and lettuce has an even higher percentage of water content. Compare the food values of the beverages given below. You might also find out the prices of each 340-g drink. It is not difficult to see that many common beverages contain only water and sugar and are fairly expensive. (See Table 7.5.)

Minerals for Body Functions

Minerals are inorganic elements needed in small amounts to assist in a variety of essential body functions. Proteins, fats, carbohydrates, and vitamins are all carbon-containing compounds (organic compounds). Minerals do not contain carbon. Minerals are readily absorbed in solution without digestion.

The minerals found, both in the body and in

Table 7.5 Comparative Nutrient Values for Some Common Beverages

Amount (mL)	Beverage	Food Energy (kJ)	Protein (g)	Carbohydrates (g)	Fat (g)	Calcium (mg)	Iron (mg)	Vitamin A (RE)	Thiamin (mg)	Riboflavin (mg)	Niacin (NE)	Vitamin C (mg)
250	milk 3.3%	660	8	12	9	306	0.1	106	.10	.42	2.3	2
250	milk 2%	540	9	12	5	315	0.1	106	.10	.43	2	2
250	milk skim	380	9	13	trace	317	0.1	106	.09	.36	2.3	3
250	apple juice	530	trace	32	trace	16	1.6		.02	.04	0.3	93
250	orange juice	530	2	30	trace	26	1.0	53	.17	.05	0.8	106
250	tea	8										
250	coffee	13										
250	cola	400		25								
250	ginger ale	300		20								
250	beer	440	1	10		13	trace		trace	.07	1.5	
50	spirits	490										

Note: mg = milligram
RE = retinol equivalent = 1 µg retinol
NE = niacin equivalent = 1 mg niacin

our food, in the largest amounts are calcium, iron, phosphorus, potassium, sodium, iodine, fluoride, and chloride. In addition, there are a number of elements found in very small or trace amounts. Even heavy metals like gold, silver, mercury, zinc, and magnesium are found in the body. We do not yet know, however, if the presence of all of these is important for good health. In most cases, heavy metals (such as lead) are toxic to the body.

The best food sources of minerals are fruits, vegetables, meats, milk, eggs, cereals, and water. Milk is a very important contributor of calcium. Liver is an excellent source of iron, but most meats and seafoods also contain good supplies of this mineral. Despite the small amounts in which most of them are found, minerals are a very important part of our diet and are crucial to good health. Many minerals are components of important molecules, such as hemoglobin, vitamins, hormones, and enzymes. Some minerals are required for the normal functioning of nerves and muscles.

Table 7.6 shows the function of the most important minerals, and the foods in which they are found. The major roles of minerals in the body can be grouped into categories:

- They control water balance.
- They regulate the acid-base balance of body fluids.
- They form part of many complex molecules, such as enzymes and hormones.
- They form a structural part of bone and cartilage.
- They help in many enzymatic activities in the body.

Table 7.6 The Minerals

Mineral	Function	Deficiencies	Sources
Calcium	Forms bone and teeth, aids blood clotting, and nerve impulse connection.	Poor calcification of teeth and bones. Osteoporosis.	Common in many foods: milk, cheese, cereals, beans, and hard water.
Phosphorus	Also required for teeth and bones, some cell reactions.	Poor development in teeth and bones.	Meats, fish, dairy products, grains; common in many foods.
Iron	Needed for hemoglobin. Helps cells obtain energy from foods.	Anemia, lack of energy, especially needed in young children, girls, and women.	Liver, heart, meats, green leaf vegetables, whole wheat bread, cereals, and nuts.
Iodine	Formation of hormones in thyroid gland.	Goiter, swollen thyroid gland.	Seafood, iodized table salt.
Sodium	Regulates water between cells and blood.	Dehydration.	Very common: table salt, canned meats, and vegetables.
Potassium	Needed for synthesis of proteins in cells.	Weakness in muscles.	Meats, cereals, milk, fruits, green vegetables.
Fluoride	Strengthens teeth, especially during development.	More rapid tooth decay.	Drinking water (natural or artificially present)

Note: This table lists the major minerals only. There are a number of other trace minerals that are very important for body functions and health.

Fibre for a Complete Diet

We can absorb most carbohydrates when they are digested and reduced to small molecular components. **Fibre** however, cannot be digested. Fibre contains cellulose, lignin, and certain polysaccharides and pectins. We lack the specific enzymes that break down these substances, even though many of them are composed of units similar to those in starch, which we *can* break down. Fibre remains in the digestive tract and is moved along into the large intestine for eventual elimination.

Functions of Fibre in the Diet
Fibre helps to hold water in the large intestine by combining with the materials present. There, it functions to conserve water and thus prevent the fecal matter from drying out. In this way, it acts to prevent constipation. Fibre also has functions which relate to cholesterol and bile salt metabolism. Diets low in fibre are also believed to increase the potential for gallstone formation and coronary heart disease.

North America has a much higher incidence of bowel cancer than do many developing countries. It may be that the high-protein, low-fibre diet in this part of the world has contributed to these problems.

Problems Known to Result from Low-Fibre Diets
Several common complaints are linked to low-fibre diets. For instance, when the feces are small and dry, the walls of the colon must provide greater muscle force to move them along. They are more difficult to transport and thus move only sluggishly. This may cause the formation of pockets (diverticula) in the bowel, causing irritation and discomfort. When diets are high in fibre, waste materials flow freely in and out of the appendix, whereas when low amounts of fibre are consumed, food tends to remain in the appendix. It then decomposes and becomes a potential source of bacterial infection. Statistics show that appendicitis is more likely to occur in individuals with a low-fibre diet. Constipation is considered an almost universal complaint in Western countries. In North America, over $250 million is spent annually on laxatives! Hemorrhoids have also been shown to be caused by fibre-deficient diets.

The problems of a low-fibre diet can be reduced by simply changing food habits. Fibre can be added to the diet by eating such foods as whole grain breads, by adding bran to morning cereals, or by eating more fruit.

Sources of Fibre
About 7 g per day of dietary fibre have been shown to produce considerable change in colon (lower intestine) functioning. The foods that contain the richest sources of dietary fibre for their mass are cereals which have not been highly processed. (See Figure 7.6.) Green vegetables and fruits provide the least amount of fibre. Fruits and vegetables do supply fibre, but must be eaten in large amounts to satisfy the body's fibre requirements.

The process by which white flour is produced removes the valuable bran components of grain. Whole wheat bread is made from flour that has not been refined as much and, therefore, has not lost its fibre content. When breads are described as being "enriched" or "restored", this usually means that some of the vitamins or minerals, which were lost during processing, have been replaced. It does not mean that any of the fibre or certain amino acids, which were also lost, have been replaced.

CHECKPOINT

1. *List four functions of water in the body.*
2. *Explain how water is lost from the body.*
3. *What three general functions are performed by minerals?*

Chapter 7 / Diet and Nutrition

Figure 7.6
A serving of any of these foods, of the amount shown, would each provide the same amount of dietary fibre (7 g) as a 30 g serving of bran.

4. List three minerals required by the body. For each of these minerals, name two foods in which they are found and state the value of these minerals to the body.

5. Why are some heavy metal elements a cause for concern?

6. What problems, or disorders, are attributed to a lack of fibre in the diet?

7.7 A Daily Food Guide

An important idea emphasized at the start of this chapter was that individuals must take responsibility for their diets and not rely on others to tell them when and what to eat. Basic rules of good nutrition can be used as a guide.

A simple system that has been adopted by The Canadian Department of Health and Welfare divides all the foods we eat into four basic groups (Figure 7.7):

- Meats
- Fruits and Vegetables
- Milk and Milk Products
- Bread and Cereals.

What Quantity of Each Group Does A Person Need?

The Milk Group
Milk and milk products provide the calcium needed to build bones and teeth. Unless an adequate amount of milk is consumed, the daily calcium requirement is difficult to meet; no other source provides this mineral in sufficient quantities.

229

Figure 7.7
The four food groups. A certain amount of food from each of these groups should be eaten each day. No single food or group of foods contains all of the nutrients that we need, not even milk or eggs.

Every day you need . . .

Milk
3 or more glasses
to drink and in foods like these —

Meat and Eggs
2 or more servings   — or some of these alternatives

Vegetables and Fruits
4 or more servings

Have one dark green   Have 1 citrus fruit.
or yellow vegetable.

Bread and Cereals
4 or more servings

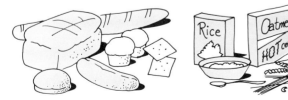

enriched or whole grain

Milk also provides proteins and some of the B vitamins, particularly riboflavin. Vitamins A and D are also added to most commercially available milk, because these are lost during processing of the milk to remove fats.

Unfortunately, many people who are anxious about their weight avoid milk, believing that it contains too much fat. Since two-percent or fat-free (skim) milk is available, there is no reason to avoid this important source of nutrients.

Teenagers need a large amount of calcium each day. Yogurt is one food from the milk group that contains many nutrients. It is a substitute selected by many who do not wish to drink two large glasses of milk each day.

The Meat Group
This group contains meat, poultry, fish, and seafoods, eggs, dried beans, peas, and nuts. These foods are the major suppliers of the proteins required for growth and tissue repair. They also supply iron and vitamins of the B complex (thiamin, riboflavin, niacin, B_6, and B_{12}).

All foods, including meat, contain stored energy. Most of these protein sources also supply fats. Even lean meat contains five to ten percent fat. The quantity and type of meat selected is important, however. A person who eats a large steak (about 350 g) will gain 6048 kJ of energy (more than half the daily total need). The fat contained in this steak would include about 320 mg of cholesterol (300 mg of cholesterol is considered the maximum advisable intake per day).

Because of their rapid growth, teenagers should try to eat at least two servings a day from the meat group, in any combination. It could be two eggs for breakfast and half a chicken breast at night, or two slices of cheese at lunch and fish for supper.

The Fruits and Vegetables Group

Fruits and vegetables are the most valuable source of vitamins and minerals. They provide 90 percent of our vitamin C and about 60 percent of our vitamin A intake. Citrus fruits provide most of the vitamin C that we use, while dark green and deep yellow vegetables are the best suppliers of vitamin A. Vitamin C is the vitamin most often lacking in the average diet.

This food group also provides some minerals, carbohydrates, and fibre. At least four servings per day from this group are required for a healthy diet. A fresh salad might equal two servings, with an apple for a snack and one serving of a vegetable. Another day's intake might include a whole banana and two vegetables at suppertime.

The Bread and Cereals Group

Breads and cereals form a much misunderstood food group. They are among the first foods to be blamed for being fattening. These foods provide energy (carbohydrates), as well as proteins, thiamin, niacin, and iron.

Commercial breadmakers have bleached and conditioned flour until it has lost much of its original nutritional value. Much white bread is now "enriched" by having the vitamins and minerals that were lost during processing replaced so that it should be as nutritious as whole wheat bread.

The recommended minimum amount for adults and teenagers is four servings per day. Two sandwiches (four slices of bread) at lunchtime often account for the whole allowance. Whole wheat brown breads are highly recommended. Granola or whole grain cereals at breakfast with toast would account for half the allowance, and spaghetti for supper would fill the quota.

Although foods from each of these groups may appear at every meal, this is not necessary. It is important, however, to ensure that the total number of recommended servings from each group is eaten sometime during the day. (See Table 7.7.)

Other Foods

Obviously, some foods have not been mentioned in this guide, items such as butter or margarine, oils, sugars, and desserts. These are not emphasized because they are common in any diet. Some of these foods also contain vitamins (oils, for instance), or furnish fatty acids; however, they primarily contribute energy.

Understanding the Daily Food Guide

Adequate servings of foods from the four groups discussed earlier can supply us with all the nutrients that we need for daily living. If you have an active lifestyle, you will need to increase the number or size of the servings. If your body mass starts to increase, you should cut back on the size of the servings, for you are eating in excess of your needs. Cut back, but don't cut out. Reduce the size of portions, but keep your diet balanced at all times.

Breakfasts and Evening Meals

Research has indicated that people who eat breakfast are more alert and productive in the morning than are those who miss breakfast. Students do better on tests and absorb information more effectively when they have had breakfast. They are less likely to tire during the day. The body goes without food all night, so it needs to be recharged with food in the morning to provide energy and nutrients for the day's activities.

Heavy meals in the evening are not necessary. When large supplies of food enter the body and are absorbed into the bloodstream, the energy contained within the food soon becomes available for use. Common evening activities such as studying, reading, or watching television do not demand much energy. As a result, this food may be stored as fat.

Table 7.7 Canada Food Guide

Milk and Milk Products	Fruits and Vegetables
Provide calcium for bones and teeth protein for tissue building and repair riboflavin for production of enzymes vitamin A for healthy skin and hair vitamin D for absorption of calcium *Servings per day* Children 4 servings Teenagers 3 to 4 servings Adults 2 servings *Sample servings* Milk – 250 mL (approximately 1 cup) Cheese – 1 slice Swiss cheese Yogurt – 250 mL Ice cream – 250 mL (equivalent to 125 mL of milk) Cottage cheese – 250 mL Some calcium is also found in: Salmon, sardines Beet and turnip greens Fresh oranges	*Provide* vitamin A for healthy skin, good night vision, bone growth, prevents infections vitamin C fights infections, promotes wound healing, cementing of cells together minerals – folic acid, helps form and break down amino acids *Servings per day* 4 to 5 including two vegetables *Sample servings* Fresh salads enough to fill a cup Cooked fruits – 125 mL (half a cup) Vegetables – 125 mL (half a cup) Fruit juice (unsweetened) – small glass One potato, carrot, or tomato Fresh apple, pear, orange, or banana
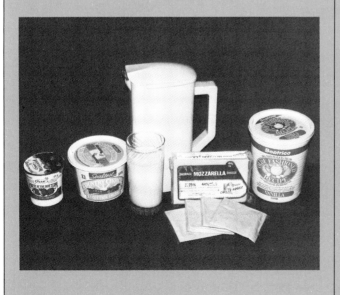	

Meat, Fish, and Alternates

Provide
proteins for building and repair of tissues and growth
iron for blood hemoglobin
riboflavin for healthy skin and eyes
niacin for growth, nervous and digestive systems
B_6 and B_{12} help with protein metabolism and the nervous system

Servings per day
all ages 2 servings

Sample servings
Beef, lamb, ham – 80-g slice, 10 cm × 6 cm
Fish – 75 g, 6 sardines, 1 fresh fillet, 4 fish fingers
Chicken – a leg or 1/2 chicken breast
Eggs 2
Cheese – 60 g, 125 mL, 2 slices of cheese, 1/2 cup of cottage cheese
Dried peas, beans, lentils – 250 mL
Peanut butter – 60 mL (4 tablespoons)
Nuts – 125 mL

Bread and Cereals

Provide
carbohydrates
proteins – building and repair of tissues
thiamin helps appetite and nervous system
niacin helps skin, tongue, nervous system
iron for hemoglobin to carry oxygen

Servings per day
3 to 5 servings

Sample servings
Bread – 1 slice
Cooked cereal – 125 to 200 mL (1/2 cup or more)
Muffins or rolls – 1
Macaroni, spaghetti, noodles – 125 to 200 mL
Ready-to-eat cereals – 125 mL (bowl)

CHECKPOINT

1. Name the four food groups of the Canada Food Guide and the numbers of daily servings recommended for a person of 16 years.

2. List four foods for each of the food groups you have named.

3. Why is it important to have a variety of food in your diet?

7.8 Being an Informed Consumer

Processed food products have at least three types of information on their packaging. First, most of the space is given over to advertising – often colourful, eye-catching pictures that make you want to buy the product. Secondly, there is information that is short, simple, and concise about how to prepare the product. Lastly, sometimes in the smallest type of all, is the important information about what is really in the product – the contents. This is the part of the label examined by a wise shopper. (See Figure 7.8.)

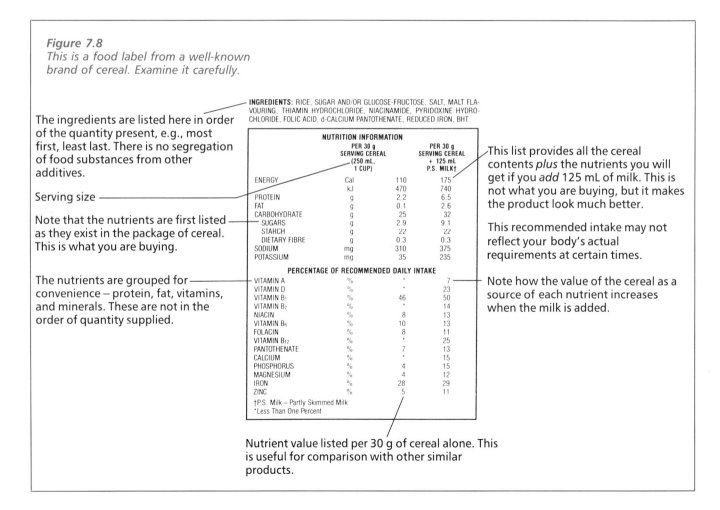

Figure 7.8
This is a food label from a well-known brand of cereal. Examine it carefully.

The ingredients are listed here in order of the quantity present, e.g., most first, least last. There is no segregation of food substances from other additives.

Serving size

Note that the nutrients are first listed as they exist in the package of cereal. This is what you are buying.

The nutrients are grouped for convenience – protein, fat, vitamins, and minerals. These are not in the order of quantity supplied.

This list provides all the cereal contents *plus* the nutrients you will get if you *add* 125 mL of milk. This is not what you are buying, but it makes the product look much better.

This recommended intake may not reflect your body's actual requirements at certain times.

Note how the value of the cereal as a source of each nutrient increases when the milk is added.

Nutrient value listed per 30 g of cereal alone. This is useful for comparison with other similar products.

When you look at the contents label on any food product, you will probably be overwhelmed by the list of chemical names and the terminology. Product producers are not trying to confuse you deliberately. They are required by law to accurately list everything the product contains, and that means using the proper terms. If we were to ask a farmer to list the contents of an apple, the result would be just as complex. (See Figure 7.9.)

Figure 7.9
What's in an apple?

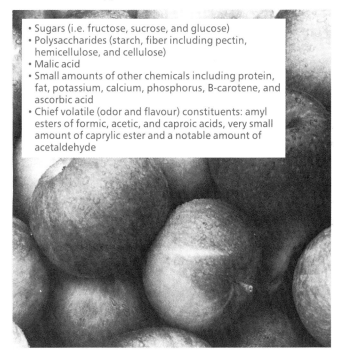

- Sugars (i.e. fructose, sucrose, and glucose)
- Polysaccharides (starch, fiber including pectin, hemicellulose, and cellulose)
- Malic acid
- Small amounts of other chemicals including protein, fat, potassium, calcium, phosphorus, B-carotene, and ascorbic acid
- Chief volatile (odor and flavour) constituents: amyl esters of formic, acetic, and caproic acids, very small amount of caprylic ester and a notable amount of acetaldehyde

A few other points to note about food labels. Not all companies separate out the products as shown in Figure 7.8. Often, only a list such as at the top of this label is given. Shoppers must know enough to sort out the important information for themselves. For example, if you wish to avoid products with a high sugar content, watch for labels listing corn syrup, fructose, lactose, sucrose, corn sweeteners, maple syrup, dextrose, and honey. It is still sugar despite these names.

Most fruit juices list their sugar content as "carbohydrates". For example, apple juice contains 11 g of carbohydrate per 100 mL. Nutrition labelling is not mandatory, but as consumers become more concerned with what they eat, major manufacturers are beginning to include a core list of nutrients along with descriptions such as "low-sodium" or "source of calcium".

Additives

Additives are substances added to food during processing. Some people think of food additives as something bad or at least a necessary evil, yet some are important and very useful. Additives make it possible for us to enjoy exotic fruits out of season, and enable us to ship products from distant towns and have them arrive fresh. Many prevent spoilage by bacteria. Some additives can even be nutritious, as vitamins and minerals also come under this general heading.

Some additives, however, have no nutritional value, being added as part of the production process. A compound might be added to stop dough sticking to baking pans or to mixer blades, or another to bleach or whiten flour, or to make sure that creams do not separate. There is always a reason for an additive being present. (See Table 7.8.)

As far as possible, food regulations try to ensure that these substances are not harmful. However, some people are allergic to certain chemical additives. Sulphite, for example, has been used to keep salads fresh and may also be in some beverages. People sensitive to sulphites experience difficulty breathing if they eat the substance. Although no longer legal for use on fresh fruit and vegetables in restaurants, sulphites could be on foods prepared before reaching the restaurant. Another problem is that, while each additive may be safe on its own, it may react with others that we take, or build up a cumulative effect, which can be very serious indeed.

Table 7.8 Common Food Additives

Reasons for Using Additives	Additives Used	Where Used
TO MAINTAIN CONSISTENCY Emulsifiers improve texture and help mixing Stabilizers and thickeners give smooth uniform texture and help increase shelf-life	Alginates, mon- and diglycerides, methyl cellulose, carrageenin, polysorbate, lecithin, agar, sodium phosphate, Guar gum	Baked goods, cake mixes, salad dressings, frozen desserts, ice cream, chocolate milk, processed cheese, chewing gum
TO IMPROVE NUTRITIVE VALUE Eliminate or prevent certain disorders of malnutrition (e.g., iodized salt has eliminated simple goiter)	Vitamin A and D, niacin, ascorbic acid (vitamin C), potassium iodide, thiamine, riboflavin, iron	Wheat flour, bread, biscuits, cereals, corn meal, macaroni, noodles, margarine, iodized salt
TO IMPROVE FLAVOUR Spices, natural and artificial flavours, etc. improve the variety of foods available and make them more attractive (e.g., ice cream flavours)	Cloves, citrus oils, carvone, monosodium glutamate, ginger, amyl acetate, benzaldehyde	Spice cake, gingerbread, ice cream, soft drinks, candy, gelatins, fruit flavoured toppings, sausage, meats, soups
TO CONTROL ACIDITY AND ALKALINITY Leavening agents used in bread Makes fruits and potatoes easier to peel for canning, neutralize sour cream in butter making	Sodium bicarbonate, lactic acid, citric acid, phosphates	Cakes, cookies, crackers, butter, chocolates, soft drinks
TO MAINTAIN APPEARANCE AND PREVENT DETERIORATION Prevent spoilage by bacteria and moulds, antioxidants prevent fats from turning rancid, used to prevent fruits from changing colour, increases shelf-life	Propionic acid, sodium and calcium salts, butylated hydroxanisole, butylated hydroxytoluene, benzoates, ascorbic acid	Bread, cheese, syrup, pie fillings, fruit juices, frozen and dried fruit, margarine, potato chips, cake mixes
TO IMPROVE COLOUR Makes foods more attractive and improves sales	Many possibilities, including annatto, cochineal, carotene, chlorophyll	Candies, baked goods, soft drinks, cheeses, ice cream, jams and jellies
OTHER FUNCTIONS Help retain moisture in some foods and in others, such as salts, keep them free flowing Binding and extending imitation sausage	Glycerine, magnesium carbonate, sorbitol, whey powder	Coconut, table salt, confectionary, sausage

Activity 7F

Food Labels and Advertising

What information is available from a food label?

Procedure

1. Collect at least three different breakfast cereal labels. Try to make them as different as you can, e.g., sugar cereals, bran cereals, rice, wheat, or oat varieties. *Make a table similar to Table 7.9, but include only those vitamins and minerals present on your labels.*

Table 7.9 Nutrient Content of Selected Breakfast Cereals

Ingredient	Sample A	Sample B	Sample C
Energy			
Protein			
Fat			
Carbohydrate			
Sugar			
Starch			
Dietary Fibre			
Salts			
Sodium			
Potassium			
Vitamins			
Vitamin B_6			
Vitamin B_{12}			
Vitamin A			
Vitamin D			
Niacin (B_3)			
Folic Acid			
Panothenate			
Thiamine (B_1)			
Riboflavin (B_2)			
Minerals			
Calcium			
Phosphorus			
Magnesium			
Iron			
Zinc			
Additives			

2. *Now complete the table by examining the labels of your selection of products and filling in the ingredients listed.* Check to see that the listings are using similar size servings. If not, try to find a way to make the figures you are comparing consistent.

Questions

1. (a) Which of the products you have used has the most sugar per serving?
   (b) Is there a big difference between your products?
   (c) Would this influence your decision to purchase the product?

2. Which product has the most protein?

3. Which product has the most starch?

4. Which product has the most dietary fibre? This is an ingredient that gives us no nutritious food value but is still important. Why?

5. Evaluate which product is the best buy for vitamins. Remember to look at both variety and quantity.

6. Evaluate which product is the best buy for minerals.

7. (a) Which product has the most salt(s)?
   (b) Is it a good reason to buy the product if it has the most salts present? Explain.

8. Which product has the most non-nutritious additives present?

9. List two good reasons for you to eat breakfast.

10. In view of your answers to question 9, which of the products you have examined is the best buy for you?

11. Consider the cost of each product, either per serving or per 100-g portion. Which is the best buy?

12. Look at the advertising on the package. List the features that try to make you buy the product. It may be anything from an endorsement by a famous sports figure to a giveaway inside. It may be special words which imply it is good – natural, fat-free, (would you expect much fat in a cereal?) low-calorie, etc.

13. Name the type of breakfast cereal you buy for your home. What reasons can you give for buying that product? Has this activity changed your mind about your future purchase of a breakfast cereal?

CHECKPOINT

1. *Why is it important to read the food content labels on food products?*

2. *Describe two ways in which food labelling can be misleading.*

3. *List two good and two poor reasons for food additives being used in food.*

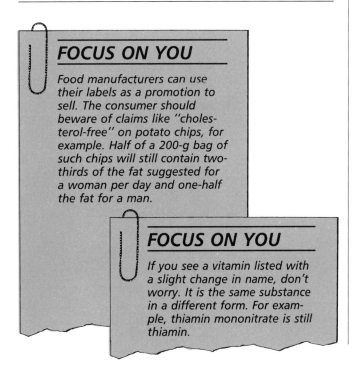

FOCUS ON YOU

Food manufacturers can use their labels as a promotion to sell. The consumer should beware of claims like "cholesterol-free" on potato chips, for example. Half of a 200-g bag of such chips will still contain two-thirds of the fat suggested for a woman per day and one-half the fat for a man.

FOCUS ON YOU

If you see a vitamin listed with a slight change in name, don't worry. It is the same substance in a different form. For example, thiamin mononitrate is still thiamin.

7.9 The Assessment of Body Size

Should I lose or gain mass? This is a question that many teenagers ask themselves, often quite without reason. Sometimes they are not satisfied with their own answers and start to ask their friends for advice. They may decide on a course of action simply because a favourite item of clothing no longer fits well, ignoring the fact that the body changes are a result of normal growth.

Although an honest appraisal of ourselves will usually give us all the information we need, it is sometimes necessary to check with a physician for a more objective standard by which to judge the amount of body fat present.

The Skinfold Test

To make sensible judgments about whether we are too fat or too thin, we sometimes need an accurate method of measurement. One simple method is to use height/mass tables, but these only compare our statistics with those of an average person. Such tables do not take into consideration individual differences, such as frame size and bone density. Sometimes the tables even ignore age and sex differences. The best method is to use skin calipers, which can provide an accurate assessment of our body fat content and where it is located. (See Figure 7.10.)

About half of the total fat in the body is subcutaneous fat, deposited just under the skin. In many parts of the body, this sheet of fat is only loosely attached to the tissues below, and can be pulled up between the thumb and forefinger, into a fold. Researchers have found that there is a direct relationship between the amount of fat deposited beneath the skin and the amount laid down around the organs of the body, where we cannot easily reach to measure; therefore, if we measure the extent of subcutaneous fat and apply this to

Figure 7.10
Skin calipers

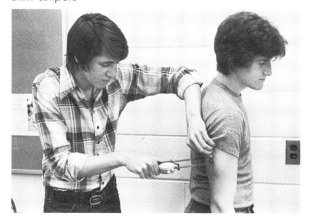

the tables produced by research, we can make estimates about the total fat in the body.

Although the total mass of skin tissue is approximately the same in men and women, the amount of subcutaneous fat varies greatly. Men's bodies have about 11 percent fat and women's bodies about 24 percent.

How Much Fat is Enough?

Your body mass should include sufficient fat to help you feel your best and supply your body needs. Specifically, you should have enough to provide adequate reserves of energy, sufficient subcutaneous fat to help insulate the body from adverse temperatures, and enough to protect the organs from injury.

Activity 7G

Your Ideal Mass

What is your ideal mass using the skinfold test?

Materials

Harpenden calipers
scales

Procedure

1. Locate the test site as in (a), (b), (c), and (d) of Figure 7.11, and mark it with a felt pen. Take a fold of skin between thumb and index finger 1 cm above the mark on the skin. Hold the flesh firmly, pulling it gently away from the body. Read step 3 before proceeding.

2. Place the jaws of the calipers at right angles to the site, over the pen mark, and carefully release the handles of the calipers. Read the measurement after the caliper pressure is fully applied and the needle is no longer drifting. *Record your result to the nearest 0.2 mm. Repeat this twice more and, if you obtain differences of more than 1 mm, take another measurement and average your results.*

Figure 7.11
Using the Harpenden calipers

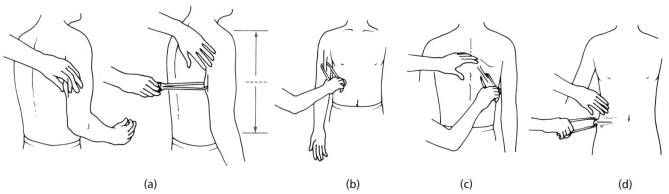

(a)   (b)   (c)   (d)

3. Make the tests at each of the following sites:

   (a) Triceps: The arm should be bent at right angles, and the site marked at the midpoint between the top of the shoulder and the tip of the elbow. The skinfold is taken parallel to the arm and the calipers applied after the subject has lowered the forearm.

   (b) Biceps: The site is marked and taken on the front of the upper arm halfway between the shoulder and the tip of the elbow. The arm should hang loosely at the side of the body. Take the skinfold parallel with the axis of the arm.

   (c) Subscapular: This site is measured about 1 cm below the lower angle of the right scapula (shoulder blade). The caliper jaws are held at 45 degrees to the spine.

   (d) Suprailiac: This site is measured just above the crest of the hip bone, with the fold at right angles to the axis of the body, parallel to the crest of the bone.

4. *To determine the amount of body fat, add together the four skinfold measurements you have taken. Refer to Table 7.10 and find the number nearest to this total in the column under skinfolds that corresponds to your age group and sex. The chart will give you the **percentage of body fat**. Record this value.*

5. *Record your body mass in kilograms. Use this to find your lean body mass (LBM). This is the mass of your body if it could be weighed without any fat present at all.*

   LBM = Present mass −
   (Present mass × percent of body fat)

6. Now *determine your ideal mass*, which should be your lean body mass plus 15 percent body fat for males or 20 percent body fat for females. *Calculate your ideal mass by multiplying your lean body mass by 1.8 if you are a male, or 1.25 if you are a female.*

Table 7.10 Body Fat and Skinfold Totals

Skinfolds (mm)	Percent of Body Fat	
	Males 17 to 29	Females 16 to 29
15	4.8	10.5
20	8.1	14.1
25	10.5	16.8
30	12.9	19.5
35	14.7	21.5
40	16.4	23.4
45	17.7	25.0
50	19.0	26.5
55	20.1	27.8
60	21.2	29.1
65	22.2	30.2
70	23.1	31.2
75	24.0	32.2
80	24.8	33.1
85	25.5	34.0
90	26.2	34.8
95	26.9	35.6
100	27.6	36.4
105	28.2	37.1
110	28.8	37.8
115	29.4	38.4
120	30.0	39.0
125	30.5	39.6
130	31.0	40.2
135	31.5	40.8
140	32.0	41.3
145	32.5	41.8
150	32.9	42.3
155	33.3	42.8
160	33.7	43.3
165	34.1	43.7
170	34.5	44.1
175	34.9	44.5
180	35.3	44.9
185	35.6	45.3
190	35.9	45.6
195	36.2	–
200	36.5	–
205	36.8	–
210	37.1	–

Example (female)
   Present Mass = 54 kg
   Total of skinfold measurements = 55 mm
   Therefore, % of body fat = 27.8%
   LBM = Present mass
         − (Present mass × % of body fat)
       = 54 kg − (54 × 27.8%)
       = 39 kg
   Ideal Mass = LBM × 1.25
              = 39 × 1.25 kg
              = 48.75 kg

7. *Determine the differences between your present body mass and your ideal mass.* For the example above, this is equal to 54 − 48.75 = 5.25 kg. This person is therefore 5.25 kg overweight.

Questions

1. Compare your calculated ideal mass with your present mass. Does this test suggest that you need to make any adjustments in your diet or exercise patterns? What do you suggest?

2. Do you feel that the test confirms your own subjective evaluation of your mass? Comment on this. (Remember that this test measures *your* body fat. It is not for making comparisons with others or the average masses of persons in your age group.)

Controlling Body Mass

In the previous pages, we have discussed the foods that we need in our daily diet and have considered the size of the portions sufficient to supply these needs. To control mass successfully, we need to balance the two sides of the equation, kilojoules of energy consumed and kilojoules expended. (See Figure 7.12.)

The Problems Associated With Being Overweight
Excess mass contributes to some very serious disorders, such as heart disease, high blood pressure, and other circulatory disorders. It adversely affects bones and muscles and crowds the internal organs. It also can have a negative effect on the individual's outlook on life. The extra mass that must be carried causes shortness of breath and constant weariness. The extra effort that must be made for every activity often results in withdrawal from activities that the individual previously enjoyed. As one becomes less active and more sedentary, the problem increases.

People who were once overweight and have lost the excess have great difficulty in maintaining their reduced mass. One of the reasons for this is that in such a person, many cells – adipose tissue cells – have been produced by the body

Figure 7.12
The balance required between the intake of food energy and the expenditure of energy on exercise, body activities, etc., to maintain a stable mass.

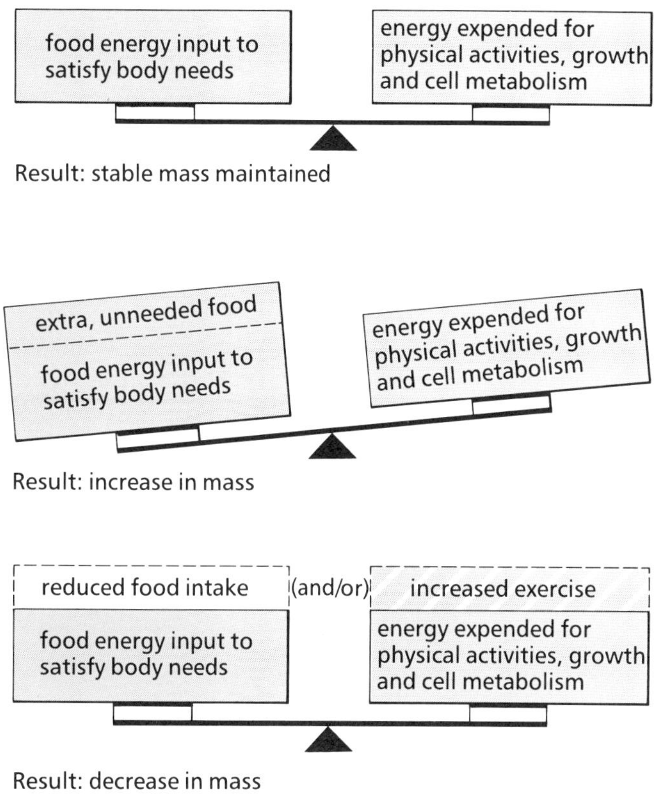

food energy input to satisfy body needs | energy expended for physical activities, growth and cell metabolism

Result: stable mass maintained

extra, unneeded food / food energy input to satisfy body needs | energy expended for physical activities, growth and cell metabolism

Result: increase in mass

reduced food intake (and/or) increased exercise
food energy input to satisfy body needs | energy expended for physical activities, growth and cell metabolism

Result: decrease in mass

to store the excess food supplies available. Once produced, these cells do not disappear for a long time. When mass is lost, these cells start to empty, but the cells are not immediately destroyed. They are still there waiting to be filled up again.

Fortunately, many of us are only slightly overweight. This is a good time to stop the pattern of increasing mass before it gets too far along. So long as we carry only a slight amount of extra mass, we can simply cut back on foods and do some exercise to work off the excess.

Working Off the Excess – By Exercise

Many people have the idea that exercise must be a strenuous, exhausting activity. It does not need to be. (See Table 7.11.) Substituting a half-hour walk to work or school for a 4-min car ride can result in a loss of 6 to 10 kg a year! The key to continued exercising is to find something that you like doing. It is no good deciding to jog if you hate jogging. You won't keep it up. It is better to find an activity that you enjoy, preferably that you do with friends. Most towns have facilities for excellent programs in swimming, basketball, squash, and other sports, as well as fitness classes. You can join an amateur team to play almost any sport. There are also many non-competitive sports – cross-country skiing, hiking, riding, and canoeing. It is important to set aside a regular time for exercise. Regular exercise usually becomes something that you will look forward to and enjoy.

Table 7.11 Food and Exercise Energy Equivalents

Food	Mass (g)	Energy (kJ)	Time in Minutes to Burn Off Kilojoules			
			Walk	Bicycle	Swim	Jog
Coke (227 mL)	240	483	20	16	12	11
White bread (one slice)	23	252	12	9	7	6
Mars bar	40	890	40	32	25	21
Peanut brittle	25	462	21	16	13	11
Popcorn, buttered (250 mL)	18	344	16	12	10	8
Chocolate chip cookie (one)	11	210	10	8	6	5
Oreo cookie (one)	12	168	8	6	5	4
Ice cream sandwich	75	873	40	32	25	21
Ice cream cone	72	672	31	24	19	16
Banana split	300	2494	114	89	71	59
Doughnut, jelly	65	949	44	34	27	23
Apple	150	365	17	13	10	9
Banana	150	533	24	19	15	13
Cheeseburger	180	1940	89	69	55	46
French fries (20)	100	1150	53	41	33	27

Note: Walking: At 3.5 to 6.5 km/h, burns off about 21.8 kJ/min.
Bicycling: At about 11 km/h, burns off 27.3 kJ/min.
Swimming: At an average rate of 30 m/min, burns off 35.7 kJ/min.
Jogging: At alternating 5-min walking-jogging, burns off 42.0 kJ/min.

LIFE · SIGNS

BODY MASS

Jean was looking at herself in the mirror, turning from side to side to view her body from all angles. Jean was not happy. Her figure seemed to her to be far too thick and she wanted to be slim. Her lack of appreciation for her appearance would have amazed her friends. They thought that she was already extremely thin and often urged her to eat with them at lunchtime. Jean always responded to such invitations with an excuse that she had an appointment to keep, or that she was not hungry, or that she would eat later.

Jean was suffering from **anorexia nervosa**, a disorder that usually affects teenage girls. Initially, such girls may be mildly overweight and start a diet program. Sometimes, when a suitable mass loss brings their body size back to an average mass, they continue to diet, convincing themselves that they still are obese.

Jean constantly devised new ways to avoid eating and also to dodge the pressures others placed upon her to eat more. Jean would try to wear bulky knit sweaters to prevent others from commenting on how thin she was becoming and whenever she could she exercised to lose still more mass.

Visits to the doctor were unhappy events as the evidence in the doctor's records showed her ever-decreasing mass. The doctor's tests showed low blood pressure and vitamin deficiencies. Jean no longer menstruated, she often became hysterical, suffered from constipation, and felt unwell. The doctor showed her some mass tables and compared her mass to the mass of other girls, but Jean could not be convinced that she was underweight, or that her symptoms were caused by her very inadequate diet.

The doctor knew from previous experiences that anorexia nervosa is a **psychosomatic** illness, a physical illness caused by the mind. After many failures to persuade Jean to eat normally, the doctor sent her to a hospital for treatment.

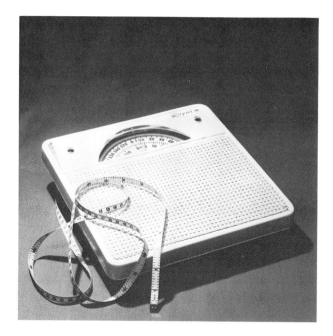

The hospital staff provided a special diet for Jean and did everything they could to encourage and gently persuade her to eat. Being in the hospital limited Jean's freedom and made it possible to keep a more accurate check on her food intake and activities.

Jean ate under protest, and then later, when the opportunity arose, she would run up and down the ten flights of stairs from her room to the hospital basement to try to burn off the food that she had absorbed. After two weeks, Jean had lost a further 2.5 kg, and the staff had to impose severe restrictions. Jean had a very strong and warm relationship with her family, and when the hospital staff refused to allow any visits by her parents or sisters until she had regained the 2.5 kg, Jean finally made up her mind to try to eat at least a small amount of food. The severe restrictions were successful and, after several months, Jean was released from the hospital.

Now, three years later, Jean is a slim young woman with a good healthy outlook on her problem. She is aware that the desire to avoid food is still there, but she is determined to overcome it. Large numbers of people who suffer from anorexia nervosa do not recover and eventually starve themselves to death.

Activity 7H

A Personal Assessment

The purpose of this exercise is to examine your own diet and discover if it is adequate and well-balanced for a person of your age, sex, size, and activity. You will try to determine if you are eating too much or too little of any particular kinds of food and discover the source of most of your kilojoules of energy. From this information you can be reassured about your food habits, or be made aware of potential problems. It will help you decide what changes you might wish to make to reduce or increase your mass as may be necessary.

This exercise is only of value if you are honest with yourself. You must record your intake faithfully, without trying to make it "look good" or hiding any energy sources.

Procedure

- Food Energy Intake

1. *Make a record of ALL the food and drink that you take in over a three-day period.* Include meals, snacks, pop, candy, etc. One of the three days should be a Saturday or Sunday, as your food and activity patterns on the weekend are often quite different from those on school days. *Keep your record on a piece of paper or in a small notebook and jot down each item, every time you take a bite or drink.* You cannot accurately remember everything you eat if you try to record it later. Before you start your record, find out how much each of the juice or milk glasses that you use regularly will hold. Fill each of them with water and pour it into a measuring jug, so that you will know what quantity of juice or milk you normally drink. *Record both the kind of food and the quantity you take.* (Note: Judging the size of portions of vegetables is not easy and rarely very accurate. The best way is to judge it in terms of "cups" – a half-cup of carrots or beans, for instance.)

For example, breakfast might be:

Toast (brown)	one slice
Butter	one pat
Jam	one tablespoon
Orange juice	170 mL (one average-sized glass)
Milk	225 mL (one average-sized glass)

2. *Prepare a table to compile your results with these headings: Food, Amount, Kilojoules, and Protein (g). Using suitable food tables, record the kilojoules and the proteins listed for each of your food items.* Try to be as accurate as possible. (See Table 7.12.) (The Department of Health and Welfare will supply more detailed tables free of charge.)

3. *Total up the number of kilojoules and the grams of protein that you consumed each day. Record these values.*

- Energy Output

Each activity in which we take part uses up energy. Some activities will use a great many kilojoules each minute, such as a game of tennis. Others will use less, such as watching television. We use energy even when we are asleep to keep the heart, respiratory organs, and other body functions working. If our intake of food energy approximately balances with our output of energy, then our body mass will probably stay fairly constant. If the intake exceeds the output, we can expect to gain mass. In this activity, you will be using an *energy factor* that has been determined experimentally for each type of activity. (See Table 7.13.)

4. *For the same days that you recorded your food and drink intake, record your activities and the time you spent on each one.* This sounds like a tall order, but you can group some of them

Table 7.12 Some Popular Fast Take-Out Foods

Food	Energy (kJ)
Hamburger (bun included)	1050
Hamburger, double	1365
Whopper	2646
Cheeseburger	1280
Cheeseburger deluxe	2520
Big Mac	2340
French fries	966
Onion rings	1260
Fried chicken – two-piece dinner	2457
Fried chicken – three-piece dinner	4135
Chili dog	1386
Fish (two pieces), cole slaw, chips	3108
Chopped steak, 112 g	1373
Chopped steak, 224 g	2742
Baked potato	1016
Salad dressing	630
Pizza (average) half of 25 cm	1953
Pizza (average) half of 35 cm	3780
Pizza (average) half of 45 cm	5040
Egg McMuffin	1310
Hot cakes, butter	1142
Milk shake	1428
Dairy Queen small cone	462
Dairy Queen medium cone	966
Dairy Queen large cone	1428
Dipped small cone	676
Dipped medium cone	1302
Dipped large cone	1890
Dairy Queen small sundae	788
Dairy Queen medium sundae	1260
Dairy Queen large sundae	1806
Hot fudge sundae	2436
Banana split	2436

(Extracted and adapted from several sources including the *Fast Food Calorie Counter* by H. Jordan, L. Levitz, and G. Kimbrell.)

together to make the task easier. For example, if you go to bed at 11:00 at night and get up at 7:20 in the morning, you can easily record it as:

Sleep 8 h, 20 min

If you have six 40-min classes in the day where you are sitting down to study, you can record:

Studying 6 × 40 min

A physical education class involves much more activity, so it will be listed separately:

Running exercise 40 min

If you walk between classes and have seven class changes, this might be shown as:

Walking 7 × 3 min

Some activities that have similar energy factors can also be grouped together. Dressing, washing, brushing teeth, making a bed, etc., might be listed as:

Getting ready for school 25 min

5. You will find that each activity has been assigned an energy factor in Table 7.13. *Place the energy factor shown in the table against each of the activities in your record.* If the particular activity is not given, try to judge the effort involved in comparison to some activity that is shown in the table. Beware of overstating exercise. Vigorous exercise is rarely continuous. You may run hard for 10 min, but then stand and rest for several minutes. A game is rarely continuous heavy exercise; adjust for these conditions.

6. *Record your mass in kilograms. Now calculate how many kilojoules you expend on each activity by using the following formula:*

Energy factor × time spent on activity (h) × your mass (kg)

Example: Studying 1.5 h
Energy factor × time × body mass
6.0 × 1.5 h × 70 kg = 630 kJ

7. When you have completed the calculations, *add up the kilojoules expended each day and compare it with the number of kilojoules you gained*

Table 7.13 Activity and Energy Factors (Energy Expended Per kg Body Mass)

Activity	Energy Factors	
	kJ/h	kJ/min
Sleeping	4.1	0.07
Sitting	5.2	0.09
Writing and studying	6.0	0.10
Standing relaxed	6.3	0.11
Singing, sewing, dressing, washing	7.1	0.13
Dishwashing	8.1	0.14
Playing cards, typing	9.0	0.15
Dusting and sweeping	10.5	0.18
Washing the car, cooking, piano-playing	11.2	0.19
Walking (3.2 km/h)	11.6	0.19
Bowling	13.6	0.23
Canoeing (1.5 km/h)	14.2	0.24
Sailing	15.8	0.26
Bicycling (3.0 km/h), walking (4.8 km/h)	16.2	0.27
Table tennis	18.0	0.30
Laundry, by hand	18.6	0.31
Walking (6.4 km/h)	20.6	0.34
Volleyball, roller skating, badminton	21.5	0.36
Dancing slow	22.6	0.38
Bicycling (15.3 km/h)	25.8	0.43
Hiking (or hunting), dancing fast, shovelling	27.0	0.45
Water-skiing, tennis, downhill skiing	36.2	0.60
Climbing stairs, running (8.8 km/h), swimming (breast stroke, 36.6 m/min)	37.5	0.62
Bicycling (20.9 km/h)	40.5	0.67
Rowing, cross-country skiing	42.0	0.70
Ice skating (vigorous)	48.8	0.81
Swimming (crawl, 45.7 m/min)	49.1	0.81
Handball	49.5	0.82
Running (12.9 km/h)	62.0	1.03
Cross-country skiing (competitive)	73.6	1.26

(Adapted from *Lazy Man's Guide to Fitness* by K.D. Rose and J.D. Martin.)

from food. Do not expect the totals to balance exactly. If they come within 400 to 800 kJ, you will be doing well. The measurement of nervous energy poses a problem in obtaining accurate results which can make a big difference in your results.

Questions

1. Go through the record of foods eaten and place a cross beside any food items that are composed mainly of carbohydrates. Many of these contain no other nutrients, e.g., cola or pop, white bread, chips, etc. Total these and see what proportion of your total intake is made up of kilojoules from these sources.

2. (a) If you wanted to gain or lose a little mass, what items could you include or omit from your diet?
   (b) If you need to reduce mass, what more nutritious food could you substitute so that you lose a little mass but don't go hungry?

3. Is the intake of protein adequate for a person of your age and size? (See Table 7.14.)

4. Does your intake vary with the output? For example, if you are going to have a strenuous day, then your breakfast should be larger, to provide the energy you will need. If you have a lazy Saturday, sleeping in until noon, your intake should be proportionately smaller. (If you have a cold or feel unwell, both your food and activity totals will probably be down.) Make a few brief statements about the days you recorded, indicating whether or not each day was a "normal" day for you. Did you have a cold, play in a team competition, or were you upset over some event or argument? List anything you think may have affected your food intake or changed your activities from your regular pattern.

Table 7.14 Recommended Daily Nutrient Intake

	Age (years)	13-15		16-18		19-35	
	Sex	*Male*	*Female*	*Male*	*Female*	*Male*	*Female*
	Mass (kg)	51	49	64	54	70	56
	Height (cm)	162	159	172	161	176	161
	Energy (kJ)	11 700	8200	13 400	8800	12 600	8800
	Protein (g)	52	43	54	43	56	41
Water-Soluble Vitamins	Thiamin (mg)	1.4	1.1	1.6	1.1	1.5	1.1
	Niacin (NE)	19	15	21	14	20	14
	Riboflavin (mg)	1.7	1.4	2.0	1.3	1.8	1.3
	Vitamin B_6 (mg)	2.0	1.5	2.0	1.5	2.0	1.5
	Folate (μg)	200	200	200	200	200	200
	Vitamin B_{12} (μg)	3.0	3.0	3.0	3.0	3.0	3.0
	Vitamin C (mg)	30	30	30	30	30	30
Fat-Soluble Vitamins	Vitamin A (RE)	1000	800	1000	800	1000	800
	Vitamin D (μg)	2.5	2.5	2.5	2.5	2.5	2.5
	Vitamin E (mg)	9	7	10	6	9	6
Minerals	Calcium (mg)	1200	800	1000	700	800	700
	Phosphorus (mg)	1200	800	1000	700	800	700
	Magnesium (mg)	250	250	300	250	300	250
	Iodine (μg)	140	110	160	110	150	110
	Iron (mg)	13	14	14	14	10	14
	Zinc (mg)	10	10	12	11	10	9

(Department of Health and Welfare Canada. Revised 1975.)

Note: mg = milligram
μg = microgram
NE = niacin equivalent = 1 mg niacin
RE = retinol equivalent = 1 μg retinol

5. Try to express your conclusions about how appropriate your diet is for you. Do you need extra or less mass? Consider your muscle tone and posture. Do you need to exercise? Try to write down any ways in which you think you could improve your health by changing your diet. Try to make your statements as positive and constructive as you can.

Many questions (and probably many answers to questions), will have occurred to you while you made the personal assessment in Activity 7H. Try to recognize that diet is a *dynamic* thing. What is an adequate diet for you today may not be tomorrow. You need to make adjustments. Your diet now, when you are involved with daily physical education classes, walking to school, and participating in sports, may be too much and may increase your mass next year, if you become less active.

CHECKPOINT

1. *Why are skin calipers a good method of judging the amount of fat on the body?*

2. *How can you make good judgments about your body size and whether or not to gain or lose weight?*

Chapter 7 / Diet and Nutrition

CHAPTER FOCUS

Now that you have completed this chapter, you should be able to do the following:

1. State the six categories of food and give the main functions of each.
2. Give the major functions of carbohydrates, fats, and proteins in the body.
3. Describe simple tests for the presence of starch, sugar, fats, and proteins in foods.
4. Name good food sources of carbohydrates, fats, and proteins.
5. List three vitamins and three minerals with sources and state why they are needed in the body.
6. Describe an experiment to determine the relative quantity of vitamin C present in a food sample.
7. State the need for fibre in the diet and name four good sources of dietary fibre.
8. Name the four food groups and the number of servings recommended each day for a person of your age.
9. Design a balanced daily menu.
10. Sensibly assess your own body mass.

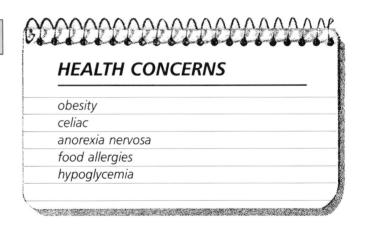

HEALTH CONCERNS

obesity
celiac
anorexia nervosa
food allergies
hypoglycemia

BIOLOGY AT WORK

MARKETING CONSULTANT
DIETICIAN
DIET COUNSELLOR
FOOD CHEMIST
CHEF
NUTRITIONIST

SOME WORDS TO KNOW

Match each of the descriptions given in the left-hand column with a word shown in the right-hand column. DO NOT WRITE IN THIS BOOK.

1. Mineral that helps build strong teeth and bones
2. Compound composed of carbon, hydrogen, and oxygen
3. A fatty compound found in animal fat, may clog blood vessels in some persons
4. Non-digestible material found in the large intestine that helps retain water
5. A reagent used to test for proteins
6. A unit of energy
7. Mineral that is needed to produce hemoglobin for red blood cells
8. Excess glucose is converted to this compound and stored in the liver or muscles
9. Substances made up of long chains of amino acids
10. Hormone involved in the control of blood sugar

A carbohydrate
B protein
C glycogen
D kilojoule
E fibre
F Biuret reagent
G Benedict's reagent
H calcium
I vitamin C
J additive
K cholesterol
L mineral
M insulin
N iron

Select any two of the unmatched words and, for each, give a proper definition or explanation in your own words.

Chapter 7 / Diet and Nutrition

SOME QUESTIONS TO ANSWER

1. What is the primary value to the body of
   (a) carbohydrates?
   (b) fats?
   (c) proteins?
2. Why is such a large part of our diet made up of carbohydrates?
3. Give four ways by which vitamins are sometimes lost from foods before they are eaten.
4. (a) List two fat-soluble and two water-soluble vitamins.
   (b) Give two examples of food that contains these vitamins which you have eaten this week.
5. What are the best dietary sources of the following minerals?
   (a) iron
   (b) calcium
   (c) iodine.
6. Why is brown whole wheat bread more nutritious than white bread?
7. (a) Of what value is fibre in the diet?
   (b) List four good sources of dietary fibre.
8. List the health hazards of being overweight.
9. List the major things that determine what *you* eat (appetite, what is available, location, type of work, social life, friends, etc.). Indicate which of these produce a "good" effect on your diet and which have a poor influence on your eating habits.
10. Make a list of the food that you ate at a recent meal. Report on how balanced the meal was by arranging the items into the four food groups. If any group was not represented, state what substitution could have been made to improve the balance.
11. Why should the amount of food eaten be varied with the amount of physical activity in which you are involved?
12. What are three useful approaches to losing mass successfully other than cutting back on food?
13. List four snack foods which would be nutritious and tasty. Now list four snacks that would not be nutritious. How could people be encouraged to choose the nutritious snacks?
14. Prepare a list of foods for breakfast, lunch, and supper for a student of your age. The day's intake should be a balance of foods and servings that meets the recommendations of the Canada Food Guide.

SOME THINGS TO FIND OUT

1. Select one of the following topics and prepare a report on it:
   "Weight Watchers"
   the school cafeteria food service
   advertising and body mass control
   food fads
   vegetarian diets
   health foods
   how much salt is enough
   vitamin supplements, costs and contents.
2. Find out what special diets are required for persons suffering from ulcers, gall bladder problems, or high cholesterol.
3. Visit a supermarket and prepare a list of the marketing techniques that are used to tempt people to buy food products. Two examples to get you started are cereals packaged with plastic toys and special displays to catch your eye while you wait at the cashier's desk.
4. (a) Discuss the nutrients in food offered by fast-food outlets.
   (b) Why are these establishments so popular?
5. Write for some brochures that advertise slimming programs, slimming "equipment", or reducing clinics. Analyse these materials and find out just what is being offered for your money. Are there any guarantees? If they claim that a person will lose a specific amount of fat, work out the cost per unit of fat lost.

DIGESTION

The system discussed in this chapter has a total length of about 9 m, greater than the width of many classrooms. It produces between 6 and 8 L of digestive juices each day. It efficiently recycles many of its own products and obtains vitamins from an active "zoo" of microscopic bacteria. All this activity serves one function – to break down the food we eat into small molecules our bodies can use.

KEY  IDEAS

- *Digestion is the process of physically and chemically breaking down food into molecules small enough to pass into cells.*
- *Digestion starts in the mouth with the action of the teeth and salivary amylase.*
- *Secretions from the liver and pancreas aid digestion in the small intestine.*
- *The small intestine is the site of final chemical digestion and nutrient absorption.*
- *The large intestine is responsible for the reabsorption of water and the production of some vitamins.*

8.1 The Process of Digestion

The process of digestion may be divided into three parts. The first stage is **physical** digestion, the mechanical process of breaking down food into smaller pieces. This starts on our plates when we use a knife and fork. It continues when we use our teeth to bite, grind, and separate the food fibres. The second stage involves **chemical** digestion; by the action of enzymes, these small food pieces are broken down into even smaller molecular-sized particles. (See Table 8.1.)

The third stage involves **absorption** of these molecules by the body. The nutrients from food broken down in the digestive tract must enter the thousands of capillaries which line the small intestine. In order to do so, the molecules that are the end products of physical and chemical digestion must be small enough to go into solution and pass through the tiny pores in the cell membranes. The blood that flows through these capillaries will then transport the nutrients to all the cells of the body, providing energy for cell processes and materials for the growth and repair of tissues.

CHECKPOINT

1. (a) What is involved in physical digestion?
   (b) What is involved in chemical digestion?
2. Give examples of both kinds of digestion.

Table 8.1 The Major Components of Food and the Final Products of Digestion After Being Broken Down into Smaller Sub-Units by the Action of Enzymes

Components of Food	Intermediate Products of Digestion	Final Products of Digestion
Carbohydrates	Disaccharides (sucrose, maltose)	Monosaccharides (glucose, fructose)
Fats	Emulsification into smaller droplets	Glycerol and fatty acids
Proteins	Short chains of amino acids	Individual amino acids
Minerals and vitamins	Require no digestion, absorbed directly	

8.2 Physical Digestion: The Teeth

Several different digestive processes begin in the mouth. The most obvious of these involves the action of the teeth. The teeth physically break down food, cutting off pieces and grinding these into smaller fragments.

If you look at your teeth in a mirror, you will see that they vary in shape and size. At the front are the **incisors**, four on the bottom and four on the top. These are chisel-shaped teeth which are excellent for biting or cutting food. On either side of the incisors are the **canines** (also called cuspids). These are more pointed in shape and are useful for tearing or shredding food. In humans, the teeth located posterior to the canines are almost square in shape. They are known as **premolars** and **molars**. These teeth, which are flattened on the upper surface, are used for grinding and chewing food, especially tough, fibrous foods, such as meat. (See Figure 8.1.)

With rare exceptions, each of us receives two sets of teeth during our lifetime. The first set, the **deciduous** or primary teeth, starts to appear at about six months of age. They then erupt at the rate of roughly one or two each month, until a full first set of 20 teeth is present. A permanent tooth appears to replace each deciduous tooth as it falls out between 6 and 13 years of age. This permanent set includes three additional teeth on each side of the jaw, which come in only once. A normal adult thus has 32 teeth – 8 incisors, 4 canines, 8 premolars, and 12 molars. (See Figure 8.2.)

Figure 8.1
The special diet that each animal enjoys is reflected in the design and arrangement of its teeth.

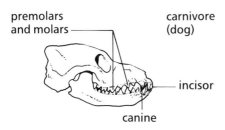

The teeth of a carnivore are specialized for catching and holding its prey. There is almost no grinding or chewing needed, as pieces of flesh can be torn off and swallowed.

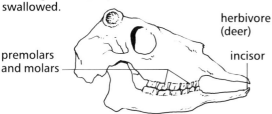

Herbivores have heavy, flat teeth specialized for grinding. Plant cells have walls made of cellulose that must be broken open to release the nutrients of the cell.

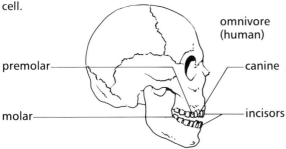

Humans belong to a group called omnivores which eat both plant and animal material. Their teeth are adapted to both bite and grind their food, having some of the abilities of both herbivores and carnivores.

Figure 8.2
The teeth and the average age of eruption through the gums

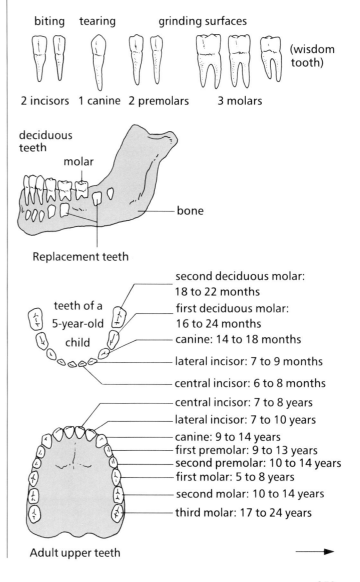

Chapter 8 / Digestion

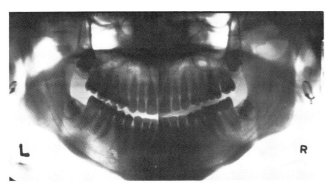

X-ray of anterior view of the mouth of a 14-year-old girl

A Close Look at a Tooth

The most obvious part of a tooth is the **crown**, which is visible above the gum. (See Figure 8.3.) It is covered with a protective coating of **enamel**, the hardest substance in the body. Unfortunately, enamel cannot be replaced naturally if it gets chipped or worn away. This loss of enamel exposes the **dentine** (den-tin). Dentine is the hard, bonelike material that forms the greater part of the tooth. Dentine is sensitive to touch, temperature, acids, and sugars. It is penetrated by minute tubes which radiate out from the pulpy interior of the tooth. These small tubes transmit information about the type of sensation received by the nerves in the pulp chamber. Although hard, dentine has a poor resistance to abrasion or damage from acid.

The **root** of the tooth is held in place by fibrous tissue which forms a firm connection between the tooth and the jawbone. The root is covered with a thin layer of **cementum**, to which the fibrous tissue connects.

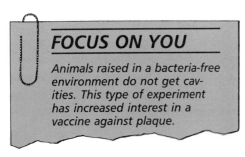

FOCUS ON YOU

Animals raised in a bacteria-free environment do not get cavities. This type of experiment has increased interest in a vaccine against plaque.

Figure 8.3
Longitudinal section through a molar tooth to show its internal structure

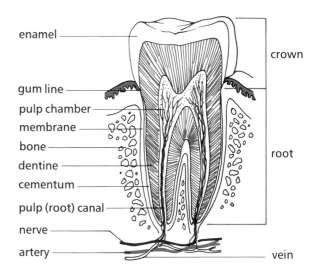

Maintaining Healthy Teeth

- **Keep them clean.** The best way to keep your teeth and gums in good condition is by regular brushing. (See Figure 8.4.) Brushing removes food and **plaque** (plak), the decay-causing bacteria. Plaque is soft and can be removed from the exposed tooth surfaces by using a toothbrush. The use of dental floss is necessary to remove plaque from between teeth. (See Figure 8.5.)

If you do not remove the excess particles of food from your mouth, they become a source of food for the plaque bacteria. If the plaque is not removed, there are two consequences. Firstly, the by-products of the bacteria include acids capable of corroding the enamel and dentine of the teeth, causing pits called **cavities**. Secondly, a hard crust, called **calculus** (or tartar), builds up over the plaque. This hard material cannot be removed by brushing. Calculus deposits irritate the gums and also make it very difficult to remove the plaque.

Chapter 8 / Digestion

Figure 8.4
Good brushing technique

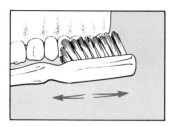 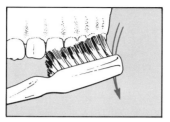 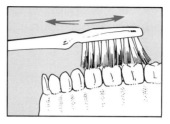 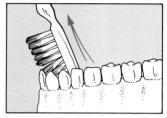

1. Begin with very small, almost vibration-in-place strokes on every tooth surface – inside, top, and outside.

2. Change the direction of movement so that the brush moves away from the gums with a rocking motion. Repeat on the insides and outsides of the top and bottom teeth.

3. Brush backward and forward across the tops of the teeth.

4. Brush the insides of the front teeth with the brush at a steep angle and brushing away from the gums.

Figure 8.5
It is necessary to use dental floss to clean between the teeth.

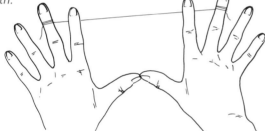

Use about 45 to 50 cm of floss.

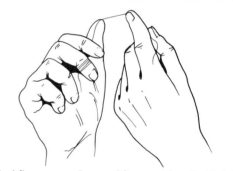

Wind floss around second fingers, then hold tight between thumbs and index fingers.

Move floss up and down sides of each tooth.

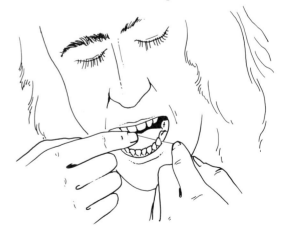

Use a different section of floss as it becomes frayed or soiled.

255

Brushing stimulates the gums, improving blood circulation in this tissue. Use a soft toothbrush that is gentle on the gums and change it every few weeks. Choose a toothpaste containing fluoride. This substance has been proven to help protect teeth from decay by making tooth enamel resistant to the acid produced by plaque.

- **Protect your teeth from injury**. Teeth need special protection if you are involved in sports, such as hockey, boxing, etc. Tooth enamel can also be damaged by trying to open bottles or cracking nuts with the teeth, as well as by habitual chewing of hard objects such as pencils.

- **Have regular dental checkups**. The dentist, or dental hygienist, must remove the accumulated calculus from your teeth. This cannot be done at home. Decay and cavities can also be detected in their early stages, permitting prompt treatment before the condition worsens.

- **A good diet is important**. A number of nutrients are particularly important for both tooth development and the maintenance of healthy teeth and gums. Calcium and phosphorus are the major structural materials in teeth, while vitamins A, C, and D help maintain the health of teeth and gums.

Foods that require vigorous chewing, such as raw fruit and vegetables, stimulate the gums and tooth sockets. Clean teeth immediately after consuming soft, sticky foods or sugary drinks, as these cling to the tooth surface and become a food source for plaque.

Activity 8A

Removing Plaque

How effective is your brushing technique?

Materials

plaque disclosing tablets, either chewable or to be dissolved in water
mirror
own toothbrush and toothpaste
paper cups

Procedure

The disclosing tablets contain a harmless dye that stains plaque bright red. You will first use the tablets to discover where the plaque is located. Then brush your teeth and re-test to discover whether or not your efforts are effective in removing the plaque.

1. On the day of the activity, brush your teeth in the morning as usual and remember to pack your toothbrush for use in class.

CAUTION!
Do not swallow the plaque disclosing solution.

2. Chew the tablet or rinse your mouth well with the disclosing solution. If you are using the solution, you should swish it around your mouth and over the teeth for about 30 s. Then rinse your mouth with water.

3. Examine your teeth in the mirror. *Note the areas where red stains indicate plaque.* Compare your own teeth with those of your partner.

4. Use either your own regular toothpaste or the toothpaste provided and brush your teeth,

making a special effort to reach the areas where the stain showed the presence of plaque.

5. Test with the disclosing solution again and re-examine the results using a mirror.

Questions

1. How effective was your morning brushing?
2. What areas are you neglecting in your regular brushing?
3. What improvements can you make in your technique to reach these areas?

Some Dental Problems

If cavities are not discovered and treated in early stages, the decay will eventually reach the tooth pulp, causing the tooth to ache as its nerves are reached and activated. The decayed material must be removed by a dentist and a patch (a filling) is used to seal the exposed pulp.

Sometimes bacteria may produce pus-filled infections at the root of the tooth. Such infections are called **abscesses** and are extremely painful. The infection may also spread into the bloodstream and cause serious problems in the body.

Peridontal diseases affect the tissues around the teeth. Symptoms include the retreat of gum tissue from the base of the teeth, tenderness, and bleeding. Good dental hygiene and proper diet are usually enough to prevent most peridontal diseases.

Halitosis, or bad breath, is an embarrassing condition which may result from a number of different causes, including smoking, some foods, the accumulation of food particles in the mouth, decay, or gum disease. In more serious conditions, throat, sinus, or digestive problems may be involved. Regular brushing and the use of a good mouthwash will usually look after the simple problems, but the advice of a dentist or doctor should be sought in more persistent conditions.

CHECKPOINT

1. How do the teeth of herbivores, carnivores, and omnivores differ?
2. What is the difference between
   (a) dentine and enamel?
   (b) plaque and calculus?
3. Describe the shape and function of each of the three kinds of human teeth.
4. Explain the connection between food particles in your mouth, bacteria, and damage to tooth enamel.
5. List four ways in which you can protect and maintain healthy teeth.
6. How are abscesses formed?

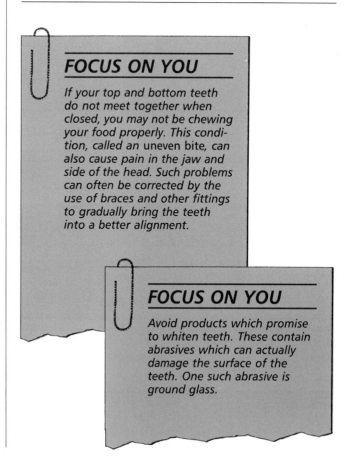

FOCUS ON YOU

If your top and bottom teeth do not meet together when closed, you may not be chewing your food properly. This condition, called an uneven bite, can also cause pain in the jaw and side of the head. Such problems can often be corrected by the use of braces and other fittings to gradually bring the teeth into a better alignment.

FOCUS ON YOU

Avoid products which promise to whiten teeth. These contain abrasives which can actually damage the surface of the teeth. One such abrasive is ground glass.

8.3 Other Aspects of Digestion in the Mouth

The Tongue

The tongue is an enormously versatile organ. It is attached not at the back of the mouth as it might seem, but rather to the floor of the mouth. It can be twisted, turned, rolled, and extended. The tongue is a very necessary part of the chewing process, for it helps to move the food onto the molars. It also mixes the food with saliva, the watery substance found in the mouth. Once the food is moist and soft, the tongue rolls it into a ball, or **bolus**, preparing it to pass into the pharynx to be swallowed. (See Figure 8.6.)

Figure 8.6
The tongue has a rough, irregular surface covered with small, raised structures called papillae. Taste buds, small specialized nerve cells which detect dissolved chemicals, are found in the furrows around the papillae.

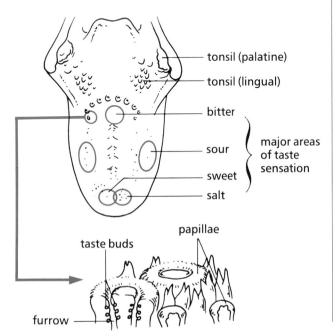

Chemical Digestion Begins: The Salivary Glands

Chemical digestion also begins in the mouth. In this process, special molecules called **enzymes** act on specific food molecules, causing them to break down into smaller units. Enzymes are produced by cells in the body and can be reused. The first enzyme to act upon a bite of food is **amylase**, contained in the watery saliva in your mouth. Amylase helps to break down starch into its component sugar molecules.

Saliva is produced by three pairs of glands which lie outside the mouth cavity and which empty their secretions into the mouth through short ducts. (See Figure 8.7.) A description of food, a whiff of an appetizing smell, or a visual image can all stimulate the production of saliva; however, it flows into the mouth, to some extent, all the time. Saliva is very slightly acidic; of the approximately 1000 mL produced per day, 99 percent is water.

Figure 8.7
The location of the salivary glands. The glands are found in pairs, one on each side of the mouth cavity.

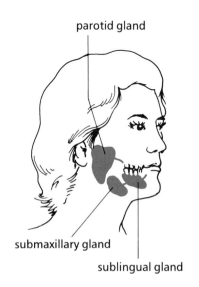

Saliva performs several useful functions, such as moistening dry food. When eating a cracker, for instance, saliva helps to bind the loose crumbs together into a smooth ball so that they will not stray into the respiratory system and cause you to cough. Saliva softens food as it is swallowed so that the rough edges of a potato chip, for example, will not scratch the walls of the esophagus. The enzyme amylase, as mentioned, begins the chemical digestion of carbohydrates such as starch.

Activity 8B

Chemical Digestion in the Mouth

Problem

What is the effect of saliva on the digestion of starch? (Saliva contains an enzyme (amylase) similar to diastase.)

Materials

25-mL test tubes
test tube rack
3 percent starch solution
diastase solution
Benedict's solution
iodine solution
wax marking pencils
hot water bath
goggles

Procedure

1. Label six test tubes from A to F.

CAUTION!
Wear goggles to protect the eyes when heating liquids.

2. Place about 20 mL of diastase solution in a test tube and boil it for 6 to 7 min.

CAUTION!
Avoid getting iodine or Benedict's solution on your skin. Rinse any spills immediately.

3. Add the different solutions to each test tube as shown under the heading "Contents" in the chart.

4. Wait 10 to 15 min, then, to test tubes A, C, and E, add a few drops of iodine. To test tubes B, D, and F, add about 1 mL of Benedict's solution and heat in a water bath.

5. *Observe any colour changes and record your results.*

Questions

1. What is the test for starch and the test for sugar?

2. What is the purpose of test tubes A and B?

3. Explain why each mixture of solutions was tested with both iodine and Benedict's solution.

Tube	Contents	Indicator	Observations	Substance Present
A	3 mL starch	Iodine		
B	3 mL starch	Benedict's		
C	3 mL starch + 6 drops diastase	Iodine	SAMPLE ONLY	
D	3 mL starch + 6 drops diastase	Benedict's		
E	3 mL starch + 6 drops boiled diastase	Iodine		
F	3 mL starch + 6 drops boiled diastase	Benedict's		

4. What conclusion can you draw about the digestive action of saliva?

5. What effect does high temperature have on the action of this enzyme?

Other Structures in the Mouth

Feel the roof of your mouth with the tip of your tongue. The front part is hard (the hard palate), but farther back is a soft part. By using a mirror to examine your mouth or by looking into the mouth of a friend, you can see that this soft palate is, at one point, formed into a piece of tissue that hangs down. This is the **uvula** (oo-view-la). (See Figure 8.8.)

Figure 8.8
This view of the mouth shows the location of the teeth, uvula, and tonsils.

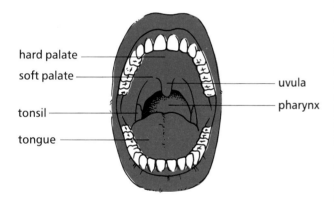

Look carefully at Figure 8.9 and you will see that the hard and soft palates separate two spaces, the nasal chamber and the mouth cavity. The palate is rather like a ceiling that separates the downstairs room from one above. You can see in the diagram that the tube leading down from the nasal chamber connects with the tube leading out of the back of the mouth. Most of us have had the unpleasant experience of having liquid go up this tube into the nose, when someone made us laugh unexpectedly while we were drinking. The tubes from the mouth and the nose come together at the pharynx. Below this junction, leading down from the pharynx are two more tubes – the trachea, which carries air into the lungs, and the esophagus, which carries the food into the digestive tract.

Swallowing

To ensure that food and drink pass directly into the esophagus and not into the lungs where they would cause us to choke or cough, special structures are present. When a bolus of swallowed food is moved to the back of the tongue, the soft palate moves upward to partially seal off the nasal passage. At the same time, the **epiglottis** (ep-i-glot-is) closes the opening into the respiratory passage (the trachea) thereby preventing the food from going into the lungs. (See Figure 8.9.) Muscular contractions of the pharynx help to pass the food down into the esophagus. The movement of food from the tongue into the pharynx is under voluntary control. The second stage, involving the epiglottis and the movement of the food into the esophagus, is involuntary.

Sometimes food can become lodged in the esophagus or the entrance to the trachea, causing the person to choke. The best way to avoid such difficulties is to cut your food into small pieces and to chew it well. Avoid talking while you have food in your mouth, as this opens the respiratory passage. What to do if someone does start to choke is explained in Chapter 6.

CHECKPOINT

1. What role does the tongue play in the digestion of food?

2. What are enzymes and how do they act in digestion? Give two examples.

3. What are the functions of saliva?

4. How does the epiglottis help when food is swallowed?

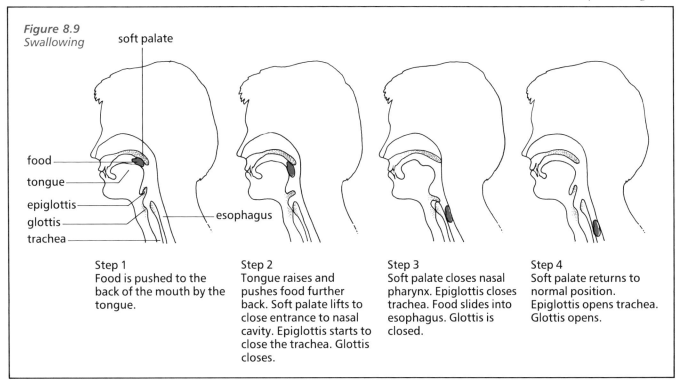

Figure 8.9
Swallowing

Step 1
Food is pushed to the back of the mouth by the tongue.

Step 2
Tongue raises and pushes food further back. Soft palate lifts to close entrance to nasal cavity. Epiglottis starts to close the trachea. Glottis closes.

Step 3
Soft palate closes nasal pharynx. Epiglottis closes trachea. Food slides into esophagus. Glottis is closed.

Step 4
Soft palate returns to normal position. Epiglottis opens trachea. Glottis opens.

8.4 Digestion in the Stomach

The Esophagus

The **esophagus** is a flexible tube, about 25 cm in length, which leads from the pharynx to the stomach. Like most of the digestive tract, its walls are composed of several layers. On the inside of the tube is a thick lining covered with a film of slippery mucus, which helps the food to pass easily. The second layer contains many glands, which produce this mucus, as well as blood vessels and nerves. External to this, there are two layers of muscle. One layer is arranged in a circular fashion and passes around the tube. The other is longitudinal, running the length of the tube. Finally, there is a thin sheet of connective tissue around the outside, which helps to anchor the esophagus to the surrounding tissues.

A bolus of food is moved down the esophagus by **peristaltic** (pair-i-stal-tik) action. This is achieved by rhythmic contractions of the circular and longitudinal muscles. (See Figure 8.10.) The esophagus eventually passes through the diaphragm and enters the abdominal cavity where the major organs of the digestive system are found. (See Figure 8.11.)

Figure 8.10
Peristalsis moves the bolus of food along the digestive tubes, including the esophagus.

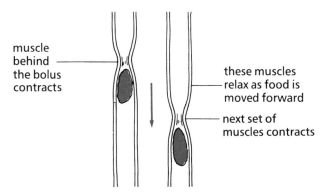

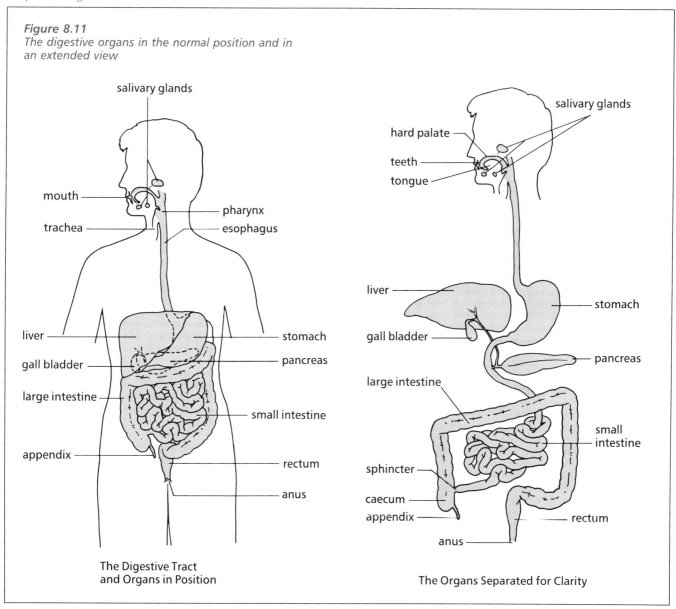

Figure 8.11
The digestive organs in the normal position and in an extended view

The Stomach

At the junction of the esophagus and the stomach, there is a ring of muscle called a **sphincter** (s-fink-ter). This ring of muscle acts like the drawstring on a bag to control the passage of materials from the esophagus into the stomach. The sphincter at the entrance of the stomach is called the **cardiac sphincter**.

The stomach is located immediately beneath the diaphragm, toward the left side of the abdomen. It forms a muscular bag that stretches as it fills with food. Its walls are made up of four layers in a manner similar to those of the esophagus, but with a few important differences. The inner wall of the stomach is folded into many

Figure 8.12
A cut-away section showing the structure of the stomach

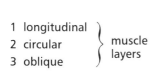
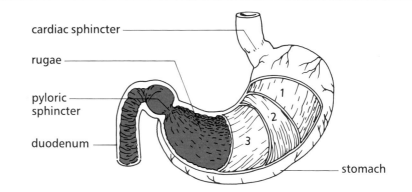

1 longitudinal
2 circular
3 oblique
} muscle layers

ridges or wrinkles. (See Figure 8.12.) At the bottom of these ridges are gastric pits or glands, through the openings of which the gastric juices are secreted. The muscle of the stomach has three layers instead of two. One runs the length of the stomach, one circles the stomach, and the other stretches diagonally around it. These layers contract rhythmically to mix the food contents with the gastric secretions.

Secretions of the Stomach
The **gastric juice** is supplied by approximately 35 million gastric glands. It contains the enzymes **pepsinogen** and **renin**, as well as small quantities of **lipase** (ly-pase).

In the cells that produce these gastric juices, carbon dioxide and water combine with salt to form **hydrochloric acid** and sodium hydrogen carbonate ($NaHCO_3$).

$$CO_2 + H_2O + NaCl \rightarrow HCl + NaHCO_3$$

The hydrochloric acid changes the pH (or acid-base balance) of the stomach contents to allow the enzymes to work efficiently. It helps kill any bacteria that may have entered the digestive tract with the food. This acid also reacts with pepsinogen to form the active protein-digesting agent, **pepsin**.

$$HCl + pepsinogen \rightarrow pepsin$$

Pepsin causes large protein chain molecules to separate into short chains consisting of 4 to 12 amino acids.

Sometimes a small part of the stomach lining may be damaged or destroyed. The protective mucous secretion (**mucin**), which coats the walls of the stomach to protect it from its own digestive enzymes, may also be inadequate. When either of these conditions occur, the enzymes and hydrochloric acid which function to digest the meat and the other proteins that we eat may start to digest the stomach lining itself. A **peptic ulcer** may result. Exactly why the lining breaks down and becomes ulcerated is not known. It is thought that excessive quantities of gastric juice in an empty stomach may irritate the lining and initiate such damage. Hydrochloric acid is usually produced in large quantities during times of stress and strong emotional pressure. For this reason, this type of ulcer is sometimes called an "executive" or "stress" ulcer, although no direct link between stress and ulcers has been proven.

The **pyloric sphincter** is located at the lower end of the stomach. This controls the flow of partially digested food out of the stomach. Food at this stage, called **chyme** (kym), is well-mixed, partly broken down, and in a semiliquid state. The time taken for food to pass through the digestive tract varies with the type of food eaten and the amount of fibre contained in the diet. The total time from entry into the body to waste elimination may vary from one to three days.

LIFE · SIGNS

ULCER

The job was getting to be too much for Mr. Babineau. Sales had spiralled down into a slump. He had recently lost an excellent assistant to a rival firm. Money problems in financing the extension to the plant had not been solved in the way that he had expected. The constant round of business meetings, business lunches, and overtime made him feel perpetually tired.

Mr. Babineau was 34 – young for his senior executive position, but he had achieved a successful career by long hours and hard work. Mr. Babineau had always been aware of having a "sensitive" stomach, and he always carried some antacid tablets to ease his discomfort.

Lately, the stomach pains seemed to be getting worse. The pains came about an hour before meal times, but disappeared shortly after he had eaten. Before he went to bed, the pains started again, but he could usually ease them by drinking a glass of milk. Today, the pain was different. He felt sick and didn't want to eat. His skin was sweaty and he looked pale.

Mr. Babineau decided to go home early and, once there, he dropped onto a chair feeling quite ill. His wife was concerned and, against Mr. Babineau's wishes, phoned their family physician, who asked them to meet her at the emergency entrance of the hospital immediately.

When Dr. Barker examined the patient, she noted that although Mr. Babineau felt sweaty, his hands and feet were cold and he was very pale. His pulse rate was 130, well above normal, and his blood pressure was low at 80/65 mm Hg. Dr. Barker gently pressed on the abdomen, but it seemed soft and normal. The doctor asked about the colour of the feces that Mr. Babineau had passed and made a note that a stool analysis must be taken. The colour of the stool had become much darker than usual and was almost black that morning.

The doctor also noted Mr. Babineau's history of using antacid tablets and the relief the patient gained from eating. She diagnosed a bleeding ulcer and prescribed immediate treatment.

Mr. Babineau received a blood transfusion to replace some of the lost blood and to help restore his blood pressure. The doctor also prescribed certain drugs to lessen his pain and discomfort. **Barium X-rays**, taken later, confirmed the doctor's diagnosis and Mr. Babineau was given some dietary advice for the future.

Once an ulcer has formed, the stomach acids irritate the ulcer site each time they are released. Food in the stomach helps to absorb the acid and give temporary relief. Antacids help to neutralize the acid condition and also ease the pain.

Bleeding from the ulcer into the digestive tract causes the feces to become much darker in colour, and it may become almost completely black.

Diet control, with smaller and more frequent meals and restriction of coffee and alcohol which stimulate the gastric secretions, is necessary. Cigarettes tend to retard the healing process, and so these also should not be used.

The objective of the treatment given to Mr. Babineau was to relieve the pain and promote the healing of the ulcer. Mr. Babineau had great difficulty in adapting to the changes the doctor prescribed. These included more regular working hours and meal times, and a reduction in work pressures. It was not long before Mr. Babineau went back to his old work habits, and two years later he had a second, more serious ulcer, which required surgery.

FOCUS ON YOU

Occasionally, the acidic gastric juice contacts the unprotected walls of the esophagus, due, for example, to overfilling of the stomach, excessive production of acid, or stomach muscle spasms. A burning pain is felt in the area of the heart, hence the name of this pain – heartburn.

Activity 8C

Protein Digestion

Problem

What is the effect of pH on enzymes and the digestion of proteins?

Materials

six 25-mL test tubes
boiled egg white in small cubes
pepsin solution
pancreatin solution
dilute NaOH (pH 8)
dilute HCl (pH 2)
incubating oven
goggles

CAUTION!
Avoid getting NaOH on your skin or clothing. Rinse any spills thoroughly in running water and report them to your teacher.

Procedure

1. Add small pieces of egg white to six clean test tubes.

CAUTION!
Wear goggles when adding NaOH or HCl to the test tubes.

2. Add 3 mL amounts of the solutions listed below to the test tubes.
   Test tube 1. Egg white + pepsin
   Test tube 2. Egg white + pepsin + HCl
   Test tube 3. Egg white + pepsin + NaOH
   Test tube 4. Egg white + pancreatin
   Test tube 5. Egg white + pancreatin + HCl
   Test tube 6. Egg white + pancreatin + NaOH

3. Incubate the tubes for about 24 h at 37°C. Prepare a table in your notebook in which to record your results. Include these headings: Tube #, Contents, Acid/Base, Observations.

Questions

1. Why was egg white selected as the test material?
2. Explain the results you obtained for each test tube.
3. Under what pH conditions does pepsin act most efficiently?
4. What conditions are most effective for the action of pancreatin?
5. Compare and comment on your results and the actual pH conditions in the stomach and small intestine.
6. (a) What controls were used in this experiment?
   (b) What other controls could be used to improve the validity of this experiment?

CHECKPOINT

1. *To what organ does the esophagus lead?*
2. *Define peristalsis and explain its function.*
3. *What tube lies in front of the esophagus?*
4. *What are the secretions of the stomach?*
5. *What types of food are digested in the stomach?*
6. *Define sphincters and explain their function.*

8.5 The Small Intestine and Associated Organs

The small intestine is a long, coiled, and looped tube about 2.5 cm in diameter, and is about 7 m in length. It fills most of the lower half of the abdomen. (See Figure 8.13.) The small intestine appears to be a jumble of tangled loops, but in fact the loops are attached to the rear wall of the abdomen by a thin membrane called the **mesentery**. This membrane serves several functions. It attaches and supports the small intestine, preventing the loops from becoming entangled. It also carries blood vessels which supply oxygen and carry away nutrients absorbed from food in the intestine.

The small intestine is divided into three parts – the duodenum, jejunum, and ileum. The first 25 to 30 cm of the intestine comprises the **duodenum**. The next part forms the **jejunum**, while the last portion is known as the **ileum**. These three sections are differentiated on the basis of their microscopic structure.

The walls of the small intestine are similar in structure to those found in other areas of the digestive tract. The outer coat is associated with the supporting mesenteries. Inside this are two layers of smooth muscle, one arranged longitudinally and the other surrounding the intestine in a circular fashion. These muscles are responsible for the peristaltic contractions of the digestive tube, which move food along inside the gut.

The inner walls of the small intestine are lined with special glandular cells. These secrete a digestive juice containing six enzymes. **Erepsin** completes the digestion of proteins, while **maltase**, **sucrase**, and **lactase** cause the final breakdown of complex sugars into single-sugar molecules. **Lipase** completes the digestion of fats, and **enterokinase** activates one of the enzymes from the pancreas after it arrives in the small intestine.

As a result of digestive processes occurring in the small intestine, the three basic food substances – carbohydrates, fats, and proteins – are broken down into molecules small enough to pass through the gut wall and enter capillaries of the circulatory system. The enzymes produced in the small intestine aid the process considerably; however, most of the digestion that occurs in the small intestine results from the activity of the liver and the pancreas, which empty their secretions into the duodenum.

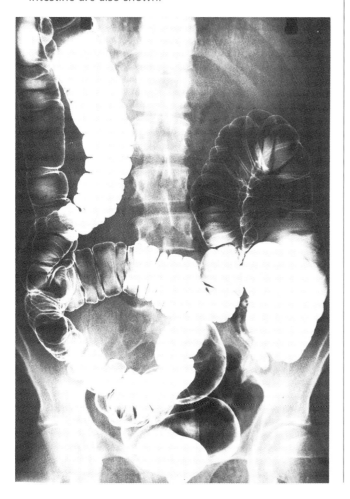

Figure 8.13
An X-ray of the small intestine. Portions of large intestine are also shown.

The Liver

The liver, which has a mass of about 1.5 kg in an adult, is the largest gland in the body. It is located high in the abdominal cavity, directly beneath the diaphragm on the right side. It partially overlaps the stomach. Two important blood vessels enter the liver. The **hepatic artery** brings blood rich in oxygen. The **portal vein** brings blood which is laden with nutrients from the intestines. The blood *leaving* the liver empties into the inferior vena cava.

The liver produces **bile** to emulsify fats. The emulsification process is similar to the action of a detergent on greasy dishes, breaking large blobs of fat into tiny droplets. Bile thus greatly increases the efficiency of fat-digesting enzymes by providing more surfaces for them to act upon. Bile is produced continuously and carried by small ducts to the **gall bladder** where it is stored until required. When fats are present and need to be emulsified, the liver can produce as much as 450 mL of bile per day. (See Figure 8.14.)

Figure 8.14
The liver, gall bladder, and pancreas. Note the ducts which deliver bile and pancreatic juices to the duodenum (small intestine).

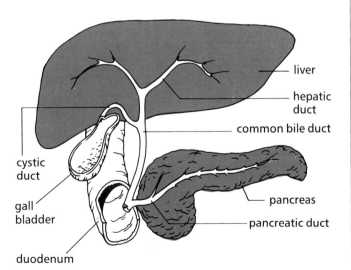

The liver is one of the most vital organs in the human body because of the many functions it performs. It acts in part as a factory producing several important chemical compounds, including the protein needed for the clotting of blood. Many of the specialized body proteins, such as gamma globulins, are also produced in the liver, as is vitamin A.

The liver also serves as a warehouse, storing such valuable supplies as glycogen, iron, and vitamins A, D, and B_{12}. When the body has an urgent need for glucose, the liver converts its glycogen store into glucose and releases it into the bloodstream. The liver can also convert protein and glycerol from fats into glucose when other sources are not available. It is believed that about 60 percent of the breakdown of fatty acids into sugar takes place in the liver, although this process can also take place in the individual cells of the body.

Additionally, the liver is also involved in waste disposal. It extracts and excretes bile pigment, urea, and other toxic substances such as alcohol and certain drugs. Some hormones are deactivated by the liver when they are no longer needed.

It is fortunate, in light of the many important functions performed by the liver, that this organ possesses an extraordinary ability to regenerate. If a part of the liver is removed, the remaining cells will gradually replace those that have been lost.

The Pancreas

The **pancreas** (pan-cree-us) is a leaf-shaped gland found close to the stomach. It measures about 20 cm in length. The pancreas has many important functions. Its primary role in digestion is to produce **pancreatic juice**, which contains about 28 digestive enzymes. The juice is alkaline and therefore neutralizes the strong acid mixture which arrives in the duodenum from the stomach, changing it from a pH of about 2 to a pH between 7 and 8, which is almost neutral. The enzymes found in this juice act on each of the three food types. Pancreatic amylase continues the breakdown of carbohydrates into sugars. Pancreatic lipase splits fats into fatty acids and glycerol. Trypsin and peptidase split proteins into small, two-unit, amino acid groups. These enzymes flow into the duodenum through a short tube called the **pancreatic duct**. (See Figures 8.11 and 8.14.)

Among the other functions of the pancreas is the production of **insulin**. Insulin plays a vital role in carbohydrate metabolism and in maintaining the balance of blood sugar.

Table 8.2 summarizes the substances secreted by the digestive system. Use Figure 8.11 to trace this process of chemical digestion structure by structure.

Table 8.2 Major Secretions of the Digestive System

Organ	Digestive Secretion	Active Digestive Agent	Action on Food
Salivary gland	Saliva	Amylase	Reduces starch to maltose
Stomach	Gastric juice	Hydrochloric acid Pepsin Rennin	Reduces a protein to short peptide chains Clots milk
Liver	Bile	Bile salts	Emulsifies fats Neutralizes acids
Pancreas	Pancreatic juice Insulin	Sodium bicarbonate Lipase Amylase Trypsin, peptidase	Neutralizes acids Reduces fats to fatty acids and glycerol Breaks down starch to maltose Continues protein breakdown to amino acids
Intestine	Intestinal juice	Maltase, sucrase, lactase	Completes digestion of sugars to glucose

Activity 8D

The Action of Lipase on Fats

Lipase, an enzyme produced mainly in the pancreas, acts specifically upon fats to split (hydrolyze) them into fatty acids and glycerol.

Materials

fresh whole milk
pancreatin or lipase
four 25-mL test tubes
phenolphthalein solution
detergent
1 mol sodium carbonate solution
water bath
hot plate
goggles

Procedure

1. Put 10 mL of fresh milk and two or three drops of detergent into each of two test tubes and shake.

2. Add a few drops of phenolphthalein indicator. If this does not produce a deep red colour, add a few drops of sodium carbonate until the red colour is distinct. Mark the tubes A and B.

BIOTECH

Hepatitis

Hepatitis is a general term that means an inflammation of the liver. Two types of viruses are usually responsible for hepatitis. **Infectious hepatitis** is most common in teenagers and young adults. Infectious hepatitis is spread when people live in crowded conditions with poor sanitation. This occasionally occurs in summer camps or in dormitories.

The first symptoms of this disease are often tiredness, poor appetite, an upset stomach, or pains in the abdomen. A fever may occur and the liver is often enlarged and tender. The most common symptom of hepatitis is **jaundice**. This is not a separate disease, but a characteristic condition in which the skin and the whites of the eyes become yellow. There is no medicine that will cure hepatitis. Bed rest and a high-protein diet will speed recovery and help to prevent permanent damage to the liver.

The other form of hepatitis, **serum hepatitis**, is transmitted by needles entering the arteries or veins. Blood transfusion equipment or the hypodermic needles of drug users can transmit this virus from one person to another. If the needle used to pierce the ears for earrings is not sterilized, it can also transmit the disease.

The best protection from hepatitis is good personal hygiene. Wash your hands at appropriate times, and never allow the use of a needle that has not been thoroughly sterilized.

CAUTION!
Use goggles while heating liquids.

3. To tube A, add a small amount of pancreatin or lipase from the tip of a spatula.

CAUTION!
Phenolphthalein is flammable. Use a hot plate rather than a Bunsen burner for this activity in case of spills.

4. Warm both test tubes in a 40°C water bath. *Note the time that you place the tubes in the bath and the time that it takes for the indicator to become colourless.*

Questions

1. What was the purpose of using the detergent?
2. What was the pH of the initial samples?
3. What was the purpose of adding the sodium carbonate?
4. What does the colour change indicate concerning the actions of lipase on the fats present in the milk?
5. Is there any significance to the 40°C temperature of the water bath and the time required for the colour change to take place?
6. What are the final products of fat digestion?

Activity 8E

pH and Enzyme Activity

Problem
What effect does the pH of a solution have on the action of a digestive enzyme?

Materials
solutions of pH 3, 6, 8, and 10
four small test tubes
test tube rack
spot plates
iodine solution
3 percent starch solution
1 percent diastase solution
goggles

Method

1. Prepare four test tubes with equal amounts of starch solution in each, about 3 mL. *Prepare a chart to record your results.*

    CAUTION! Wear goggles while mixing liquids.

2. To test tube A, add 3 mL of pH 3.
   To test tube B, add 3 mL of pH 6.
   To test tube C, add 3 mL of pH 8.
   To test tube D, add 3 mL of pH 10.

3. One student from each bench should watch the time and signal his or her partner at 1-min intervals after the diastase enzyme has been added. To test tube A, add 10 drops of diastase solution. Shake gently to mix the contents and start the timing. After 1 min, tip a few drops from the test tube into one of the depressions in the spot plate and immediately test with iodine.

4. After every 1-min interval, add a few more drops of the starch and enzyme mixture to another depression in the spot plate and test with iodine. Continue to test until you no longer get a positive test for starch. *Record the number of minutes required to digest the starch.*

5. Repeat the experiment using test tubes B, C, and D. If time is short, the teacher may ask different rows of benches to test a different pH effect.

Questions

1. Which pH gave the best digestion time?
2. Relate your results to the digestion of starch in the body. Use your text to review the pH of the duodenum. You may wish to test the pH of the mouth with litmus paper or Hydrion paper. Which of the pH solutions comes closest to the pH of your mouth?
3. Draw a graph of your results, plotting pH against time taken for digestion.
4. Explain why the rates of digestion vary.

CHECKPOINT

1. Name the three sections of the small intestine.
2. (a) Describe the structure of the walls of the small intestine.
   (b) What is significant about this structure?
3. Which vein takes blood rich with nutrients to the liver?
4. (a) What is the function of bile?
   (b) Where is it stored?
5. Name two functions of the liver, other than bile production.
6. Name four secretions of the pancreas and state what food substances they digest.
7. (a) Why is pH important for digestion in the stomach and small intestine?
   (b) How does the pancreas help in creating the proper pH conditions?
8. What is the function of insulin?

FOCUS ON YOU

Certain drugs can affect the strength of peristoltic contractions and so change the rate of movement of materials through the digestive system. For example, codeine, present in some cough medications and pain relievers, temporarily slows and weakens peristoltic contractions. Constipation may result.

8.6 Absorption of Nutrients

The surface lining of the small intestine has small projections which improve its ability to absorb the end products of digestion. These small, fingerlike structures are called **villi**. (See Figure 8.15.) The purpose of the villi is to increase the internal surface area of the small intestine so that nutrients can pass through the lining cells and enter the circulatory blood vessels more efficiently and rapidly. Each villus contains many blood and lymph vessels. These vessels collect the nutrients and transport them to locations where they are needed. The cells that make up the wall of the villi are themselves equipped with even smaller microvilli on the exposed surface of epithelial cells. These further improve the absorptive abilities of the intestine.

Some substances taken into the body are composed of small molecules that dissolve readily. Alcohol and aspirin, for example, are not digested; their molecules are already small enough to pass through the wall of the stomach and be absorbed directly into the capillaries without entering the small intestine.

Digestion is an active process. Food passes along the digestive tract as a result of peristalsis, which also serves to constantly churn and mix the food. The villi wave about with the movement of food and even this small action contributes to the mixing process. The amount of time required for food to pass through the digestive tract will vary with the individual and the type of food eaten. Figure 8.16 gives an idea of the average time taken to reach particular points along the route.

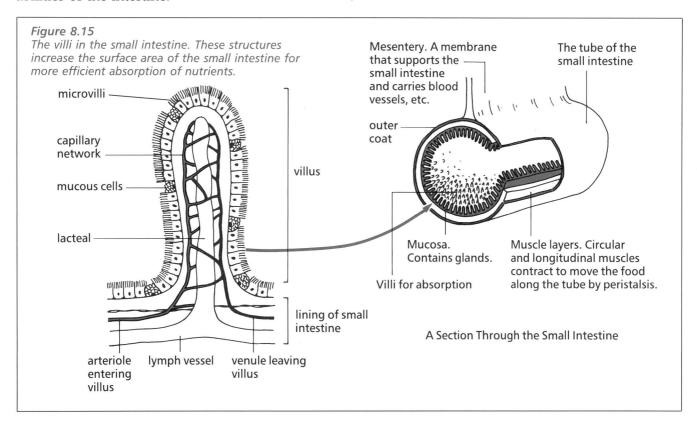

Figure 8.15
The villi in the small intestine. These structures increase the surface area of the small intestine for more efficient absorption of nutrients.

A Section Through the Small Intestine

Figure 8.16
Approximate times taken for food to pass through the digestive tract. The time taken varies greatly, depending on the type of diet (especially the amount of fibre present).

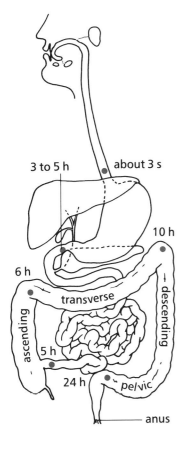

Times taken for food to pass along the digestive tract. The times indicate how long the first part of a meal would take to reach the different parts of the digestive tract. Food taken in during the last part of a meal may take from 3 to 5 h to leave the stomach.

CHECKPOINT

1. (a) What is meant by absorption?
   (b) Why is it necessary?

2. What structures are present in the small intestine to improve the efficiency of absorption?

BIOTECH

Alcohol and the Human Body

Ethyl alcohol or ethanol (C_2H_5OH) is the active ingredient in beverages such as beer, wine, whiskey, gin, and rum. Beer and wines contain between 2.5 and 14 percent alcohol while spirits and liqueurs contain between 35 and 50 percent alcohol.

Alcohol is absorbed from the stomach wall (20 percent) and from the intestine (80 percent) and goes into the bloodstream. Milk and fatty foods impede its absorption, whereas the addition of mixes such as carbonated "soft drinks" hastens absorption. The alcohol level in blood reaches its maximum within one-half to two hours after consumption. Alcohol enters various organs, including the brain, lungs, and kidneys.

In the liver, an enzyme changes alcohol to another compound that will then be broken down into carbon dioxide and water.

Body mass and the amount of food in the stomach primarily determine the effects of alcohol concentration in the body. After consumption of two bottles of beer or 75 mL of whiskey or 150 mL of wine, the average-sized male of about 75 kg will have a blood alcohol concentration of 0.05 percent. Consumption of one more drink could bring the blood alcohol level to the legal limit of 0.08 percent. The body must chemically change this alcohol for the level to return to zero

percent. It takes the body about 1 h to reduce the blood alcohol level of 0.025 percent to zero. For some people, a reading of 0.03 percent could start impairment, but impairment is evident in everyone at a reading of 0.08 percent. (In the courtroom, this amount is expressed as 80 mg percent.)

Alcohol is a depressant of the central nervous system. All kinds of motor performance, for example, the control of speech, eye movements, ability to stand, and nerve reflexes, are affected; therefore, the ability to control an automobile is adversely affected by alcohol.

Alcohol affects the gastrointestinal tract. Frequent complaints from people who drink excessively are distension of the abdomen, belching, and "burning stomach". Excessive consumption of alcohol can lead to erosion of the stomach lining, causing ulcers, and promotes the enlargement of the liver, leading to cirrhosis. The normal liver functions are slowed down greatly. The pancreas can also be affected, leading to obstruction of pancreatic enzyme production.

Alcohol has also been known to affect the unborn. Scientists have found that children born to women who drink excessively or even moderately while pregnant may have a pattern of physical and mental birth defects called the fetal alcohol syndrome. Most affected youngsters have smaller brains, narrow eyes, and low nasal bridges. No one knows how much alcohol is too much, so pregnant women should not drink at all.

The statistics in Canada relating to alcohol-related traffic infractions and accidents are staggering. Over 1000 people are killed in alcohol-related traffic accidents in Canada each year. In the province of British Columbia alone, approximately 25 000 charges for impaired driving per year have recently been made. The total cost in terms of necessary law enforcement, hospitalization, and lost productivity in that province is estimated at about $130 million per year.

Police throughout Canada are using the breathalyzer analysis to test suspected drinking drivers. If a person exhales over 0.08 percent blood alcohol, a charge for impaired driving will be laid under Section 236 of the *Criminal Code of Canada*. It is a criminal offense to drive with a reading of 80 mg percent or 0.08 percent blood alcohol.

The effects of alcohol on the brain

Parietal Lobe
Sensory control is affected by 0.10-0.30 percent blood alcohol. Senses are dulled or distorted. Difficulty in writing well, speech may be slurred, technical abilities reduced.

Frontal Lobe
Reason and self-control are affected by blood alcohol levels of 0.01-0.10 percent. The alcohol removes the inhibitions we usually have and weakens self-control. There is usually a feeling of well-being and a false confidence. Judgment becomes impaired, a person usually talks more and listens less.

Occipital Lobe
Visual perceptions are impaired by 0.20-0.30 percent blood alcohol. Colours may be distorted, double vision may occur. Poor judgment of distance and speed is evident.

Cerebellum
The co-ordination of muscles is affected by 0.15-0.35 percent blood alcohol. Difficult to walk or turn. Balance difficult to maintain.

Thalamus and the Medulla
Autonomic nervous system affected by 0.25-0.50 percent alcohol. Tired, apathetic attitude, respiration and circulation depressed. Eventually a lowering of body temperature, stupor, shock, and possibly death.

Chapter 8 / Digestion

8.7 Elimination: The Large Intestine and Associated Structures

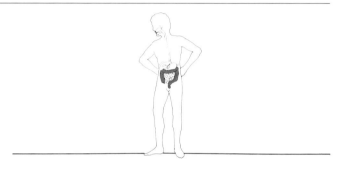

The junction of the small and large intestines is located in the lower right side of the abdomen. (See Figure 8.17.) Another sphincter, the **ileocaecal valve**, occurs at this junction. This valve controls the rate at which the contents of the small intestine pass into the large intestine. These contents consist of undigestible food remains and water. The water is needed by the body and must be reabsorbed. The food remains are waste and so are eliminated from the body.

The Caecum and Appendix

Immediately beyond the sphincter is a small pouch called the **caecum**. Extending from this pouch is a blind tube about 6 to 8 cm long which is called the **appendix**. (See Figure 8.17.) Although humans cannot break down cellulose, it is still an important dietary component that

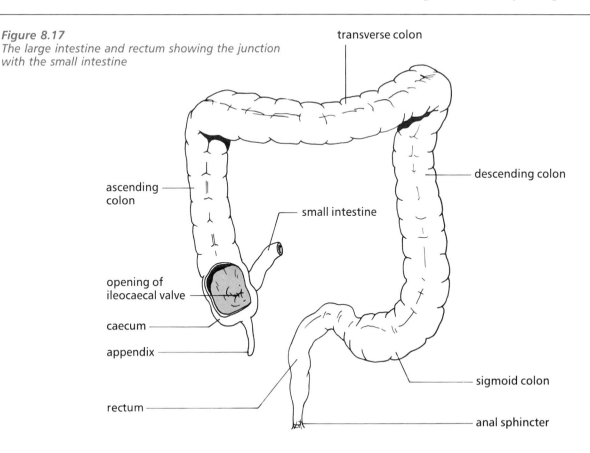

Figure 8.17
The large intestine and rectum showing the junction with the small intestine

adds bulk to digested food and thus helps to prevent constipation. In herbivores, micro-organisms in the caecum and appendix break down the cellulose. In humans, these structures are not functional.

Unfortunately, food material can easily get into the blind-ended appendix sac and may lodge there, unable to get out. (This situation is similar to a backwater along a small stream, where the water becomes stagnant because there is no current.) If pathogenic (disease-causing) bacteria lodge in this sac, where moisture, food, and warmth provide excellent growing conditions, the appendix may become infected. This disorder is known as **appendicitis**. Surgical removal of the appendix may then be necessary.

Should an infected appendix burst (which can happen if it does not receive proper medical attention), the contents of the appendix, including the bacteria, may spill out into the abdominal cavity. This then causes a much more serious condition called **peritonitis**. Prompt action to obtain medical help can avoid such dangers.

The Large Intestine

The large intestine is about 1.5 m in length, 7.6 cm in diameter, and makes up almost one-fifth of the total length of the digestive tract. It forms an inverted U-shape that fills the lower abdomen. It is divided into three sections. The first section, the **ascending colon**, rises up the right side of the abdominal cavity. The second portion, called the **transverse colon**, passes across the middle of the abdomen. The **descending colon** follows down the left side of the cavity and bends toward the centre of the lower body. (See Figure 8.17.)

The large intestine functions mainly to reabsorb water, which is always in demand for cell metabolism. As this water is reabsorbed, the undigested food residues present in the large intestine are dried into a suitable consistency for defecation. As the colon takes up this water, some dissolved inorganic salts are also absorbed.

The large intestine usually contains cultures of bacteria and other micro-organisms which do us no harm. Some of these organisms operate on the waste material to produce vitamins B and K which are then absorbed for body use. Others reduce the volume of waste to be discarded.

As the contents of the large intestine are being dried, they are also being mixed by periodic, sluggish contractions of the colon walls. This gradually propels the contents toward the rectum. The waste material, or **feces**, which are formed in the colon, are made up of about 60 percent solid material and 40 percent water. Although the walls of the large intestine possess muscle layers much like those found in other parts of the digestive tract, this organ does not make the same regular, peristaltic contractions. Instead, several times a day, there is a strong series of wavelike constrictions, which move the entire contents of the large intestine along a short distance.

The Rectum

The **rectum** is the last section of the digestive tract. (See Figure 8.17.) It ends at the **anus** with a ring of sphincter muscle. When the rectum is sufficiently distended and full, nerve endings in the walls send out messages which produce a mild feeling of discomfort, an indication that the feces are ready for elimination.

The rectal veins are found near the anal opening. These veins sometimes become enlarged and distorted. This increase in size restricts the rectal tube and may make the passage of feces difficult and painful. These enlarged bulges in the rectal veins are known as **hemorrhoids** or piles.

If the diet is lacking in fibre, waste products become drier and more compacted. When sufficient fibre is present, the waste holds more water and is much softer in texture and therefore passes more easily through the rectum.

CHECKPOINT

1. Where, exactly, are the caecum and appendix found?
2. What function does the appendix serve in herbivores?
3. (a) Compare the contents of the digestive system in the small intestine and the large intestine.
   (b) What process accounts for any differences?
4. Give two functions of the large intestine.
5. What is the function of the rectum?

8.8 Living with Your Digestive System

Many proprietary medications for common ailments associated with the digestive system, such as indigestion or constipation, are to be found in any drug store. These products generally treat the symptoms and not the cause of the problem. Relying on the use of these products can mean a relatively minor condition could become severe.

For example, constipation in young people is usually the result of poor dietary habits. Increasing the amount of fluids and high-fibre foods and

taking regular exercise are far better methods of solving constipation than the regular use of laxatives. Similarly, with indigestion, antacid tablets can provide relief, but eating smaller amounts of food, eating more slowly, reducing the quantity of very spicy or acidic foods, and eating in a relaxed situation offer a more positive solution to the problem. Any persistent or recurring problem should be treated by a physician.

At times of tension or temporary stress, as, for example, before an exam, almost everyone experiences "butterflies" in their stomach, loss of appetite, or even a bout of mild diarrhea. Maintain your intake of fluids and realize this reaction to tension is natural. If, however, the symptoms continue or become painful, you may need to see your physician for advice on how to handle the stressful situation in a more relaxed manner. Paying attention to such "body language" can help prevent more serious problems later.

BIOLOGY AT WORK

DENTIST
DENTAL ASSISTANT
ORTHODONTIST
DENTAL HYGIENIST
PERIODONTIST
NURSING AIDE

HEALTH CONCERNS

diabetes mellitis
peptic ulcers
food poisoning
gingivitis
cirrhosis
diarrhea
gastritis
hiatus hernia
hepatitis

Chapter 8 / Digestion

CHAPTER FOCUS

Now that you have completed this chapter, you should be able to do the following:

1. Describe the types, structure, and function of human teeth.
2. Describe and be able to apply good oral hygiene.
3. Outline the cause, symptoms, and treatment of selected disorders of the digestive system.
4. Describe the location and function of the salivary glands.
5. Describe the mechanics of swallowing and the movement of food by peristalsis.
6. List the major digestive secretions and their action on specific food types.
7. State how pH affects the action of specific enzymes.
8. List the parts of the digestive system and describe the structure and function of each.
9. Describe how food is absorbed into the bloodstream.
10. Explain the process of elimination.

SOME WORDS TO KNOW

Match each of the descriptions given in the left-hand column with a word shown in the right-hand column. DO NOT WRITE IN THIS BOOK.

1. A leaf-shaped gland secreting many enzymes into the duodenum
2. Substance secreted by the liver
3. Active digestive agent that breaks down proteins in the stomach
4. Rhythmic movements of the small intestine that help to move food along the gut
5. Small fingerlike structures that aid in absorption
6. A hard material found in teeth that has poor resistance to acids
7. Rings of muscle that restrict the passage of materials along a tube
8. The thin sheets of membrane that hold the loops of the small intestine to the back wall of the abdominal cavity
9. Helps prevent food from entering the larynx and trachea
10. One of the functions of this organ is to reabsorb water

Select any two of the unmatched words and, for each, give a proper definition or explanation in your own words.

A absorption
B plaque
C stomach
D large intestine
E small intestine
F pancreas
G calculus
H liver
I bile
J peristalsis
K pepsin
L incisor
M amylase
N villi
O dentine
P saliva
Q epiglottis
R sphincter
S mesentery
T mucin

SOME QUESTIONS TO ANSWER

1. What is the difference between physical and chemical digestion?

2. Which of your three kinds of teeth would you use to
   (a) bite off a piece of celery?
   (b) chew some tough meat?
   (c) pull off a piece of toffee?
   Explain how the shape of each kind of tooth helps you in the given action.

3. What is the difference between digestion and absorption?

4. (a) Describe the pH of the stomach and small intestine contents during digestion.
   (b) Explain what produces these conditions and why they are necessary.

5. Describe how animals with different diets are adapted to the type of food they eat.

6. If, for some reason, a person lost the ability to produce saliva, what adjustments to their eating habits would they have to make in order to duplicate all the functions of saliva?

7. What techniques of good oral hygiene can you use to promote healthy teeth?

8. (a) Using a diagram, explain how the villi increase the surface area of the small intestine.
   (b) Why is this important?

9. If the alkaline secretions from the pancreas failed to raise the pH of the stomach acids as they passed into the small intestine, what would happen to the effectiveness of the enzymes there?

10. (a) What control mechanisms prevent the stomach contents from moving into the small intestine too soon or in too large a quantity?
    (b) Where else in the digestive system are such controls located?

11. Make a table for the digestive system, using the following headings: Secretion, Food Digested, End Product of Digestion. List the organs in the sequence in which they act down the left-hand side. Complete your table in point form.

12. What is peristalsis, why is it needed, and how does it work?

13. What is the value of fibre in the diet?

14. You have just finished eating a hamburger and a glass of milk. Explain how the foods you have eaten will be digested in your body. Be specific about the organs involved and the secretions they will use.

SOME THINGS TO FIND OUT

1. The role of the large intestine is to reabsorb water. What happens when a person is constipated, or has diarrhea? Can you explain why the speed at which the wastes pass through the large intestine is important in both these conditions?

2. Find out some of the causes of food poisoning and what controls the government uses to try and protect us.

3. What regulations are required in your school cafeteria for food preparation, washing of dishes, personnel health, and the control of cockroaches and other organisms?

4. Humans cannot digest cellulose, but herbivores, such as cows, sheep, and rabbits, can. Find out what differences there are between the digestive tracts of these animals and that of humans.

5. Find out what diet changes are required for a person suffering from an ulcer or diabetes.

6. Read the labels on as many antacid products as you can. Prepare a chart on the contents of the various brands and the cost per tablet. With the help of your teacher, design an experiment to determine which tablet is the most effective in neutralizing an acid of specific strength.

9

THE EXCRETORY SYSTEM

An archeologist digs in the soil of a site once used by tribes of early humans, searching for information on what they ate. Carefully, she sorts out pieces of shells, dried grains, and fragments of animal bone. Microscopic examination of the soil itself reveals pollen grains from the plants these people collected for food thousands of years ago. A lot of information can be found in what people throw away.

In a similar way, urine, the final waste produced by your excretory system, can reveal by its composition a great deal about your body. Urinalysis, the chemical analysis of urine, is a routinely used and powerful medical tool, giving information on overall health and specific body functions. This is possible because the excretory system is responsible for maintaining the chemistry of the blood.

KEY IDEAS

- Wastes are removed from the body by the lungs, skin, and digestive system, as well as the kidneys of the excretory system.
- Excretion removes the wastes produced by various cell activities before they accumulate to dangerous levels.
- The millions of nephrons in the kidneys are responsible for filtering out the wastes carried in the blood.
- Nephrons concentrate wastes into urine. At the same time, usable products are recycled into the blood.

9.1 What is Excretion?

Every body activity uses energy and generates wastes. If waste products were not removed, they would quickly accumulate in harmful proportions. Some wastes are toxic (poisonous) and pose a serious threat to health if they are not removed promptly. The process of getting rid of metabolic wastes is called **excretion**.

Fortunately, atoms and molecules do not wear out, but may be rearranged and used over and over again. Many of the end products of various cell activities can be recycled and used in other processes. As a result, the amount of waste which actually needs to be discharged from the body is very small in relation to the amount of work done.

Excretion Through the Lungs

There are some substances which must be excreted quickly before they have time to build up and become toxic. Carbon dioxide, which is one of these, is excreted through the lungs. The lungs are not often thought of as an excretory organ. Any structure, however, that enables us to eliminate a metabolic waste product plays a part in the excretory process. The lungs can also eliminate alcohol, which, since it is made up of small molecules, can be rapidly passed from the bloodstream into the alveoli of the lungs and exhaled. This is why the breathalyzer test is an effective and simple way of determining the amount of alcohol that people have taken into their systems.

Carbon dioxide and water are the major waste products of the energy-forming reactions that are constantly taking place inside the body. When carbohydrates and fats are used to supply fuel to muscles and organs, carbon dioxide and water are released. The water produced is not really waste, since the body needs all the water it can get. This "waste" water is thus used in many of the other important processes and activities carried on by the body. Not all of the water can be saved; some water vapour is lost from the lungs and some water must be used to dissolve the wastes we excrete as urine. (See Figure 9.1.) Water also evaporates from the skin. (See Table 9.1.)

Table 9.1 Daily Water Gain and Loss by the Body

Water Intake (mL)		Water Output (mL)	
Liquids	1200	Urine	1500
Food	1000	Feces	150
Oxidation of food	300	Lungs	300
		Skin	550
	2500		2500

Figure 9.1
The loss of water through the lungs can easily be seen on a crisp winter morning.

Chapter 9 / The Excretory System

Excretion Through the Skin and Anus

Salts, as well as some urea and water, are excreted through the pores of the skin. If you taste the sweat on your skin, you will find it has a distinctly salty flavour.

Bile and other substances are secreted into the digestive tract to help in digesting food. These substances will eventually be removed from the body in the feces, along with the food materials, such as fibre, that we cannot digest. Solid waste excretion (egestion) is discussed at the end of the chapter on digestion. (See Table 9.2.)

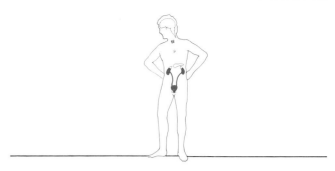

Figure 9.2
The structures of the urinary system

Table 9.2 The Products Excreted by Various Organs of the Body

Excretory Organ	Products Excreted
Kidneys	Nitrogenous wastes resulting from protein metabolism, such as urea, poisons, water, mineral salts
Lungs	Carbon dioxide, water
Intestine	Waste residues of digestion (mostly cellulose), metabolic wastes, such as bile pigments
Skin	Water, mineral salts

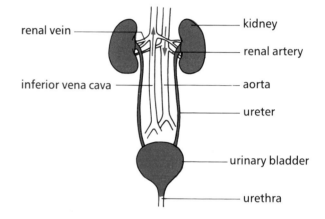

Figure 9.3
X-ray of lower abdomen showing the ureters and bladder

9.2 The Organs of the Excretory System

The excretory system consists of two **kidneys** and two **ureters** (you-ree-ter), which carry the urine produced in the kidneys to the **urinary bladder**. (See Figures 9.2 and 9.3.) Another tube, the **urethra** (you-ree-thrah) carries the urine out of the body. In males, the urethra passes through the penis. In females, the urethra lies between the pubic bone and the front wall of the vagina.

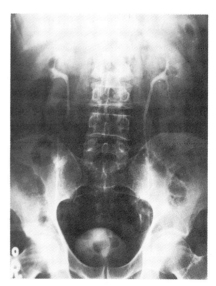

The Kidneys

The kidneys are the major excretory organs of the body. They are found in the posterior wall of the abdominal cavity, one on either side of the spine, just below the level of the ribs. They are snugly secured in a packing of fat and are provided with a rich supply of blood, which comes directly from the aorta. This blood enters the kidney through the **renal arteries** (ree-nul), which carry about 25 percent of the blood pumped out during each contraction of the heart. The **renal veins** carry blood back from the kidneys to the inferior vena cava.

The Functions of the Kidneys
The major function of the kidneys is the elimination of nitrogenous and other dissolved wastes. Wastes released from cells throughout the body are dissolved in water and carried by the circulatory system to the kidneys. The kidneys filter these wastes, making them more concentrated in preparation for excretion and, in the process of doing this, retaining as much water as possible; thus, the second major function of the kidneys is maintenance of water balance.

The kidneys also play an important role in regulating the acid-base balance of the body. Excessive amounts of either acidic or basic (also called "alkaline") materials are constantly being extracted and excreted by the kidneys. Because a normal diet includes more acid-producing than alkaline-producing foods, neutralizing extra acid is a problem for the body. As well as being able to excrete such unwanted chemicals, the kidneys can also manufacture ammonia, an alkaline substance that acts to neutralize acids, thereby removing their harmful effects.

The Structure of the Kidney
The kidney is shaped like a lima or brown bean and is about the size of a small fist. It has a dark-coloured core which is densely packed with blood vessels. These vessels form a network of capillaries and small tubes known as the **medulla**. (See Figure 9.4.)

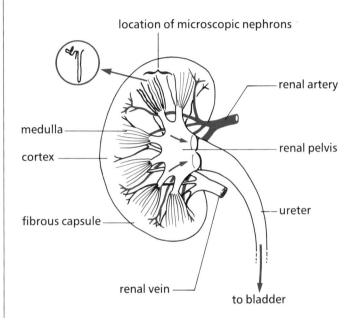

Figure 9.4
A longitudinal section through one kidney

The filtration of blood takes place in minute structures called **nephrons** (nef-rons), found in the **cortex**, which is the outer layer of the kidney. These structures are very small; there are estimated to be more than 1 million nephrons in each kidney. In spite of their microscopic size, the nephrons filter a total of more than 180 L of blood every 24 h. Most of this filtered fluid is reclaimed by the body, and only about 1.5 L of urine is passed each day. Sometimes, in addition to the wastes we would expect to find in the urine, some usable substances are present. Physicians use the analysis of a urine sample as a common diagnostic technique. The presence or absence of a particular substance in the urine may direct the physician's attention to a problem or provide information about the malfunction of a particular organ.

Chapter 9 / The Excretory System

CHECKPOINT

1. (a) Which other systems excrete waste products besides the excretory system?
   (b) What waste products do these systems excrete?
2. List four ways by which the body loses water.
3. List the major organs and tubes of the excretory system.
4. Which vessels carry blood to and from the kidneys?
5. What are the microscopic filtering units of the kidneys called?

9.3 The Nephron: Site of Filtration

Each nephron consists of a series of tubules and a ball of capillaries called the **glomerulus** (glom-air-you-lus) which lies inside a cup-shaped structure known as the **Bowman's capsule**. (See Figure 9.5.)

Blood arrives at the kidney through the renal artery. It then flows into increasingly smaller vessels until it is eventually delivered into an afferent arteriole leading into a glomerulus. After passing through the ball of capillaries in the glomerulus, the blood flows through a net of capillaries surrounding the tubules and the loop of Henle, and leaves through the efferent arteriole and the renal veins. Follow this path in Figure 9.5.

Figure 9.5
The nephron, the microscopic filtering unit of the kidney. There are more than 1 million of these units in each kidney.

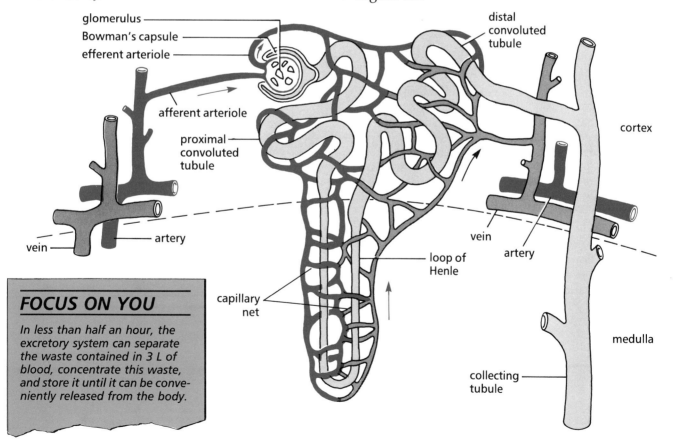

FOCUS ON YOU

In less than half an hour, the excretory system can separate the waste contained in 3 L of blood, concentrate this waste, and store it until it can be conveniently released from the body.

Table 9.3 How the Nephron Works

- Each of the 1 million nephrons in each kidney filters a tiny amount of the blood plasma. The combined action of the nephrons results in the processing of 120 mL of plasma each minute.

1. The pressure of the blood in the supply arteries to the kidney causes plasma to pass through the capillary walls of the glomerulus into the Bowman's capsule.

2. Filtered plasma enters the tubules. The filtrate contains substances still needed by the body, such as glucose, salts and amino acids, as well as waste products. Blood cells and large proteins are too large to enter the Bowman's capsule.

3. Capillaries around the proximal tubule reabsorb glucose, amino acids, and potassium ions by diffusion. Wastes, and some useful substances yet to be reabsorbed, are left in the tubule.

4. Sodium ions must be moved from a dilute solution in the tubule to a more concentrated one in the cells around the tubules. This requires energy to work the sodium "pumps". Cells around the tubules take the sodium ions from the filtrate and pump them into the capillaries.

5. As the sodium ions are pumped out, chloride ions follow and combine with them to form sodium chloride (salt). This salty solution draws water from the tubules for reabsorption by the capillaries.

6. About 20 mL of the initial 120 mL of filtrate remains in the distal tubule. Here the anti-diuretic hormone (ADH) determines how much water will be reabsorbed to balance the body's needs.

7. The reabsorbed substances, needed by the body, are returned to the veins and enter the circulatory system. The wastes, dissolved in small amounts of water, remain in the tubule and form urine. The urine produced by all the nephrons enters the collecting tubules and eventually reaches the bladder.

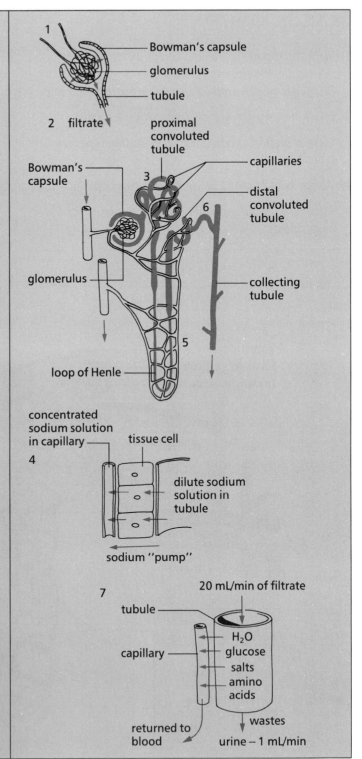

The blood flowing into the glomerulus is rich in oxygen and contains all of the normal constituents of blood – blood cells, proteins, amino acids, glucose, nutrients of all kinds, salts, and wastes. The nephron's function is to remove the wastes, while retaining the other useful substances. It is rather like trying to clean out a drawer with many different items in it. It is difficult to pick out just the junk items. It is easier to tip out the entire contents, replace the things you wish to keep, then throw away what is left. The nephrons work in a similar fashion. Both wanted and unwanted substances cross into the Bowman's capsule. Other parts of the nephron then remove the useful materials and pass them back into the blood. The remaining substances – the wastes – are delivered to the bladder for disposal as urine.

Although each nephron only produces a minute quantity of filtrate, the amount produced by 2 million nephrons together is considerable. It amounts to more than twice the individual's body mass each day! From 80 to 85 percent of the water in the filtrate is reabsorbed by the capillaries, however. It has been estimated that more than 1 kg of sodium and 180 g of sugar are reabsorbed with the water. This is indeed an impressive recycling job! The functions of the nephron are summarized in Table 9.3.

9.4 Kidney Transplant: Sharing for Life

Each year, more than 500 Canadians receive a kidney **transplant**, a surgical procedure which replaces a non-functional kidney with a healthy one. This technique has saved the lives of many people suffering from some kind of kidney failure. The healthy kidneys are obtained from donors. A donor may be a close blood relative (parent, brother, or sister). One kidney of the pair can be removed without harm to the donor.

The other source of organs are donations from recently deceased individuals. Canadians receive an organ donor card with their driver's licence. It is a legal document which indicates that the person who signed the card is willing to donate his or her organs for transplantation or research upon death. (See Figure 9.6.) Time is vital if a donation is to be made. A driver's licence is often one of the first documents inspected at the scene of an accident. The donor card will thus give immediate information about the wishes of the victim to offer this lifesaving gift to others in need.

Figure 9.6
By signing a card like this, or the one attached to your driver's licence, you give permission for your organs to be used to help someone else in the event of your death.

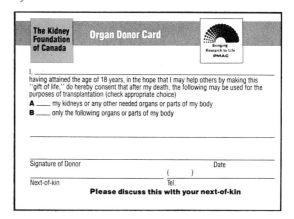

A major problem with any transplant is to prevent the donated kidney from being rejected by the recipient's immune system. This system tries to destroy any foreign protein that enters the body, regardless of whether the invader is a bacterium or a vitally needed transplant organ. To prevent rejection, great care must be taken to see that there is a good tissue match. For genetic reasons, the best matches, and most successful transplants, are with organs from close relatives. The present one year survival rate is 97 percent, if the kidney comes from a blood relative. The rate is 90 percent if the kidney comes from a donor that is not a blood relative, but provides a good tissue match.

BIOTECH

The Artificial Kidney

The artificial kidney machine is used to purify blood when the kidneys fail to do their work. This most commonly happens when kidneys become diseased.

In many cases, a permanent plastic "shunt" is implanted between the artery and vein in the arm or leg of the patient (see illustration at the end of this box). During the **dialysis** or blood cleansing, arterial blood is led to the artificial kidney and returned to the vein.

In this machine, a semipermeable membrane such as cellophane tubing is used to do the filtering normally done by the kidney. The entire apparatus is immersed in a tank containing a *dialysing fluid* (a liquid with a concentration of salts and other substances the same as that of blood).

The basic feature of all kidney machines is a semipermeable membrane with tiny pores from 0.0004 to 0.0008 μm in diameter. The membrane pores are of a size that will allow water, acid, ammonia, and salts to pass through, but will hold larger molecules such as glucose, fatty acids, amino acids, and proteins. Sodium and potassium will pass freely into the dialysing fluid. These salts are essential to the body, so their levels as well as others must be kept constant. It is difficult to prevent the movement of these salts out of the blood into the fluid. That is why the dialysing fluid contains the same concentration of these salts as the blood does. The migration of these salts can go both ways across the membrane. The hemodialysis takes from 3 to 6 h and is usually done three times a week.

Another kind of dialysis is now possible for patients with kidney failure. It is called continuous ambulatory peritoneal dialysis (CAPD for short). It has made life much easier for patients with kidney failure because it does not require a large machine.

This method uses the body's own abdominal lining called the peritoneum as the semipermeable membrane for filtering. A fluid with the same salt concentrations as the blood is introduced through a tube (called a catheter) directly into the abdominal wall. A person who uses this method has a catheter implanted into the abdominal cavity. The catheter is simply capped when not used. First, the fluid is inserted through the catheter. Then, the peritoneum acts as a filter so that diffusion and osmosis take place

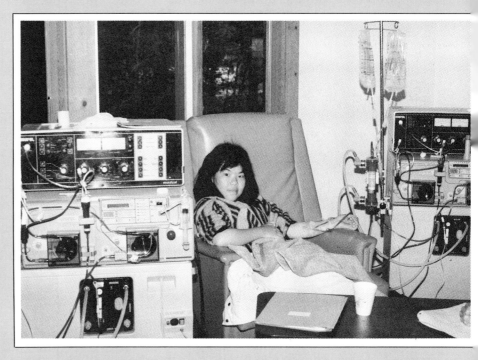

The dialysis machine operates as an artificial kidney for this girl.

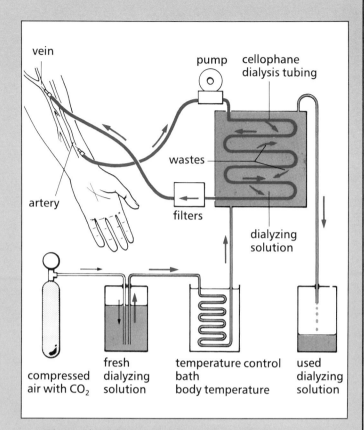

The artificial kidney machine

between body fluids and blood supplying the stomach and intestines.

Patients on this peritoneal dialysis have to replace about 2 L of the fluid (the dialysate) four times a day and replace it with fresh dialysate. The whole process takes about 30 min and can be done almost anywhere.

Research is being carried out by engineers and physicians to test a permanently wearable artificial kidney.

CHECKPOINT

1. (a) What cells and substances are found in the blood entering the kidney?
   (b) What substances or cells in the blood do not cross into the Bowman's capsule?
   (c) What kinds of substances are found in the filtrate of the Bowman's capsule?

2. What substances needed by the body are reabsorbed from the proximal tubule and loop of Henle?

3. Why is careful tissue matching necessary before a successful organ transplant can be made?

9.5 Urine and the Antidiuretic Hormone (ADH)

The **antidiuretic hormone** is produced by the pituitary gland and released into the bloodstream. This hormone affects the permeability of the collecting tubules of the kidneys. The purpose of this action is to control the amount of water being reabsorbed and match this amount to the water requirements of the body. In doing so, it also helps control the blood-water concentration.

If the water content in the body is high, ADH slows down the reabsorption of water. The volume of urine increases and the wastes the urine contains are diluted. If the body needs more water, ADH permits more water to be reabsorbed. The amount passed into the urine is reduced.

The Composition of Urine

Normal urine is an amber or yellow, transparent fluid, which is usually rather acidic. These characteristics can vary a great deal and still be within the normal range, depending on a number of factors. If we eat a standard mixed diet, acid-producing foods predominate and the urine will be acidic. If we consume a vegetable diet, the urine will be alkaline. The quantity of urine also varies with the temperature and the quantity of liquid ingested. The average amount of urine passed each day is about 1.5 L. About 95 percent of this urine is water and the rest is made up of about 60 g of dissolved solids. Table 9.4 gives a summary of the characteristics of normal urine.

Table 9.4 The Physical Characteristics of Normal Urine

Amount (per 24 h)	1500 mL. This varies with intake of fluid, temperature, amount of sweating, etc.
Colour	Straw-coloured or amber. If only small amounts are passed, the colour is darker. Changes with diet (red from beets).
Clarity	Clear or transparent, but cloudy after it is left standing
Odour	After standing for awhile, it has a characteristic "ammonia" odour
Acid-Base balance	Normal range pH 4.8-7.5, usually about pH 6. May be less acidic if diet is mainly vegetables. Acidic on a high protein diet

Urinalysis

The substances found in urine come from what we take into our bodies as well as from processes occurring within our bodies. This is why **urinalysis**, the analysis of urine composition, is such a powerful medical tool. (See Figure 9.7.)

Figure 9.7
Modern urinalysis equipment

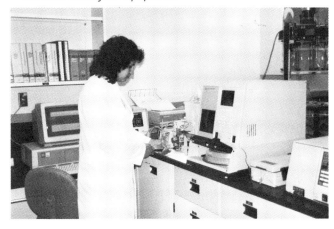

Table 9.5 lists some of the common substances found in urine. In addition, urine normally contains a few epithelial and blood cells.

Table 9.5 Some of the Common Substances Found in Urine

Some Substances Found in Urine	Dietary Source
Water	Fluid intake and metabolism
Uric acid	Breakdown of proteins
Sulphates	Breakdown of proteins
Creatine	Product of muscle action
Ammonia	Breakdown of amino acids
Salts	Diet intake

Some substances are not normally found in urine. These include glucose, proteins, large amounts of blood, and albumen (blood protein). The presence of these substances may have harmless causes. For example, some glucose or protein might be present after a heavy intake of these compounds in food. The body excretes what it cannot immediately use.

Several serious disorders also cause distinctive changes in the composition of urine. (See Table 9.6.) For example, glucose in urine is also a warning of diabetes mellitus. If the results of a urinalysis include any unusual substances, the physician will know to immediately have other tests performed to confirm the diagnosis. With early detection, treatment can begin before the disorder worsens.

Table 9.6 Some Disorders Which Can Be Detected by Urinalysis

If Found in Urine	May Indicate
Amino acids (protein)	Nephritis
Glucose	Diabetes mellitus, liver disease, endocrine disorders
Urobilinogen (in large amounts; from breakdown of red blood cells)	Jaundice, hepatitis, hemolytic disease
Uric acid (crystals) (in large amounts)	Kidney stones, gout (uric acid crystals in joints)
Large numbers of cells	Renal disease

CHECKPOINT

1. (a) What hormone controls the amount of water reabsorbed by the kidney?
   (b) How does it act?

2. List the characteristics of normal urine.

3. List three substances found in normal urine and three substances not normally present.

4. (a) If a person's urine was found to contain glucose, amino acids, and large amounts of water, what might a physician want to find out?
   (b) What disorders might a physician suspect?

9.6 The Collection and Release of Urine

The Ureters

Once urine has been formed by the nephrons, it drains into the collecting tubules and then empties into the renal pelvis. (See Figure 9.4.) From the pelvis, the urine flows into the ureters (one from each kidney) and is carried down to the urinary bladder. (See Figure 9.8.)

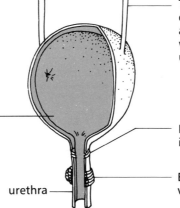

Figure 9.8
The urinary bladder and its associated tubes

The ureters are long thin tubes which deliver urine from the kidneys. There are smooth muscles in the ureter walls which contract regularly to move the urine down to the bladder.

Internal sphincter muscles are under involuntary control.

External sphincter muscles are under voluntary control.

urethra

Smooth muscle coating in the walls of the bladder allows it to expand as it fills. It also contracts periodically to expel urine through the urethra. The urinary bladder is a hollow muscular sac which acts as a reservoir for urine.

The ureters, which are 25 to 30 cm long in adults, become somewhat larger and thicker-walled as they descend. Their inner cell lining secretes mucus which provides protection from the urine. Layers of muscle in the ureter walls contract in peristaltic fashion to move the urine along.

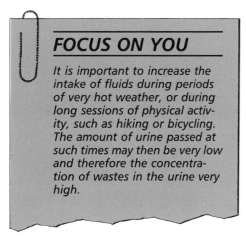

FOCUS ON YOU

It is important to increase the intake of fluids during periods of very hot weather, or during long sessions of physical activity, such as hiking or bicycling. The amount of urine passed at such times may then be very low and therefore the concentration of wastes in the urine very high.

The Urinary Bladder

The urinary bladder is a hollow organ found in the pelvic cavity, just behind the pubic bone. In men, it is directly in front of the rectum and in women it is in front of, and under, the uterus. It is loosely held in place by folds of the abdominal lining (**peritoneum**).

At the base of the bladder is the opening to the **urethra** surrounded by circular muscle fibres forming an internal sphincter. Just below this is another ring of skeletal muscle, the external sphincter. (See Figure 9.8.)

Urination

The quantity of urine contained in a full bladder may vary from 200 to 400 mL. When the bladder is full, stretch receptors in the walls of the bladder send impulses to the spinal cord making an individual conscious of the need to urinate. The brain then sends messages to the external sphincter when an opportunity to empty the bladder occurs. Although emptying the bladder (**micturition**) is basically under reflex control, the external sphincter muscles are under voluntary control. The external sphincters over-ride the reflex action and we can control the need to urinate by conscious effort.

9.7 Some Disorders Affecting the Excretory System

Urinary Tract Infections

Infections in the urinary tract are among the most common problems treated by doctors. These infections are caused by bacteria invading the kidneys, bladder, or urinary tract. Normally urine is sterile, but when an infection develops, bacteria can be detected in the urine. Some examples of urinary infections are cystitis, bladder and kidney infection, and nephritis. Most infections are limited to the bladder.

Common symptoms of urinary infections are pain or a burning sensation during urination and a sense of needing to pass urine frequently, although the volume passed may be small. The urine may be cloudy, foul smelling, or even bloody at times. Sometimes these symptoms are accompanied by fever and chills or lower back pain. It is also possible to have urinary infections without any symptoms at all. If the doctor determines that you have a urinary infection, an antibiotic will usually be prescribed and possibly some medication to reduce the pain and discomfort.

Hypertension and the Kidney

Hypertension, or high blood pressure, is common among people in all types of occupations and with all types of personalities. Hypertension does not mean, as many people believe, being nervous, irritable, high-strung, or impatient. In Chapter 5, we discussed blood pressure and how it can be checked with a blood pressure cuff. What has hypertension got to do with the kidney? When the arteries of the kidney are narrowed, as by high blood pressure, there is an increase in the production of **renin** by the kidney. Renin activates another substance in the bloodstream called **angiotensin**, which further constricts the blood vessels, causing another increase in blood pressure. If this cycle is not checked, the kidneys can be seriously damaged.

Kidney Stones

Sometimes, some of the salts found in the urine form small crystals that are hard and insoluble, called kidney stones. They can form almost anywhere in the urinary tract, but are usually found in the collecting points of the kidney. When kidney stones are painful and obstruct the flow of urine, they may have to be surgically removed. Ultrasound is also used to shatter the stones, so that the fragments can be passed with urine.

Another method uses flexible fibre optics. A tube is passed up through the urethra or through a small hole in the kidney. The optic fibres enable the surgeon to see into the kidney and locate the stones, which can then be removed by suction.

CHECKPOINT

1. How is the emptying of the bladder controlled?
2. (a) Through which tube does urine leave the body?
   (b) What other system uses this tube in the male?
3. What are the symptoms of a urinary tract infection?
4. (a) What is hypertension?
   (b) How does it affect the kidneys?
5. (a) What are kidney stones?
   (b) Name two methods which doctors use to remove these obstructions.

BIOLOGY AT WORK

DRUG ANALYST
LABORATORY TECHNOLOGIST
BIOCHEMIST
UROLOGIST
LASER TECHNICIAN

HEALTH CONCERNS

jaundice
cystitis
kidney stones
nephritis
gout
renal disease

Chapter 9 / The Excretory System

CHAPTER FOCUS

Now that you have completed this chapter, you should be able to do the following:

1. Name the structures and functions of the parts of the excretory system.
2. Describe the process by which the nephron separates wastes from the blood entering the kidney, and reabsorbs nutrients and other needed substances.
3. Name and describe the function of the hormone controlling the concentration of urine.
4. List the characteristics of normal urine and name some of the common substances found in urine.
5. Explain the usefulness of urinalysis in diagnosing illness.

SOME WORDS TO KNOW

Match each of the descriptions given in the left-hand column with a word shown in the right-hand column. DO NOT WRITE IN THIS BOOK.

1. The tube leading from the kidney to the urinary bladder
2. Microscopic filtration units in the kidney
3. The name of the artery carrying blood to the kidney
4. Minute "balls" of capillaries in the nephron
5. The major organ of excretion
6. The tubule carrying filtrate from the Bowman's capsule to the loop of Henle
7. The tube that carries urine away from the bladder
8. The fluid that passes from the glomerulus to the Bowman's capsule
9. A ring of muscles that can close off a tube
10. A waste product, high in nitrogen, that is found in the urine

Select any two of the unmatched words and, for each, give a proper definition or explanation in your own words.

A Bowman's capsule
B glomerulus
C loop of Henle
D nephron
E ureter
F urethra
G filtrate
H proximal tubule
I distal tubule
J urea
K renal artery
L sphincter
M proteins
N urine
O kidney
P micturition

SOME QUESTIONS TO ANSWER

1. Draw a sketch of the excretory system and label the major parts.
2. Why are there so many nephrons in each kidney (about 1 million)?
3. What causes the filtrate to cross from the capillaries of the glomerulus to the Bowman's capsule?
4. Why are large proteins and blood cells not found in the filtrate?
5. What is the function of the antidiuretic hormone?
6. How does the composition of urine differ from that of the filtrate?
7. Why does the doctor sometimes ask a patient for a urine sample as part of a physical examination?
8. Why is it recommended that a person drink six to eight glasses of fluids a day?

9. During an assault, especially if the victim is on the ground and trying to keep from being kicked, the kidneys often suffer serious damage. This is due to their position in the body. Explain why this is so.

10. How would the volume and characteristics of urine change
    (a) for a field hockey player after playing for several hours in hot weather?
    (b) for a person who had been drinking pop or alcohol for several hours at a party?
    Explain your answers.

SOME THINGS TO FIND OUT

1. Find out about organ transplants. Who can donate? Do parents have to be consulted? What time problems are involved?

2. Make a table in which to compare the main components of plasma, filtrate, and urine. Note both differences and similarities.

3. Urinalysis is becoming a common part of international sport competition. Why has this become necessary? Discuss the following: testing will be part of all sports soon; urinalysis will be popular for a while but won't stop the problem of drug abuse; people should have the option of refusing or requesting testing of their urine.

4. Sweat is mainly water, but it also contains inorganic and organic substances. Research the substances that are excreted in sweat. Some substances that are harmful to us can also be absorbed through the skin. Try to find out what these substances are.

GENETICS: BLUEPRINT FOR LIFE

Before a machine or building is constructed, a blueprint is prepared to describe the materials to be used as well as where they are to be placed.

Imagine a blueprint able to describe the formation of a new, unique human being. This blueprint would have to include instructions on growth and development as well as structure. It would have to provide information for the working of every cell. If this human "blueprint" were translated into words on a page, it would fill 1000 volumes of an encyclopedia! Yet, within your body, the genetic blueprint of who and what you are is contained in 46 microscopic chromosomes, found in the nucleus of your cells.

KEY IDEAS

- *DNA is the basic molecule of life, controlling the assembly of proteins, all cell activities, and the transmission of hereditary information.*
- *The genetic code involves sequences of nitrogen bases along the DNA molecule.*
- *The inheritance of chromosomes and their complement of genes determines the sex of a child, the appearance of identical and fraternal twins, and the presence of genetic disorders.*
- *The solving of simple inheritance problems allows us to predict the probability of a trait appearing in the next generation. This has considerable importance for genetic counselling.*

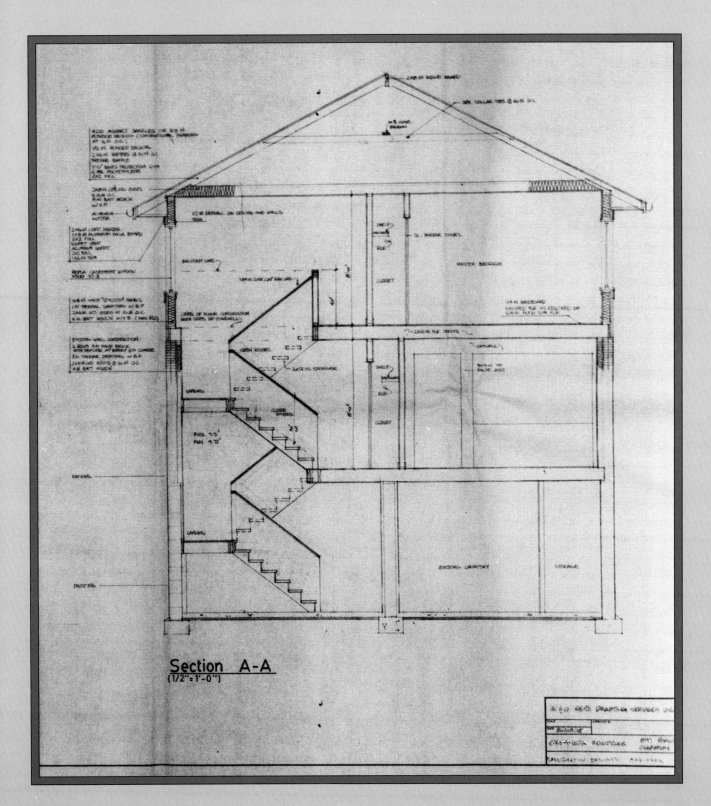

10.1 Information for Development

We are all aware that we have characteristics in common with other members of our family. It may be eye or hair colour that we share in common with one parent, or perhaps a blood type or tallness that we share with the other. Brothers and sisters may possess several characteristics in common and be so alike that other people can see the family resemblance without previously knowing that a relationship existed.

The "blueprint" for the development and appearance of each individual is contained within the chromosomes of our cells. This information is so efficiently organized and tightly packed together that every detail of our body structures, the chemical processes of each system, and even some of the elements of our personalities are contained within the 23 pairs of chromosomes located inside the cell nucleus. In spite of the enormous amount of information they contain, these chromosomes are so small that it takes a high-powered microscope to see them.

What Are Chromosomes?

Chromosomes are made of complex molecules called **DNA (deoxyribonucleic acid)**. The general structure of this molecule has been understood for some time. Imagine a ladder with flexible (instead of rigid) sides that can be twisted around and around into a spiral. The sides of the DNA molecule are made up of alternating deoxyribose sugar and phosphate molecules. The "rungs" of the DNA ladder, stretched between the sugar molecules, are formed of pairs of molecules called **nitrogen bases**. (See Figure 10.1.)

The nitrogen bases are very important molecules, for the sequence in which they occur in the DNA molecule determines the "code" for all the genetic information. There are four nitrogen base molecules – adenine (A), thymine (T), guanine (G), and cytosine (C). These molecules always occur in the same linked pairs in DNA. Adenine pairs with thymine and cytosine pairs with guanine because of the physical size, shape, and the nature of the bond formed between these molecules.

The Genetic Code

An example of another code will help explain what is meant by a "genetic code". You probably know that the Morse code consists of two symbols, a dot and a dash. With this code, any word can be communicated by using these two symbols. For example, the word "black" can be written

B	L	A	C	K
— • • •	• — • •	• —	— • — •	— • —

The genetic code has four symbols – adenine, thymine, guanine, and cytosine. All the information needed for building all body structures, proteins, hormones, and other molecules is coded into sequences using these four chemical substances. With four symbols, an enormous number of variations is possible in the arrangement of these bases along the DNA chain, for in each position, there are four possibilities.

Our knowledge of the chemical nature of the chromosome is relatively recent and there is still much to be discovered; however, the microscopic image of the chromosomes has been quite thoroughly examined. We know that chromosomes are formed from granular material in the nucleus just before cell division and, when stained, appear as short dark threads in the nucleus. They may be counted and recognized as distinctive shapes, similar to each other, yet slightly different from all the other chromosomes. We know that each species of organism has a specific

Chapter 10 / Genetics: Blueprint for Life

Figure 10.1
The Watson-Crick model of the DNA molecule. This is the molecule that has the unique ability to replicate. It forms the blueprint in which all genetic information for each cell is stored. The sequence of the nitrogen bases acts as a code to convey the chemical messages of heredity.

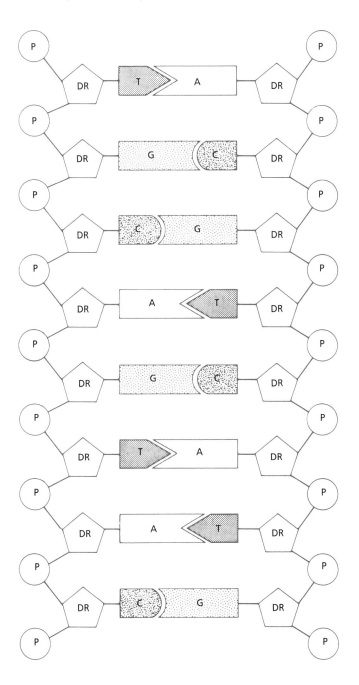

number of chromosomes. Humans have 23 pairs, fruit flies have 4 pairs, and horses have 33 pairs.

Chemical studies have shown that along the length of the chromosomes are short sections of DNA, with coded information giving instructions for a particular characteristic or characteristics. These short sections are called **genes**. There are estimated to be hundreds of genes along each chromosome. Each gene carries the information for a characteristic, such as hair colour, from one generation to the next.

The four nitrogen bases

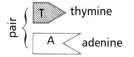

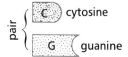

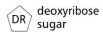

The "ladder" formation of the DNA molecule is twisted into a spiral with the nitrogen bases forming the rungs.

299

BIOTECH

Cracking the Genetic Code

New techniques in genetic engineering allow scientists to decode and modify tiny sections of the DNA molecule. Here is what we know so far.

A massive international program has just begun which will involve tens of thousands of scientists for the next 15 years. Their aim is to map every one of the 3 billion nucleotides that link together and form the long twisted coils of DNA in each human cell. It is known as the **"Human Genome Project"**. The many combinations of nucleotides located along the DNA molecule form the 100 000 genes which determine every characteristic of our bodies – our height and hair colour, what we can eat and digest, what diseases we can resist, how we repair torn tissues, and even components of our intelligence. Already, the short sections of DNA that have been unravelled and mapped are yielding up their secrets about what causes some birth defects. Of the more than 4000 hereditary diseases, the genes of some 170 have been identified.

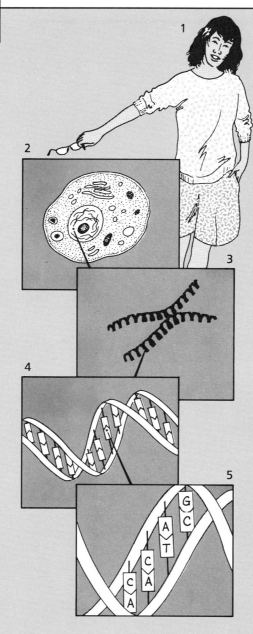

1. The human body holds about 10 trillion cells. Large molecules called proteins determine the structure and function of each cell.

2. A cell nucleus contains 23 pairs of chromosomes. The chromosomes control the manufacture of proteins by the cell.

3. A chromosome consists of two long, twisted strands of DNA, the chemical that carries genetic information from parents to offspring. Human DNA is divided into about 100 000 clusters called genes.

4. A gene determines a human characteristic such as height, eye colour, or disease resistance. Genes are composed of thousands of nucleotides, the smallest genetic unit.

5. Nucleotides come in four different shapes called A, C, G, and T, arranged in pairs along the strands of DNA. About 3 billion nucleotides make up the human genome, the blueprint of a human being.

Each pair of chromosomes contains specific information about many characteristics in the body. One chromosome pair could have information about hair colour, liver structure, body build, etc., while another chromosome pair might have information about entirely different characteristics. You received one-half of each pair of your chromosomes from each of your parents. Two chromosomes with information about the same characteristics make a **homologous** (hoe-mol-uh-gus) **pair**.

How Genes Operate in the Cell

Genes carry the information on how to build each and every protein needed by the body, from enzymes to the structural components of muscle. This information is coded in the sequence of bases, on a small part of the DNA molecule. This part of the DNA molecule unravels and makes copies of itself called **messenger ribonucleic acid (mRNA)** and **transfer ribonucleic acid (tRNA)**. The protein information has now been transcribed onto the mRNA and tRNA strands. The mRNA and tRNA move out of the nucleus into the cytoplasm. The mRNA locates on the ribosomes associated with rough endoplasmic reticulum, the site of protein synthesis. (See Figure 10.2.)

Building a Protein
There are 22 different amino acids which must be arranged into a particular sequence to form a specific protein. Each amino acid associates with a particular three-unit group of transfer RNA. Energy delivered by ATP activates a bond between the tRNA and the amino acid. Once activated, the tRNA carries the amino acid to the mRNA "plan" and finds a site where the pairs of bases will fit.

Once the tRNA has found a matching site on the mRNA, bonds (peptide) form between adjacent amino acids, and a protein is formed. The bonds between the amino acids and the tRNA are broken and the protein is free to leave. The tRNA is also free to leave and seek new amino acids to fill other sites.

The sequence of nitrogen bases in the chromosome (DNA) is thus responsible for determining the final sequence in which the amino acids are joined to form a certain protein.

CHECKPOINT

1. (a) What do the initials DNA stand for?
   (b) Why is this molecule so important?
2. Describe the structural shape of the DNA molecule.
3. Name the four bases present in a molecule of DNA.
4. Why is the sequence of nitrogen bases along a chromosome so important?
5. (a) What is a gene?
   (b) How are genes involved in the building of proteins?
6. (a) What is a homologous pair of chromosomes?
   (b) Why are chromosomes present in pairs?

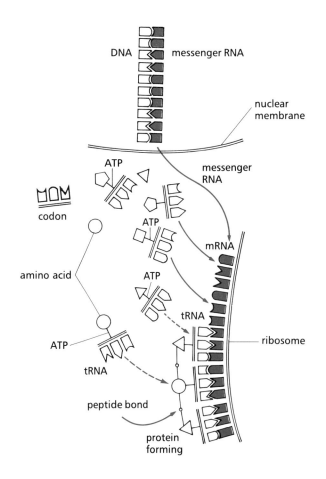

Figure 10.2
Building a protein from the cells' genetic code

10.2 Chromosomes and the Production of Gametes

Mitosis is the way in which cells multiply to allow growth and replacement of worn out or damaged cells and tissues. In **mitosis**, cells identical to the parent cell are produced. They have the same number of chromosomes and identical genetic information as the parent cell. (See Figure 10.3.)

Figure 10.3
Mitosis – the division of one parent cell into two identical cells

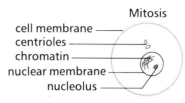

Interphase

Nuclear membrane and nucleolus visible. Chromatin present as granular mass. The cell is active, involved in its normal functional activities. Just prior to entering prophase, the chromosomes duplicate.

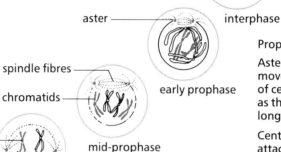

Prophase

Aster rays appear. Centrioles move toward opposite poles of cell. Chromosomes appear as thin threads. Nucleolus no longer visible.

Centrioles form spindle fibres attached to chromosomes now seen as two chromatids. Centrioles reach the poles of the cell. Nuclear membrane no longer visible.

Chromatid pairs start to migrate toward the equator of the cell.

Metaphase

Chromatid pairs line up across the middle of the cell. Separation of chromatid pairs occurs as the centromere splits. Each chromatid is joined by a spindle fibre from the centromere to the aster.

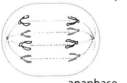

Anaphase

Two complete sets of chromosomes now are drawn toward the opposite poles of the cell. The contraction of the spindle fibres causes this movement.

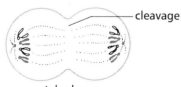

Telophase

Nuclear membrane reappears and surrounds chromosomes. Nucleoli reappear. Chromosomes become less distinct. Division of the cytoplasm occurs. Cell membrane starts to indent.

Daughter Cells Enter Interphase

Cells now resemble original parent cell, containing an identical set of genetic material. Cleavage and division of the cytoplasm is complete.

A special kind of division takes place in sex cells, also called **gametes**. This process is called **meiosis**. (See Figure 10.4.) In meiosis, the cells that are produced are not identical to the parent cell. Instead, they have only half the number of chromosomes present – one of each homologous pair. It is important to understand why the process of meiosis is necessary in sex cells.

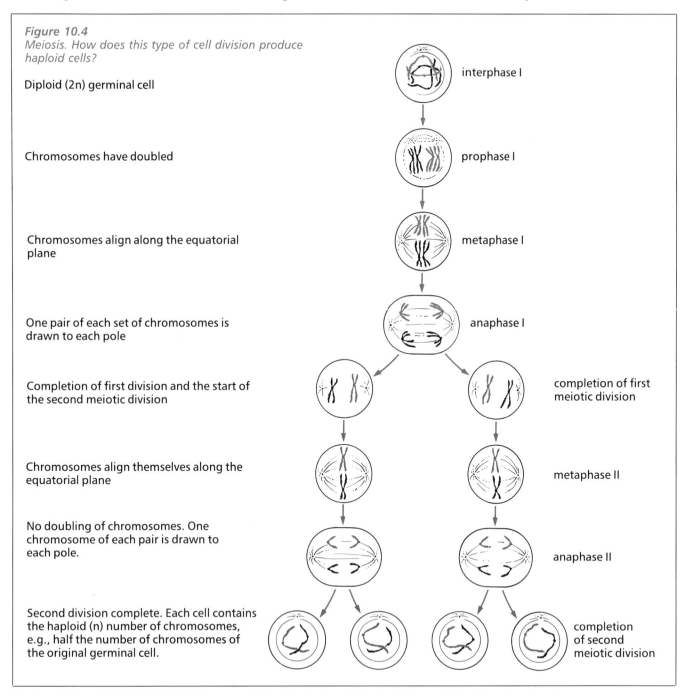

Figure 10.4
Meiosis. How does this type of cell division produce haploid cells?

Diploid (2n) germinal cell — interphase I

Chromosomes have doubled — prophase I

Chromosomes align along the equatorial plane — metaphase I

One pair of each set of chromosomes is drawn to each pole — anaphase I

Completion of first division and the start of the second meiotic division — completion of first meiotic division

Chromosomes align themselves along the equatorial plane — metaphase II

No doubling of chromosomes. One chromosome of each pair is drawn to each pole. — anaphase II

Second division complete. Each cell contains the haploid (n) number of chromosomes, e.g., half the number of chromosomes of the original germinal cell. — completion of second meiotic division

If a sperm or ovum had the full complement of 23 pairs (46 chromosomes), when they joined together at fertilization, there would be 92 chromosomes in each of the cells of the embryo. In each succeeding generation, the number would then double. To prevent doubling during the development of the gametes, the number of chromosomes is halved during the process of meiosis. Only one chromosome from each pair is present in each sperm or ovum cell.

Cells that contain 23 pairs or a total of 46 chromosomes are called **diploid** cells (2n). Cells with half the number of chromosomes – one of each kind – are **haploid**, 23 in number (n). (See Figure 10.4.) The only haploid cells in the human body are the egg and sperm cells.

Sperm Production

The seminiferous tubules of the testes are lined with special germinal cells called **spermatogonia**, which, like all body cells, have the diploid (2n) chromosome number. At puberty, these cells start active meiotic division, which continues throughout a male's lifetime. The cells produced by this division are, therefore, haploid (n) cells. These cells become specialized as sperm. (See Figure 10.5.)

Egg Production

A similar process of meiotic division takes place in the ovary; however, when the germinal cells called **oogonia** divide, only one cell eventually matures into the haploid egg (ovum). The other three small units, called **polar bodies**, that are produced during meiosis break down and disintegrate. See Figure 10.5 to compare ovum

Figure 10.5
The development and meiotic divisions which produce the gametes

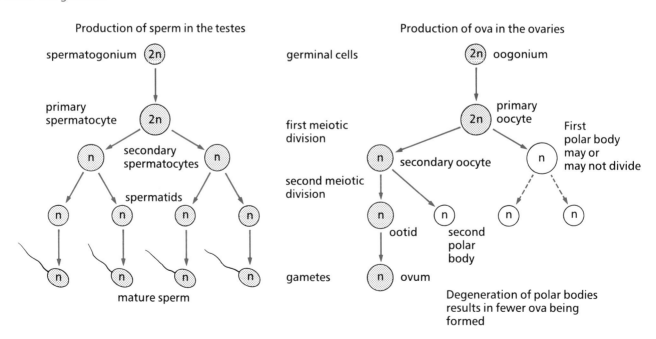

formation and sperm formation. Unlike sperm production, which occurs throughout the male's lifetime after puberty, eggs are produced before a female is born. Normally, one egg completes development with each menstrual cycle. This is very significant when the risk of long-term exposure to adverse environmental effects, such as radiation, is considered.

CHECKPOINT

1. How does mitosis differ from meiosis?
2. (a) What types of cells undergo meiosis?
   (b) Why is this process necessary?
3. (a) How many chromosomes are there in a human diploid body cell?
   (b) How many chromosomes are there in a human sex cell?

10.3 Inheritance

Inheritance is the transfer of characteristics or traits from one generation to another. As the pairs of chromosomes are pulled apart during meiosis, chance alone will determine which member of each homologous pair ends up in the same cell. The following example will show the importance of this "chance" event. To keep the possible combinations simple, we will use only three pairs of chromosomes. (See Figure 10.6.)

Eight combinations (2^3) in a sperm or egg are possible with three pairs of chromosomes; however, humans have not just three, but 23 pairs of chromosomes. Therefore, the possible combinations are enormous. It is calculated that 2^{23} or over 8 million different combinations are possible in each sperm or ovum.

This is only part of the variation that is possible when an egg is fertilized, for now any one of those sperm can unite with an ovum with equally as many possible combinations. The combinations possible in the fertilized egg (**zygote**) are greater than 70 trillion! No wonder it is impossible to find an exact double for any one human being.

Activity 10A

The Laws of Probability

By observing the results of the random flipping of coins, you can understand why the inheritance of traits can be predicted, based on the laws of probability.

Materials

two coins of the same value per student pair

Figure 10.6
Possible combinations of chromosomes

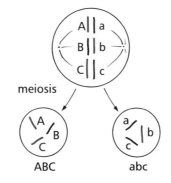

Pairs of chromosomes at the second metaphase ready to pull apart. One of each pair moves to each end of the cell.

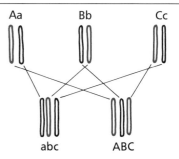

Two random chance selections are illustrated. Another six are possible. Can you work out what they will be?

Procedure

1. Place a coin on your thumb, without looking to see if it is heads or tails. Use the same technique for each toss. Flip the coin and, *in a table in your notebook similar to the one below, record whether the coin comes down heads or tails.*

2. Do this 20 times and *total your results.*

Trial	Heads	Tails
1		
2		
3		
⋮		
9		
10		
11		
12		
⋮		
20		

3. *What ratio of heads to tails did you obtain?* (Note: To find the ratio, express the larger number as an approximate multiple of the smaller term.)

   For example: In 20 trials, the results were
   (a) 7 heads and 13 tails
       $7/13 \doteq 1/2$ or 1:2 ratio
   (b) 15 heads and 5 tails
       $15/5 \doteq 3/1$ or 3:1 ratio

4. The class will write all the results (as totals) on the chalkboard. *Combine these values and calculate the ratio as in step 3.*

Questions

1. (a) Did you get a 1:1 ratio in the sample you made with 20 trials?
   (b) How close to this ratio were your results?

2. What were the largest variations among the class results?

3. Did the class results come much closer to a 1:1 ratio than your own results? Why or why not?

4. What does this suggest about the expectation of results in small and large samples of numbers?

Chromosomes in Combination: How Genes Act

When the information inherited from both parents is the same, e.g., both chromosomes carry the same gene for a particular characteristic, the chromosomes are said to be **homozygous** (hoe-moh-zy-gus). When the information inherited differs, with each chromosome of the pair carrying different instructions about a characteristic, the chromosomes are referred to as **heterozygous** (het-e-row-zy-gus).

Consider the example in Table 10.1 concerning the inheritance of curly or straight hair. Although there are usually many genes involved in the coding of a trait, in this example we will assume that the trait is due to the presence of one gene and that only two expressions, *curly* and *straight*, are possible. In cases A and C, the chromosomes are homozygous for the particular trait, carrying the same gene. Whichever characteristic is present will be expressed. When the chromosomes carry different genes for a characteristic as in case B, the chromosomes are heterozygous. The gene from one parent calls for curly hair and the gene from the other parent calls for straight hair.

Table 10.1 The Inheritance of Curly or Straight Hair

	Case A	Case B	Case C
Information contained in the gene	curly hair curly hair	curly hair straight hair	straight hair straight hair
Person will have	curly hair	curly hair	straight hair

When two different sets of instructions are present at the same time, both characteristics cannot be expressed. The hair cannot be both straight and curly; a person has one or the other. In most cases, one set of genetic instructions is "stronger" than the other. This set is called **dominant** and will be the characteristic which shows up in the offspring. The other gene remains hidden and is called **recessive**. (See Table 10.2.)

By tradition, capital letters are used to represent dominant genes and lower-case letters are used for the recessive genes. Returning to our example, curly hair has been found to be dominant over straight hair in humans. If we now assign letters in our example, we get the results in Table 10.3.

The term **phenotype** is used to describe the visible characteristics of an individual that are produced by genes. The gene combination that produces the characteristics is the **genotype**. Table 10.3 shows the phenotypes and genotypes that are possible for curly or straight hair. What is your phenotype for this characteristic?

Table 10.2 Some Human Traits that are Due to Dominant or Recessive Genes

Dominant Genes	Recessive Genes
brown eyes	blue or grey eyes
hazel or green eyes	blue or grey eyes
congenital cataract	normal
far-sightedness	normal vision
short-sightedness	normal vision
astigmatism	normal vision
dark hair	blond hair
non-red hair	red hair
curly hair	straight hair
long eyelashes	short eyelashes
normal hearing	congenital deafness
blood groups A, B, and AB	blood group O
normal blood clotting	hemophilia (sex-linked)
normal red cells and hemoglobin	sickle cell anemia
Rh blood factor, positive	Rh blood factor, negative
ability to curl tongue	cannot roll tongue
free ear lobes	attached ear lobes

Table 10.3 Using Letters to Follow the Inheritance of Traits

	C = curly hair	c = straight hair	
	C C	C c	c c
Phenotype of the person	curly hair	curly hair	straight hair
Genotype of the person	homozygous CC	heterozygous Cc	homozygous cc

Activity 10B

More Laws of Probability

Materials

two coins of the same value per student pair

Procedure

1. Work in pairs. This time, toss your coin at the same time your partner does. *Record both results in a table as shown below.* Again, make 20 trials and *total your results in a column.*

Trial	Heads/Heads	Heads/Tails	Tails/Tails
1			
2			
3	SAMPLE	ONLY	
4			
⋮			

2. Record the class data in a large chart on the chalkboard. *Total all the results and determine the ratio.*

Questions

1. Assume that "heads" represents a gene for growing tall and "tails" represents a gene for remaining short. Also assume that if heads showed up either as heads/heads or as heads/tails, heads would be the gene actually expressed. (You will encounter the concept of dominant and recessive genes later in this chapter.) Add together the totals for heads/heads and heads/tails to determine the number of "tall" offspring. Do this for your own results and also with the chalkboard totals. What ratios does this produce?

2. A pioneer in the field of genetics, Gregor Mendel, predicted that a ratio like the one you calculated in the previous question should be 3:1. How do your results compare with this prediction? Another way of saying this is to show that each coin has one side out of two with a head on it (1/2). Therefore, since $1/2 \times 1/2 = 1/4$, one offspring in four will have the chance of expressing the recessive trait, and so, three offspring in four will express the dominant trait.

3. What might have happened if your answers are not exactly 3:1? Consider the size of the samples used. How does the number of trials affect the accuracy of the answer?

4. As an extra challenge, explain why the results, after four crosses between two heterozygous parents, do not always produce offspring with a 3:1 ratio of the traits.

CHECKPOINT

1. *Explain what is meant by dominant and recessive genes.*
2. *Explain the terms homozygous and heterozygous.*
3. *Give an example of a human trait phenotype.*
4. *If the trait in question 3 was recessive what would the genotype be?*

10.4 The Transmission of Genes

To show how genes are transmitted from one generation to another, let us continue our example from the previous section. Assume that, in a family, the father is homozygous for curly hair (CC). The mother is homozygous for straight hair (cc). We can now determine precisely what type of hair will be inherited by their children.

As the same information is present in each chromosome of the homozygous pair in the father, each sperm can carry only a gene for curly hair. Since the mother has homozygous genes for straight hair, every ovum will carry only a gene for straight hair.

When a sperm and ovum from this couple unite, the fertilized cell that results will have a curly hair gene (C) and a straight hair gene (c). It will be heterozygous (Cc).

Since the curly gene (C) is dominant over the straight gene (c), each of the children will have curly hair (Cc), the heterozygous condition. This means that the straight gene will be present, but its effects will not be seen in the children.

C = curly hair

c = straight hair

Father   Mother
CC   ×   cc

Gametes   Possible ♂ male gametes

		C	C
Possible ♀	c	Cc	Cc
female gametes	c	Cc	Cc

Genotype: All offspring are heterozygous

Phenotype: All have curly hair

If one of these children, when an adult, mates with another heterozygous (Cc) curly-haired individual, the children of such a union could have the following possible combinations:

Father   Mother
Cc   ×   Cc

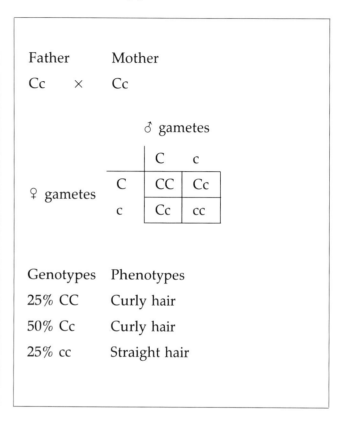

Genotypes	Phenotypes
25% CC	Curly hair
50% Cc	Curly hair
25% cc	Straight hair

When both parents are heterozygous (Cc), the chances of having a child with curly hair are three times as great as having a child with straight hair.

Although one child would be homozygous curly and the other two heterozygous curly, you would not be able to tell the difference by looking at them. Only by examining the parents and their future children would it be possible to determine whether the genes present are heterozygous or homozygous; however, if a child has straight hair, the child must be homozygous for the recessive gene. In other words, for a recessive trait to be expressed, both genes of the pair must be recessive.

The Inheritance of Two Characteristics

When chromosomes are pulled apart during cell division and meiosis, it is a matter of chance which chromosome of each pair is drawn to a particular pole.

If two characteristics controlled by genes on two different chromosomes are considered, you will see how different combinations of characteristics are possible.

We will use, as an example, suspended ear lobes which are dominant over lobes attached to the head, and a straight little finger which is dominant over a bent little finger. (See Figure 10.7.)

Figure 10.7
Which of these characteristics are dominant?

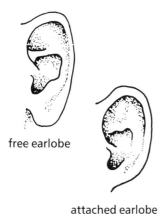

free earlobe

attached earlobe

Consider a union of a man with *suspended ear lobes* (L) and *straight little fingers* (S) and a woman with *attached lobes* (l) and *bent little fingers* (s). If the father is homozygous for suspended ear lobes and straight little fingers and the mother is homozygous for attached ear lobes and bent little fingers, we can predict that the children will all be heterozygous and will show the dominant traits of suspended ear lobes and straight little fingers.

Parental genotypes　　　LL SS × ll ss

Possible male gametes (LLSS)

		LS	LS
Possible female gametes (llss)	ls	LlSs	LlSs
	ls	LlSs	LlSs

If both parents are heterozygous individuals (LlSs), then the predicted ratios of the offspring would be found as follows:

The possible combinations of gamete types:

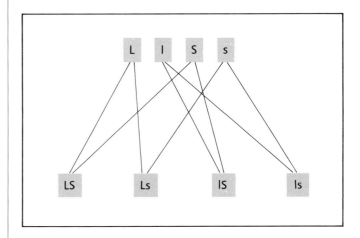

Possible male gametes (LlSs)

		LS	Ls	lS	ls
	LS	LLSS	LLSs	LlSS	LlSs
Possible female gametes (LlSs)	Ls	LLSs	LLss	LlSs	Llss
	lS	LlSS	LlSs	llSS	llSs
	ls	LlSs	Llss	llSs	llss

Ratio of phenotypes from the heterozygous cross:
9 suspended ear lobes, straight little fingers
3 suspended ear lobes, bent little fingers
3 attached ear lobes, straight little fingers
1 attached ear lobes, bent little fingers.

Activity 10C

Hereditary Traits in Humans

There are thousands of traits in humans that conform to the simple laws of genetic inheritance. A few are given in Table 10.4. In this activity, you will examine some of the traits that you have inherited and try to determine if they are present as heterozygous or homozygous conditions.

Table 10.4 Some Human Traits

Trait	Dominant (D)	Recessive (d)
ear lobes	suspended or notched	attached to head
tongue rolling	can curl into "U" shape	cannot roll into "U" shape
finger hair	hair between second and third joints, on fingers 3, 4 and 5	no hair present
freckles	present	no freckles present
hairline	pointed at front	straight across at front
little finger	last joint bends in	last joint is straight
thumb joint	bends back less than 50 degrees	can be naturally bent back more than 50 degrees

Note: Certain characteristics may be indistinct. If any finger hair is present at all, consider yourself dominant for that characteristic. If the thumb or little finger is recessive in one hand, consider it recessive in both.

Procedure

1. *Make a table in your notebook with the headings shown and list the traits in Table 10.4 in the first column.* Use the table to determine which traits are dominant or recessive. Your lab partner should help you determine your trait.

Trait	My Trait's Phenotype	My Probable Genotype	Mother's Phenotype	Mother's Possible Genotype	Father's Phenotype	Father's Possible Genotype
freckles/ no freckles	freckles	FF or Ff (✓)	freckles	FF or Ff	no freckles	ff

2. At home, determine the phenotypes of your parents for these traits. If only one parent is available, you might try to trace a pattern between yourself, a parent, and a grandparent, or choose to work with a partner with a large family. *Record these results in the proper column.*

3. When a trait is present in the dominant condition, you cannot tell by looking at the person if they are homozygous or heterozygous; however, by examining the parent genotypes, or your genotypes, or those of your brothers and sisters, you can often work out their genotypes. For example, if you have freckles (a dominant trait, F), your genotype could be FF or Ff. Your mother also shows the dominant trait, so her genotype is FF or Ff. If your father shows the recessive trait (no freckles), you know his genotype must be ff, since recessives are expressed only in the homozygous condition. Therefore, in this case, your genotype must be Ff for that trait.

Try, in the columns marked genotype, to work out as many of the answers as you can. (Note: If there are some results that do not fit exactly, it may be because the genes have not been completely expressed. This is a simplified exercise and exceptions are possible.)

Incomplete Dominance

In the examples we have been considering, we have made the assumption that a trait has only two expressions, one of which is dominant and the other recessive. In reality, there are often several genes involved for each trait and variables exist to modify the basic rules we have studied. Some plants, for example, show **incomplete dominance**, where crosses between red and white flowers result in pink blooms or striped red and white flowers. Many human characteristics, such as height or hair colour, produce an almost continuous range of variation.

Our Genetic Past: Determining a Pedigree

A very useful way of showing how genes are inherited and how they may be traced back through several generations is to use a **pedigree** diagram. (See Figure 10.8.)

STEP 1

If we examine this pedigree and know that the gene for short-sightedness is dominant over the normal vision gene, then, where a person has normal vision, they must have two recessive genes. Let S = short-sightedness, s = normal vision. We can easily fit the genotypes of all the individuals who have normal vision beneath the symbols. (See Figure 10.9.)

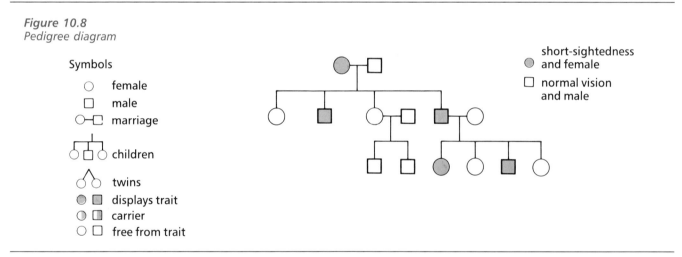

Figure 10.8
Pedigree diagram

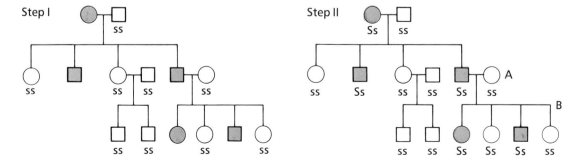

Figure 10.9
Determining genotypes

STEP 2

We know that if the dominant characteristic of short-sightedness is expressed, then at least one of the genes for this trait is present; thus, we can now add S to all the shaded symbols. (See Figure 10.9.) If we now look at the parents who are short-sighted and their offspring, we can determine whether the genotype is heterozygous or homozygous in each case. For instance, at A the mother has double recessive genes, so each ovum must carry a recessive gene and the children must, therefore, all carry at least one s. The child at B, who is short-sighted, must be heterozygous (Ss).

Activity 10D

Tracing a Trait

Select a trait that is fairly common in your family and produce a pedigree including as many family members as you can, such as brothers and sisters, parents, aunts, uncles, and grandparents. A telephone call or a note to them should supply the information you need. If you cannot determine the phenotypes of some individuals, show them in your pedigree and mark them "unknown" for the trait.

> **FOCUS ON YOU**
>
> Long before the molecular basis for genetics was understood, plant and animal breeders kept track of desirable characteristics in their breeding stock using pedigrees. In this way, they increased the probability of those characteristics appearing in later generations — resulting in the varieties of domestic plants and animals we have today.

> **FOCUS ON YOU**
>
> In the case of the inheritance of a recessive trait, generations may pass without that characteristic being expressed as a phenotype. The child must receive the gene from both parents for the characteristic to appear.

10.5 An Example of Inheritance Due to Multiple Gene Forms: Blood Type

When there are several alternative forms of a gene, they are known as **alleles**. The inheritance of a particular blood type is determined by an allele that controls the production of a specific antigen. Recall that the human blood types are A, B, AB, and O. Blood group A carries the A antigen, group B the B antigen, AB carries both A and B antigens, and O blood carries neither antigen. In this case, A and B have equal dominance to each other; both are dominant over type O.

- Type A blood individuals have an allele that produces the A antigen.
- Type B blood individuals have an allele that produces the B antigen.
- Type AB blood has alleles that produce both antigens.
- Type O blood produces neither antigen.

Table 10.5 illustrates this complex inheritance of blood types and shows the possible blood types produced by parents of differing genotypes. If we know whether the parents are heterozygous or homozygous, we can determine the possible blood types of their children.

Chapter 10 / Genetics: Blueprint for Life

Table 10.5 The Inheritance of Blood Types

Phenotypes (Blood Groups)	Type A Blood	Type B Blood	Type AB Blood	Type O Blood
Possible genotypes	A ∥ A or A ∥ O	B ∥ B or B ∥ O	A ∥ B	O ∥ O

Example

Parental phenotypes   Possible genotypes

B   A

	BB × AA			BO × AO			BO × AA			BB × AO	
	A	A ♂	or	A	O ♂	or	A	A ♂	or	A	O ♂
B	AB	AB		B	AB	BO		B	AB	AB		B	AB	BO
B	AB	AB		O	AO	OO		O	AO	AO		B	AB	BO
♀				♀				♀				♀		

offspring phenotype (blood types)

All offspring would have AB blood – 100%

25% AB blood
25% A blood
25% B blood
25% O blood

50% AB blood
50% A blood

50% AB blood
50% B blood

Consider a father with type O blood and a mother with type A blood. The possible number of blood types in the children is then quite small.

Possible Genotypes of the Parents	Possible Blood Types in the Offspring
Mother AA or AO Father OO	(AO) → A (OO) → O

Sometimes an understanding of blood type inheritance can be used to sort out a possible mix-up of babies in a hospital or provide evidence in a paternity suit. In the previous example, if the baby in question had either B or AB blood, then the man shown could not have been the father.

It is important to note that if the blood type does match, it would not prove that the man was the father, only that he *could* be the father.

Any other male with the same blood type could be the father. Many other factors that are found in blood are also inherited. The Rh factor is one. These and other factors can also be taken into consideration in such cases.

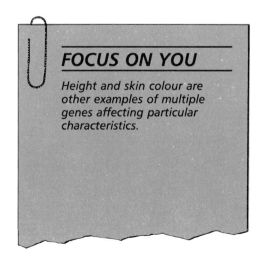

FOCUS ON YOU

Height and skin colour are other examples of multiple genes affecting particular characteristics.

314

CHECKPOINT

1. What ratio of offspring will carry the recessive phenotype when two heterozygous parents are crossed?

2. Draw a simple pedigree using the symbols for male, female, marriage, offspring, etc.

3. What blood types are possible in the children if the father has B blood (homozygous) and the mother has type O blood?

4. Is it possible for a child to have AB blood if one parent has O blood and the other parent has type AB blood? Explain your answer.

10.6 Special Information Carried by the Chromosomes

Sex Determination

If we examine the chromosomes of a cell under a microscope, we see that there are 22 pairs which can be matched together by their physical appearance. The 23rd pair has chromosomes that are quite different in shape. These are the **sex chromosomes**. The two types of sex chromosomes are called "X" and "Y", so called because they have the shape of these letters. In females, two X chromosomes are present. In males, there is one X and one Y chromosome.

During the development of sperm and ova, when the cells undergo meiosis, women (who have two X chromosomes) produce ova with a single X chromosome in each. Men, with one X and one Y chromosome, will produce sperm, 50 percent with an X chromosome and 50 percent with a Y chromosome present. It is, therefore, the male that determines the sex of a child, depending upon whether the particular sperm that fertilizes the ovum carries an X or a Y chromosome. (See Table 10.6.)

Sex-Linkage

Each chromosome carries its own complement of genes. Genes carried on the X and Y chromosomes are called **sex-linked**. Since the X chromosome is larger than the Y, it contains a number of genes that are not present on the shorter Y chromosome. (See Figure 10.10.) This is important in the transmission of sex-linked traits.

Some sex-linked disorders are inherited much more commonly in males than in females. Red-green colour-blindness and hemophilia are two examples of sex-linked traits that are transmitted to the offspring by genes that are on the sex chromosomes.

Table 10.6 Sex Determination

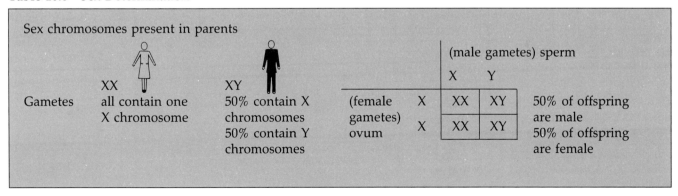

Figure 10.10
Diagrammatic representation of the X and Y chromosomes showing that homologous genes are present in only some portions of the two chromosomes. Sex-linked genes are located in the non-homologous portions of the chromosomes.

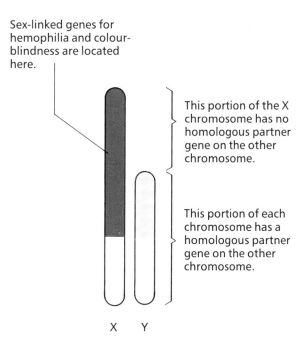

The most common sex-linked traits are recessive. For example, red-green colour-blindness is recessive (c), and normal colour vision is dominant (C). If a female has one normal X chromosome (C) and one X chromosome with the gene for red-green colour-blindness (c), she will simply "carry" the trait (Cc). The dominant normal "C" gene is expressed. If a male has an X chromosome with the gene for red-green colour-blindness, he will be colour-blind. Why? Because his other chromosome, the Y chromosome, due to its shorter length, has *no* gene, normal or otherwise, to pair with the defective recessive gene. With no normal colour-vision gene to dominate the recessive defective gene, he is colour-blind. The following example shows how colour-blindness could be inherited by males in a family.

- Example: Colour-blindness in Male Offspring

Mother carries red-green colour-blindness trait

Father has normal vision

$X^C X^c$ —— × —— $X^C Y$

	X^C	X^c
X^C	$X^C X^C$	$X^C X^c$
Y	$X^C Y$	$X^c Y$

One-half of the females produced would be normal ($X^C X^C$) and one-half would be carriers ($X^C X^c$). One-half of the males would be red-green colour-blind ($X^c Y$). One-half would be normal ($X^C Y$).

Can a female be colour-blind? Yes, she can, but only if her father is and her mother is a carrier! If she gets a "c" gene from her mother (a "carrier") and a "c" gene from her father (who would be colour-blind), she would be homozygous recessive "cc" and, thus, colour-blind.

- Example: Colour-blindness in Male and Female Offspring

Mother

Father

$X^C X^c$ —— × —— $X^c Y$

Carrier for red-green colour-blindness

Red-green colour-blindness

	X^C	X^c
X^c	$X^C X^c$	$X^c X^c$
Y	$X^C Y$	$X^c Y$

The expected proportion of affected offspring would be one-half of the females and one-half of the males with the red-green colour-blindness trait.

CHECKPOINT

1. *Which parent will carry the greatest proportion of sex-linked genes? Why?*

2. *In a recessive sex-linked disorder, which offspring will display the disorder most often? Why?*

3. *Give three examples of sex-linked disorders.*

10.7 Errors in the Genetic Blueprint

We have seen that the information that determines many of our characteristics is coded in the chromosomes that we receive from our parents. If we receive a gene that is responsible for some abnormality, it may be passed on to our children. Whether the defect is expressed or not will depend on such factors as whether the trait is caused by a dominant or recessive gene.

In North America, it is estimated that about one in 14 children is born with a serious mental or physical defect, that is, in about seven percent of all live births. In addition, there are thousands of spontaneous abortions, stillbirths, and miscarriages each year. These problems are sometimes caused by a severe abnormality that interrupts the fetal development. About 20 percent of all human birth defects can be traced to genetic factors. Another 20 percent are caused by environmental factors that affect the baby while it is developing in the uterus. Still other defects are caused by the interaction of both hereditary and environmental factors.

Mutations

Sometimes there is an alteration in the DNA structure of a chromosome, and a variation appears which has not been evident in previous generations. Such changes are known as **mutations**.

Most mutations are harmful, and a few cause such drastic changes that offspring cannot survive; however, an occasional mutation may prove useful by helping to improve an organism's ability to survive. For example, a mutation that alters a digestive enzyme so that a new food source can be more easily digested can give an animal an advantage in a competitive environment. It has a better chance of survival and a greater chance of transmitting its new gene to the next generation.

Mutagenic Agents
There are a number of agents that can upset the normal passage of genetic information from one generation to another. **Radiation** of several types – X-rays, gamma rays, and ultraviolet light – has been shown to release subatomic particles that can penetrate cells and cause chromosomal changes. Mutations are also induced by several chemicals. Mustard gas used in World War I, nitrous acid, dimethyl sulphate, coal tar derivatives, and a number of medical and non-medical drugs have been found to produce chromosomal changes.

Here are the major points about mutations:

1. Mutations alter an organism's genetic code.

2. Mutations are unusual events.

3. Mutations are more often harmful than beneficial.

4. Mutations happen randomly (by chance).

5. Mutation defects cannot at present be cured.

6. Mutations are transmitted to the next generation.

7. There are many known causes of mutations.

8. Mutations in bacteria and other organisms can be experimentally manipulated by scientists in a process known as gene splicing.

BIOTECH

Karyotypes

It is sometimes necessary, for diagnosis or for genetic counselling, to examine a person's chromosomes. White blood cells from that person are cultured in the laboratory and allowed to divide. When cell division reaches metaphase, a chemical (colchicine) is added to stop the cell from dividing further. Special stains are added to prepare the materials for microscopic examination. The prepared slides are then photographed and enlarged. Each chromosome in the photograph is examined for size, shape, and banding, and sorted into matching pairs. The pairs of chromosomes are then taped to a chart as shown. This chart is that person's **karyotype**.

From the karyotype, an extra or missing chromosome can be quickly identified. Figure 10.11 shows a karyotype of a Down Syndrome baby with an extra chromosome 21 and a missing chromosome 5. Compare this with the karyotype of a normal male shown here.

When concern over the health of a fetus warrants amniocentesis, a sample of amniotic fluid is withdrawn from the uterus and centrifuged to separate any fetal cells which are present. The cells are then cultured and examined as described above.

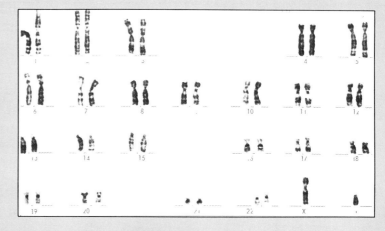

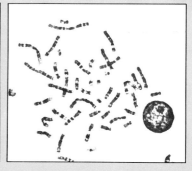

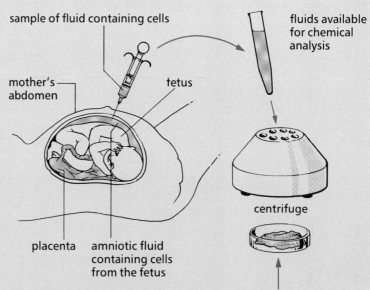

Cells that settle out during centrifuging can be cultured, used for sex determination, or examined for chromosomal defects.

Genetic Factors

Some genetic defects arise during the process of cell division when an extra chromosome or too few chromosomes might be produced. Chromosomes can also be broken and perhaps reattach in another place, deleting or duplicating the information they carry. Normally, chromosomes are present in the nucleus of each cell in pairs, but in some birth defects, the nuclei contain an extra chromosome in addition to the normal pair. This condition is known as **trisomy**.

Trisomy

The extra chromosome (or chromosomes in some cases) in the cells of the embryo are the result of an error that occurs during cell division of the egg or sperm from which the embryo is formed. The most common cause of trisomy is the failure of a pair of chromosomes to separate during the production of sperm or ovum cells. This lack of separation is known as **meiotic nondisjunction**. Normally, a protein filament pulls each chromosome of a pair away from its partner, drawing them to opposite ends of the cell. If one filament is broken, both chromosomes will be drawn to the same end. As a result, a sperm (or ovum) will be formed with two identical chromosomes present. If an embryo results from the union of this sperm with an ovum, that embryo will have trisomy. Another sperm from this male will lack this particular chromosome entirely.

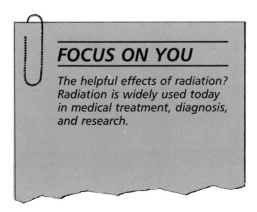

FOCUS ON YOU

The helpful effects of radiation? Radiation is widely used today in medical treatment, diagnosis, and research.

When trisomy occurs in chromosome 21, it results in a condition known as **Down Syndrome**. People with this defect are mentally retarded. The face is typically rather broad and flat, the eyes are slanted, and the tongue appears larger than normal. (See Figure 10.11.)

Figure 10.11
Trisomy. Karyotype of a person with Down Syndrome. Chromosome 21 shows trisomy and a number 5 chromosome is missing. The XX chromosomes show this is the karyotype of a female.

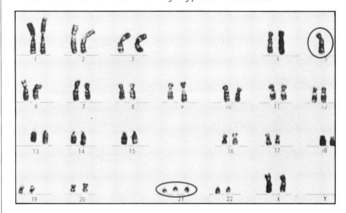

The Inheritance of Diseases and Defects

When a disease is transmitted from one generation to another, it is known as a **hereditary disease**. There are at least 4000 hereditary diseases, including such ailments as muscular dystrophy and Huntington's chorea (which attacks nerve cells in the brain). Cystic fibrosis is an often fatal disease of childhood. The disease affects the pancreas and the bronchioles of the lungs. Cystic fibrosis is a recessive trait represented in Figure 10.12 by the letter c. The gene for a normal pancreas and bronchioles is represented by C.

Detecting Hereditary Abnormalities

People with family histories of certain hereditary abnormalities can seek genetic counselling.

Expert advice can free some potential parents from the concern that their child might have some deformity when there is no need to worry. **Genetic counsellors**, after a close study of each family pedigree, can often give accurate assessments of the predicted ratios for normal and defective children in a particular family. This will give them a good idea of what the chances are that they will produce a child with a particular condition. (See Figure 10.12.)

Figure 10.12
The inheritance of diseases and defects

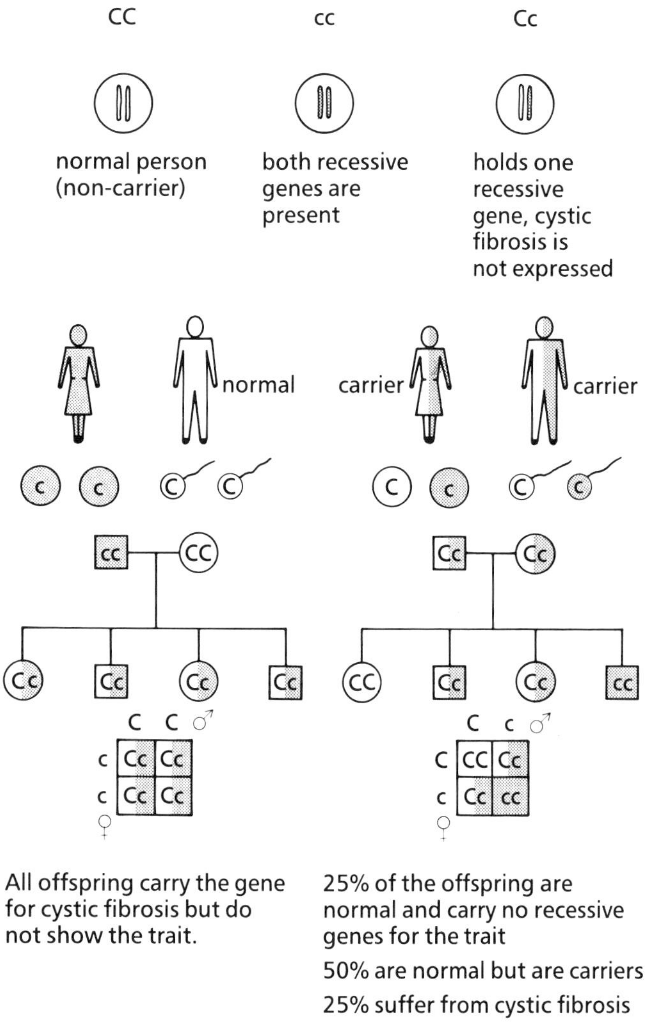

CC — normal person (non-carrier)
cc — both recessive genes are present
Cc — holds one recessive gene, cystic fibrosis is not expressed

All offspring carry the gene for cystic fibrosis but do not show the trait.

25% of the offspring are normal and carry no recessive genes for the trait
50% are normal but are carriers
25% suffer from cystic fibrosis

After the third month of pregnancy, a sample of the cells from the amniotic fluid surrounding the fetus in the uterus can be withdrawn and analysed by a process called **amniocentesis**. The potential for the development of many abnormalities can be recognized at this early stage by examining the cells in this fluid. Abnormalities caused by the presence of the wrong number of chromosomes or by their abnormal appearance can be detected by amniocentesis. Chorionic villi sampling and certain blood tests are also used for early identification of some disorders, while chemical analysis may be used to recognize many enzyme-deficiency diseases. Early detection can, in some cases, save the life of the baby by alerting physicians to the need for special treatments.

CHECKPOINT

1. Define a mutation.
2. Give four examples of mutagenic agents.
3. How can amniocentesis help to determine if a fetus carries a birth defect?
4. What can people who carry the genes for a genetic disorder do to try and prevent their children from having the same trait?

BIOLOGY AT WORK

AGROLOGIST
GENETIC COUNSELLOR
BIOTECHNOLOGIST
INDUSTRIAL PHARMACOLOGIST
AGRICULTURAL TECHNOLOGIST

LIFE · SIGNS

GENETIC COUNSELLING

Barbara and Jim had been married for six years and had a three-year-old son John. Their son suffered from a disorder that produced a number of physical and mental defects. John had small folds of skin in the inner corner of each eye and a large tongue which protruded slightly from his mouth. His hands were small and his fingers stubby. He had a heart defect and he was mentally retarded as a result of the malformation of his brain. These characteristics are typical for people with the disorder known as Down Syndrome.

Barbara and Jim had no illusions as to the future of their son. This was not a condition from which he could recover.

Barbara and Jim wanted to have another child but were concerned that their next child might also have Down Syndrome. After many months of indecision, they visited their physician and explained their concern. The physician said that they were fortunate to be able, in their case, to benefit from known information about the disorder. The physician explained that genetic counselling was available for cases such as theirs.

In advance of their first visit to the counsellor, a form was sent to Barbara and Jim which requested information about their own medical history and that of members of their families.

The visit to the genetic counsellor was a long, informative one. The counsellor, Mary Willet, was a knowledgeable and sympathetic person to whom Barbara and Jim were quickly able to relate. Mary told Barbara and Jim that the presence of John's extra chromosome number 21 was probably not caused by a gene that either of them carried. She referred to the medical histories that they had provided and noted that there were no other cases of the disorder in either family. This suggested that it was very unlikely that Jim or Barbara carried a defective gene. It was more likely that their son's Down Syndrome was caused by a spontaneous event in egg or sperm cell before fertilization. In fact, in more than 90 percent of Down Syndrome cases, the parents do not carry a defective gene. The "mistake" arises unexpectedly and for an unknown reason.

Women over 40 years of age are much more likely to have children with Down Syndrome than younger women. As Barbara was still below the age at which the incidence of Down Syndrome becomes more common, the counsellor said that, in Barbara's and Jim's case, the chance of a second child suffering from the same disorder was as small as that for other parents of their ages.

Barbara and Jim were delighted and very relieved. Mary Willet explained that if Barbara became pregnant, amniocentesis could be carried out early in the pregnancy and, in the unlikely event that the baby did not have the normal number of chromosomes, the pregnancy could be terminated if she and Jim so wished.

Before they left the office, Mary gave them information on various care centres and assistance programs that were available in their area for children with Down Syndrome. Barbara and Jim went home, feeling that they had finally learned what they needed to know. They were excited about the prospect of having another baby and began planning for the future.

Almost a year later, Barbara gave birth to a normal, healthy baby.

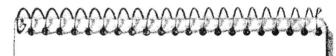

HEALTH CONCERNS

Turner's Syndrome	hemophilia
Down Syndrome	sickle cell anemia
Huntington's chorea	muscular dystrophy

CHAPTER FOCUS

Now that you have completed this chapter, you should be able to do the following:

1. State that DNA is the basic hereditary chemical of life and that it acts by directing the assembly of amino acids and proteins in the cell.
2. Explain the significance of the following terms: chromosome, gene, phenotype, genotype, homozygous, heterozygous, mutation, allele, dominant and recessive.
3. State that the genetic code is the basis of inheritance.
4. Demonstrate examples of inheritance by drawing Punnett squares and solving simple problems, using specific human characteristics.
5. Explain the role of chromosomes in sex determination, fraternal and identical twins, and genetic disorders.

SOME QUESTIONS TO ANSWER

1. Draw a simple diagram of a DNA molecule and label the parts involved. Be sure to show the proper pairing of bases.
2. (a) What are the functions of a chromosome?
   (b) What are the functions of a gene?
3. Explain the following terms:
   (a) homozygous and heterozygous
   (b) homologous chromosomes
   (c) phenotype and genotype.
4. (a) What is the purpose of meiosis?
   (b) How many chromosomes are there in the normal human body cell?
   (c) How many chromosomes are there in a sperm or ovum?
5. Explain how the sex of the offspring is determined genetically.
6. (a) What is a mutation?
   (b) List four different mutagenic agents.
7. (a) List some of the causes of birth defects.
   (b) How can the risk of birth defects be reduced?
8. What is the value of amniocentesis?
9. A woman has cataracts. This is a dominant trait and she is homozygous. The father is homozygous and normal. What will be the predicted phenotypes and genotypes of the children? Use C for cataract, c for normal.
10. In humans, curly hair is dominant over straight hair. If a straight-haired man marries a curly-haired woman, what are the *possible* genotypes and phenotypes of their children? Give the predicted ratios in each case.
11. A pedigree for grey eye colour is given below. Shaded symbols indicate the presence of the trait.

    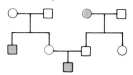

    (a) Is the trait dominant or recessive? Explain how you can tell.
    (b) Give the genotypes of as many of the individuals in the pedigree as you can.
12. If a man with type O blood, whose parents both have type B blood, marries a woman with type AB blood, what will be the theoretical percentage of children with type B blood?
13. In a paternity lawsuit, a woman with type AB blood sues a man with type O blood. The child has the same blood type as the mother. Could this male be the father? Explain your answer.
14. Hemophilia in humans is controlled by a recessive sex-linked gene.
    (a) Could both a father and son be hemophiliac?
    (b) Explain why hemophiliac mothers always have hemophiliac sons.

SOME WORDS TO KNOW

Match each of the descriptions given in the left-hand column with a word shown in the right-hand column. DO NOT WRITE IN THIS BOOK.

1. Two different genes or traits on two homologous chromosomes
2. The appearance or presence of a trait in an organism
3. A unit of heredity and a location on a DNA molecule responsible for a specific characteristic
4. Rod-shaped bodies in the cell nucleus
5. Two similar chromosomes carrying information about the same characteristics
6. A change in a chromosome producing a new inheritable characteristic
7. The presence of an extra chromosome in a cell
8. Two successive cell divisions with only one duplication of chromosomes producing haploid cells
9. Process of removing some amniotic fluid for culture and analysis
10. A single set of chromosomes such as are found in gametes

Select any two of the unmatched words and, for each, give a proper definition or explanation in your own words.

A diploid
B genotype
C phenotype
D gene
E amniocentesis
F dominant
G recessive
H trisomy
I karyotype
J haploid
K chromosome
L homologous
M mutation
N meiosis
O mitosis
P homozygous
Q heterozygous
R allele

SOME THINGS TO FIND OUT

1. Select one of the following topics and prepare a report for presentation to the class:

   the importance of genetic counselling
   karyotyping and its value
   genetic engineering
   gene splicing
   cloning
   hybridization; producing new varieties of plants and animals
   chromosome mapping
   human genome project.

2. What are some of the practical advantages of being able to detect a carrier of a disorder?

3. Many magazines carry articles related to new discoveries and new techniques for using genetics. For example, by changing the genes in specific bacteria, they can be made to produce certain important antibiotics and insulin for us. Select one of these articles and prepare a short summary of what you read.

REPRODUCTION: PRODUCING A NEW INDIVIDUAL

Sex and reproduction. The attraction between male and female. These are concepts which advertisers and moviemakers alike realize will sell their product, whether toothpaste or a thriller.

Yet, as a biological system, reproduction has to be considered in terms of its function. Through the combining of sex cells from two different individuals (fertilization), genetic material is mixed and re-combined to form a new, distinct person. This system is so well-designed that the human population of this planet is expected to double to 12 billion in the next 39 years, including the children and possibly grandchildren of individuals in your class.

KEY IDEAS

- *Male and female reproductive systems are controlled by hormones in the bloodstream.*

- *Menstruation is a natural process in which unfertilized eggs and the lining of the uterus are released from the body each month.*

- *Pregnancy involves the implantation of a fertilized egg in the uterus and the growth of this single cell into a complete human being.*

- *Understanding the human reproductive system lessens your risk of unwanted pregnancy or of contracting a sexually transmitted disease.*

11.1 The Male Reproductive System

The male reproductive system consists of two testes, an arrangement of excretory ducts and accessory glands, and an organ for sperm transfer to the female. The ducts are called the epididymis, vas deferens, and the ejaculatory ducts. The accessory glands and structures include the prostate gland, seminal vesicle, Cowper's gland, and the penis. (See Figure 11.1.)

The Testes

The **testes** begin developing early in the embryonic growth of a male child. They appear initially high in the abdomen, close to the kidneys, then move downward and appear outside the body shortly before birth, in a small sac called the **scrotum**. The external placement of the testes is important for production and development of sperm. Sperm require a lower temperature than other body cells for maturation and development. If the testes fail to descend and remain in the abdominal cavity, the ability to produce viable sperm is reduced and the male may be sterile.

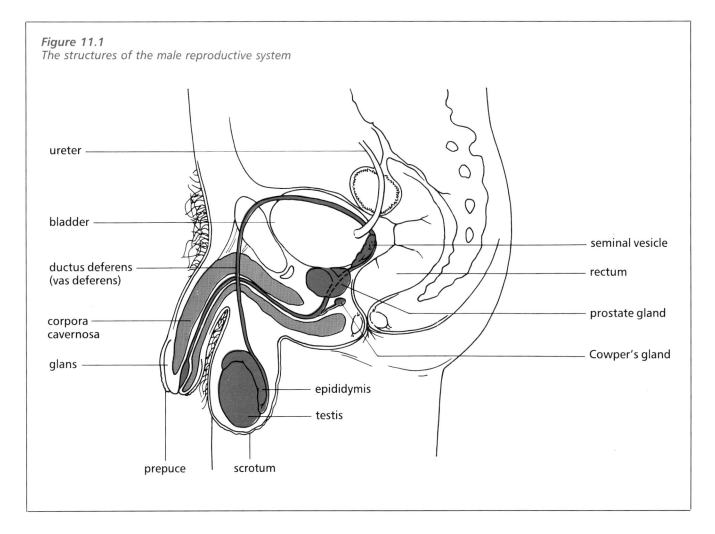

Figure 11.1
The structures of the male reproductive system

The Structure of the Testes

The adult testes are oval bodies about 4 cm in length and 2.5 cm in diameter. They are divided into many compartments containing the tightly packed coils of the **seminiferous tubules**. (See Figure 11.2.) The tubules are lined with germinal epithelial cells. The word germinate means "start to grow". These cells divide by mitosis to form two similar cells. One of these cells undergoes meiosis to produce the haploid gametes, or sperm. (You will recall that only one of each chromosome pair is transferred to each tiny sperm cell.)

The Production of Sperm

There are estimated to be more than 400 seminiferous tubules in each testis, each of which is about 0.5 m in length. It is here that sperm production takes place. Four hundred million sperm commonly leave the male's body in a single ejaculation. The system of sperm production is highly efficient, and furnishes very large numbers of gametes.

The sperm cells begin their development in the germinal epithelium. As they grow, they gradually mature and take on the appearance of the typical sperm. (See Figure 11.3.) They then move slowly toward the centre of the seminiferous tubule, ready to pass down the hollow tube toward the epididymis. **Interstitial cells** located between the coils of the seminiferous tubules secrete the male hormone **testosterone**.

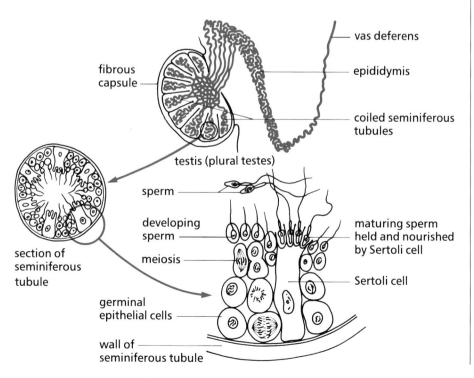

Figure 11.2
The testes and seminiferous tubules. The seminiferous tubules are the site of sperm production.

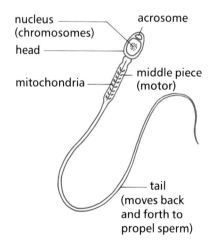

Figure 11.3
The sperm is specially adapted to transport its 23 chromosomes to the ovum and penetrate the protective ovum membrane.

acrosome
– contains an enzyme which helps the sperm to penetrate the egg

head
– contains the nucleus and chromosomes

Hormones and the Male Reproductive System

At puberty, **FSH (follicle stimulating hormone)** supplied by the pituitary gland stimulates the production of sperm in the testes. **ISCH (interstitial cell stimulating hormone)** stimulates the interstitial cells to produce testosterone and thus promotes the development of male organs and secondary sex characteristics. These characteristics, such as a deep voice, facial and body hair, and a muscular frame, are the outward signs of sexual maturity in the male.

The Ducts of the Male Reproductive System

When sperm leave the testes they pass through four distinct ducts:

- the **epididymis**, a long narrow structure on the back of the testes containing many coiled tubes where the sperm continue to mature;
- the **vas deferens**, which loops over the pubic bone and bladder, conveying sperm from the epididymis to the ejaculatory duct;
- the **ejaculatory duct**, which links tubes from the accessory glands and passes through the prostate gland to enter the urethra;
- the **urethra**, which, in the male, carries both urine and semen through the penis.

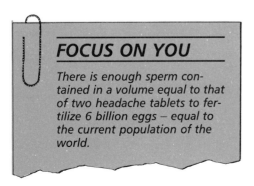

FOCUS ON YOU

There is enough sperm contained in a volume equal to that of two headache tablets to fertilize 6 billion eggs — equal to the current population of the world.

Accessory Structures

There are several glands associated with the male reproductive system, which produce secretions that surround and nourish the sperm. These are the **seminal vesicles**, and **prostate** and **Cowper's glands**. These glands secrete a thick, yellow substance which contains a rich supply of sugars and citric acid. The secretions provide a fluid medium for the sperm, provide them with nourishment, improve their motility, and neutralize the acidity of the ducts, as well as the vagina, through which the sperm must pass. **Semen** is a mixture of sperm and all the secretions of the accessory glands. A single ejaculation contains about 400 million sperm in about 3 to 4 mL of semen.

The prostate gland is a firm muscular organ about 4 cm across. It surrounds the urethra and parts of the ejaculatory duct. Contractions of smooth muscles in the prostate gland help to push the semen out explosively during ejaculation. Sometimes the prostate becomes harder and less flexible and squeezes the urethra that passes through it. Any constriction, or blocking, of this tube prevents the passage of urine. It is quite common for older men to undergo surgery to relieve this constriction. (See Figure 11.4.)

Figure 11.4
Front view of the male reproductive system. Note that, with the exception of the prostate gland, the other glands and structures are all paired.

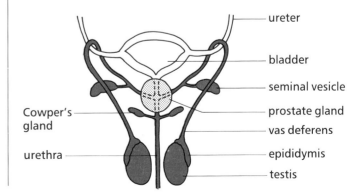

The Penis

The **penis** is the external organ of the male system. It is made up of three masses of spongy tissue, held together by bands of elastic connective tissue. This spongy tissue, the **corpus cavernosum**, forms two cylinders lying side by side. Below these, a third cylinder of spongy tissue carries the urethra through its centre. (See Figure 11.5.) This tissue forms a blunt cone at the end of the penis called the **glans** which is covered with a fold of skin called the **foreskin**. This fold may be removed soon after birth in a process known as *circumcision*.

Figure 11.5
A section through the penis showing the spongy corpus cavernosum

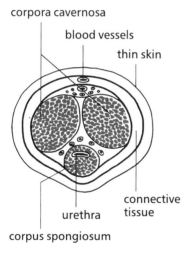

Normally, the penis is soft and relaxed. At this time, the tissues of the spongy corpus cavernosum contain very little blood. During periods of sexual excitement, the blood vessels which supply the penis expand and more blood flows into the spongy chambers producing a higher pressure within the tissues. The tissues then become enlarged, compressing the veins that usually carry blood away from the penis and preventing them from doing so. This building of blood pressure makes the penis firm and erect (erection) and capable of transferring sperm into the vagina.

Ejaculation

Ejaculation consists of a sequence of two events. The first occurs when smooth muscles in the reproductive glands and vessels move semen into the urethra, at a point close to the prostate gland. The second, and main event, involves contraction of a small muscle that propels the semen through the urethra and out through the opening in the penis. After leaving the penis, sperm live for 24 to 72 h.

CHECKPOINT

1. Name the organs of the male reproductive system.

2. (a) Where do the testes originate?
   (b) How do they get down into the scrotum?

3. What is the location of sperm production?

4. Describe the structure of a sperm cell.

5. Name, in order, the ducts that form a pathway for sperm to travel from the seminiferous tubules to the exterior of the body.

6. What functions do the accessory structures perform?

7. Describe how the tissues of the penis are adapted to achieve erection.

11.2 The Female Reproductive System

The female reproductive system consists of a pair of ovaries, two Fallopian tubes, a single uterus, and a vagina. (See Figure 11.6.)

The Ovaries

The two **ovaries** are the primary sex organs of the female. The ovaries, which resemble almonds in shape and size, are located in the back of the abdominal cavity near the rim of the pelvis. They are held in place by ligaments. During pregnancy, as the uterus and the fetus enlarge, the ovaries are pushed aside. The uterus must, therefore, have some flexibility in its attachment. To allow for these changes during pregnancy, the broad ligaments and the ovarian ligaments hold the ovaries loosely to the uterus, and the suspensory ligaments suspend each ovary from the wall of the pelvis.

The Development of the Follicles

Each ovary is contained within an outer layer of special epithelium (germinal epithelium). Inside the ovary, there is a network of connective tissue, which contains small groups of cells called **follicles**. Within each follicle is an egg or **ovum** (*ova* plural) formed by meiosis. At birth, there are some 400 000 tiny follicles present in each ovary. Each ovum, with its surrounding follicular cells, is called a **primary follicle** and is only partly developed at birth. It will remain in this state until a pituitary hormone restarts the development of the follicles at puberty. Only about 400 of these primary follicles will ever reach full maturity and be released, one per month, during the reproductive life of the female. The other follicles will gradually degenerate.

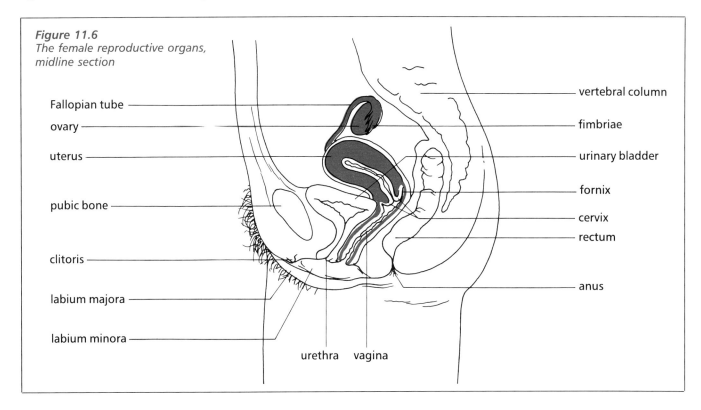

Figure 11.6
The female reproductive organs, midline section

Puberty

Very little development of the primary or secondary sexual features of a female occurs before puberty. At **puberty**, which normally occurs between 10 and 14 years of age, the secondary sexual characteristics develop. The breasts, uterus, and vagina increase in size and mature. Pubic and axillary hair (under the arms) becomes noticeable, and the general contours of the body become more rounded by deposits of fat beneath the skin. Puberty ends with the onset of the first menstrual period, which may occur at anywhere from 10 to 18 years of age. The age at which menstruation begins also varies with different cultures and with such factors as diet. In western cultures, the age of females at first menstruation has been getting steadily younger over the last half century. In both sexes, puberty is due to increased pituitary secretions, although what triggers these secretions is not known.

After puberty, under the influence of FSH (follicle stimulating hormone), a few primary follicles develop into **secondary follicles**. The follicular cells surrounding the ovum multiply and produce fluid which accumulates in these cells. The ovum also changes during the development of the follicle. By the process of meiosis, the number of chromosomes present is halved, so that each cell contains one chromosome of each original pair of chromosomes. One cell remains, the other degenerates.

Although a number of follicles may start to develop each month, usually only one will reach maturity. The mature follicle is greatly enlarged and causes a bulge in the side of the ovary. Eventually, this mature follicle ruptures the ovarian membrane in the process known as **ovulation** and begins its journey along the Fallopian tube. (See Figures 11.7 and 11.8.)

Figure 11.8
Developing ovum in the ovary

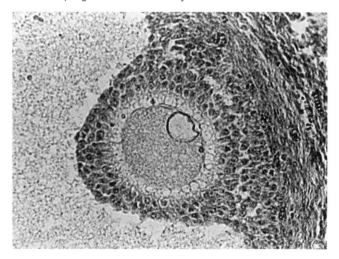

Figure 11.7
The sequence of follicle development, ovulation, and changes to the corpus luteum and corpus albicans

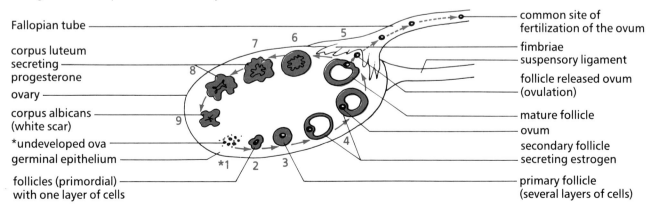

Female Hormones Produced by the Ovaries

As the follicular cells develop, they produce the hormone **estrogen** which is secreted into the bloodstream for transfer to other organs. After ovulation, some of the follicular cells are left behind in the ovary. Under the influence of another pituitary hormone, LH (luteinizing hormone), these remaining cells become organized into a yellow mass, called the **corpus luteum**. The corpus luteum produces the female hormone **progesterone** and some estrogen, which influence maintenance of the endometrial lining in the uterus and pregnancy.

If fertilization of an ovum occurs, progesterone will continue to be secreted for two to three months. After ten days, if the ovum is not fertilized, the corpus luteum shrinks and degenerates until it is just a small white patch of scar tissue, known as the **corpus albicans** (*corpus* means body and *albicans* means white).

Activity 11A

Examination of Reproductive Tissues and Cells

Using prepared material, look for the special structures associated with reproductive tissues and cells.

Materials

prepared slides of rat or human testis, sperm cells, rat or human ovary with Graafian follicle and corpus luteum

Procedure

1. Examine the cross-section of a seminiferous tubule of the rat testis under the low power of your microscope. Turn to high power and examine a section of the tubule wall. Refer to Figure 11.2. Sketch any four specialized cells. Repeat the above procedure with the human sperm cell. *Sketch and label the acrosome, head, middle piece, and flagellum.*

2. Observe the slide of the rat/human testis under low power. Observe the cells between the seminiferous tubules. These are interstitial cells. *What is their function?*

3. Examine the slide of the rat or human ovary at $100\times$ magnification. There is no need to use your high power. *Sketch a mature follicle with its oocyte (future egg) and label it.* Examine the corpus luteum under low power. Refer to Figure 11.7.

Questions

1. (a) Which cells are haploid?
   (b) Which cells are diploid?

2. What happens to the spermatids as they migrate into the centre of the tubule?

3. What is the function of the acrosome?

4. What is the function of the follicle cells?

5. When does the corpus luteum appear in the female menstrual cycle and what is its primary function?

6. What would be the relative level of each hormone just prior to ovulation?

The Fallopian Tubes

The **Fallopian tubes** (*oviducts*) conduct the ovum from the ovary to the uterus. They are attached to the upper part of the broad ligament and are 10 to 12 cm in length. The end nearest to the ovary has a funnel-shaped opening surrounded by a fringe of tiny projections called **fimbriae**. (See Figure 11.9.)

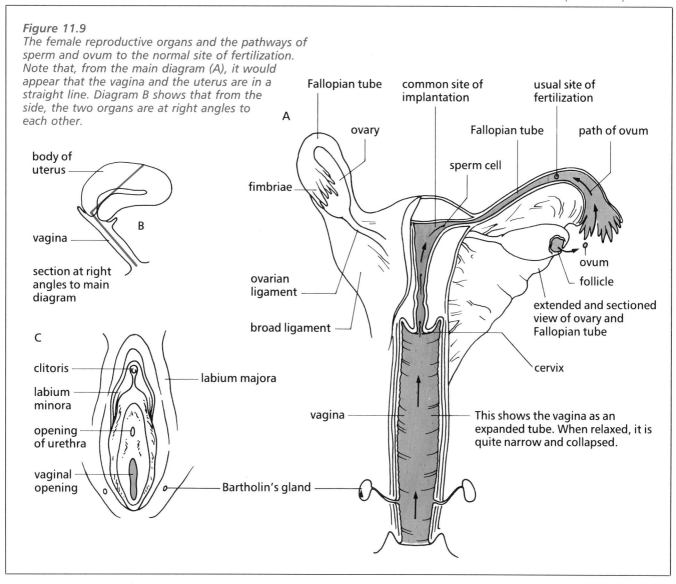

Figure 11.9
The female reproductive organs and the pathways of sperm and ovum to the normal site of fertilization. Note that, from the main diagram (A), it would appear that the vagina and the uterus are in a straight line. Diagram B shows that from the side, the two organs are at right angles to each other.

The ovary is not directly attached to the Fallopian tube. The fimbriae almost entirely enclose the ovary and when a mature follicle is released, these fimbriae "catch" it and, with wavelike movements, sweep it into the Fallopian tube. Since the ova have no means of propulsion, they are moved along the tube by peristaltic movements of the tube muscles. There is also a current of fluid moved along by the sweeping action of tiny cilia which line the Fallopian tubes. Many secretory cells present in the walls of the tubes help to produce the fluid in which the follicle is carried. The other end of the Fallopian tube leads directly into the uterus.

The Fallopian tube is the usual site of fertilization. The follicle containing the ovum will be carried along the tube over a period of three to five days. If the ovum is not fertilized it will normally die within 24 h; however, the life span of an ovum varies considerably in different women.

The life span of the ovum is so short that normally the sperm must reach it while it is still travelling along the first third of the Fallopian tube to ensure that fertilization takes place. Furthermore, the fertilized ovum must reach the uterus within a fixed period or it will be rejected, because progesterone levels will have started to decline.

The Uterus

In women who have not borne children, the **uterus** is a hollow, thick-walled, and muscular organ, about the size and shape of a large pear. The upper part of the uterus is free and movable. It rests on the top of the urinary bladder. The bottom part of the uterus is embedded in the floor of the pelvis, between the bladder and the rectum. The uterus is securely attached to the walls of the abdomen. During pregnancy, however, the ligaments that perform this function are able to adjust to the increasing size of the uterus. The ligaments support the uterus firmly so that any unusual or abrupt body movements do not endanger the safety of the developing fetus.

> **FOCUS ON YOU**
>
> The cervix is a common site for cancer in women. Like many diseases, if treatment is started when the disease is in its early stages, considerable success in effecting a cure can be achieved. The Pap smear is a simple process. It involves the physician wiping off a few cells from the surface of the cervix and examining them under a microscope. If cancer cells are present, they are quickly identified. With regular examinations, the disease can be detected and treated while in its very early stages.

The upper part of the uterus, or body, is much larger than the narrow lower end which forms the **cervix**. The cervix, which is like the neck of a bottle, opens into the vagina. The cavity inside the uterus forms the shape of a wide capital T. At the top of the uterus, between the openings of the two Fallopian tubes, is an area known as the **fundus**. This is a common site for the implantation of a fertilized ovum.

The uterus contains three layers – an outer layer of connective tissue (the **peritoneum**), a middle layer of smooth muscle tissue (**myometrium**) which makes up much of the uterine wall, and an inner layer (**endometrium**).

The Vagina

The **vagina** is a muscular, collapsible tube, about 8 cm in length. The upper end of the vagina encloses the cervix, which opens into the uterus where the cervix extends into the vagina. The upper end of the vagina forms a narrow circular recess around the cervix and between the walls of the vagina. At the open end of the vagina is a thin folded membrane called the **hymen**. The hymen partially covers the vagina, leaving a small central opening.

The lining of the vagina is composed of smooth muscle and connective tissue. It is covered with a mucous membrane which has many folds. These folds permit the enlargement of the vagina during childbirth, when it serves as the birth canal. The vagina is the organ which receives the penis during sexual intercourse.

The External Genital Organs

These features have a minor role compared to the primary organs already described. The external genital organs consist of the following parts:

- the **mons pubis**. This is a rounded pad of fatty tissue in front of the pelvic bones of the pubic symphysis. It has a thick covering of skin and, after puberty, is covered with hair.

- the **labia majora**. These are the two fatty folds of skin which surround the opening to the vagina. These folds contain many sebaceous sweat glands.

- the **labia minora**. These are two small folds of skin inside the labia majora. At the top, the folds come together and form a small hood (prepuce) that partly covers the clitoris. The folds of these labia enclose two openings – the urethra from the urinary bladder and the opening of the vagina.

- the **clitoris**. This is a small structure made of erectile tissue, richly supplied with blood vessels and nerve endings. It is important in the sexual stimulation felt by the female.

CHECKPOINT

1. Describe the development of a follicle within the ovary.
2. What hormones act on the ovary?
3. What happens when ovulation occurs?
4. (a) What hormones are produced by the ovary?
   (b) Where do these hormones have their effect?
5. What tubes carry the ovum to the uterus?
6. Describe the shape, structure, and location of the uterus.

11.3 The Menstrual Cycle

Within the uterus, the endometrium forms a soft protective "bed" where a fertilized ovum can become attached and develop. It provides support and nourishment for the tiny embryo in its initial stages. If the ovum is not fertilized, this lining is not needed. As it breaks down, it is released in the menstrual flow.

The menstrual cycle has three distinct phases. The **menstrual phase** covers the time of menstrual flow, during which the endometrial wall is shed. The **follicular phase** begins when the pituitary gland releases follicle stimulating hormone into the bloodstream. FSH stimulates the development of a new follicle in the ovary. As the follicle develops, it releases estrogen into the bloodstream which, in turn, stimulates the lining of the uterus and other organs to prepare for the possible arrival of a fertilized ovum.

In the **luteal phase**, the ovum has been released, the endometrium lining has been prepared, and progesterone, a hormone from the ovary, is produced to maintain the condition of the endometrium until it is determined if the ovum is fertilized or not. Progesterone and other hormones are produced by the cells of the corpus luteum left behind in the ovary after the ovum is released. (See Figure 11.10 and Table 11.1.)

Hormonal Feedback

Progesterone causes the endometrium to develop a support system for the ovum, with a rich supply of blood vessels and glands to provide nutrients. If the ovum is not fertilized, high levels of estrogen and progesterone are recognized by the pituitary gland as a signal that luteinizing hormone is no longer required. As the supply of LH diminishes, the corpus luteum dries up and forms a pale scar, called the corpus albicans, and the supply of progesterone is cut off. FSH is also suppressed by high levels of estrogen. This is another example of hormonal feedback.

If a fertilized ovum is received and implanted in the uterus, the cells around the developing ovum produce another hormone called **chorionic gonadotrophin**. This hormone enables the corpus luteum to maintain progesterone and estrogen secretions until the embrionic placenta starts its own hormone production. Chorionic gonadotrophin passes into the bloodstream and some appears in the urine. Pregnancy tests can detect this hormone about 10 to 14 days after the first missed menstruation.

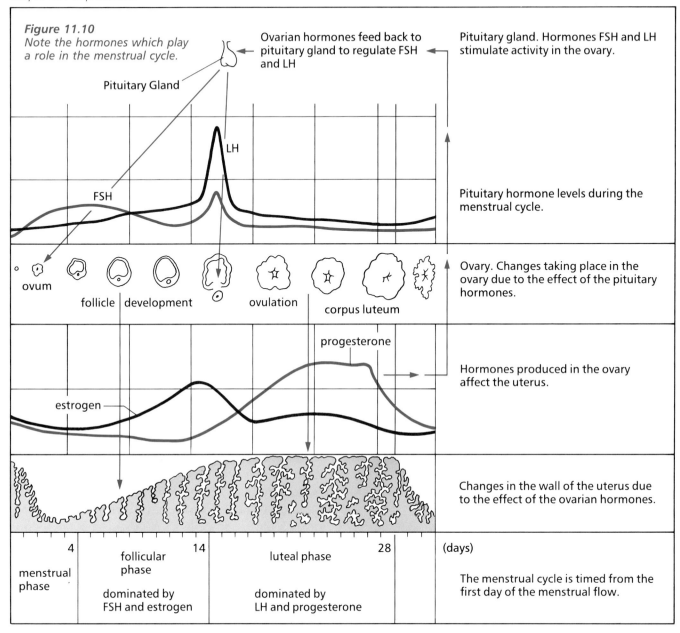

Figure 11.10 Note the hormones which play a role in the menstrual cycle.

CHECKPOINT

1. Three different parts of the body are involved in the menstrual cycle and are connected by hormone messengers. Name these three parts.

2. What is the function of the endometrium?

3. What are the three phases of the menstrual cycle?

4. What effect do estrogen and progesterone have on the endometrium of the uterus?

Table 11.1 Summary of the Organs and Hormones Involved in the Menstrual Cycle

Organ	Organ Function, Hormones Produced	Organ Affected	Effect of Hormone
Pituitary gland	FSH (follicle stimulating hormone)	ovary	Stimulates the growth of a new follicle in the ovary
	LH (luteinizing hormone)	ovary	Controls the development and secretions of the corpus luteum
Ovary	Produces ovum		
	Estrogen from the follicular cells	uterus	Initiates the thickening of the endometrial lining to receive the ovum
	Progesterone and estrogen by the corpus luteum	uterus and other organs	Maintains the endometrium
Uterus	Prepares soft "bed" to receive ovum		Quantity of progesterone and estrogen in blood direct pituitary to continue secretion of LH if ovum is fertilized, or discontinue secretion if ovum is unfertilized
	Site where fetus develops		
	Provides feedback to pituitary whether or not ovum is fertilized	pituitary	

11.4 Copulation and Fertilization

If fertilization is to take place, sperm must be deposited in the vagina close to the time of ovulation. **Fertilization** is defined as the union of a sperm and ovum. It normally occurs in the Fallopian tube (within 24 h of ovulation) when the ovum is about one-third of the way along the tube. Sperm are transferred to the vagina and then to the Fallopian tube by copulation. When the male becomes sexually excited, blood vessels supplying the penis enlarge and fill the spongy sinuses, thereby causing the penis to increase in size and become erect. When erect, the penis is able to penetrate the vagina.

Back and forth movements of the penis inside the vagina increase sexual tension in the male until ejaculation occurs. The walls of the male ducts contract and push the semen rapidly through the vas deferens, into the urethra, and from the penis into the vagina.

Sexual excitement is evident in the female as in the male. Touch and sensory stimulation arouse nerve endings in the female genitalia, especially in the clitoris, which becomes erect. Impulses to the walls of the vagina cause the secretion of lubricating fluids, which facilitate sexual intercourse. Reflexes in a woman initiate emotional and muscular reactions similar to those occurring in the male. This response is referred to as **orgasm** (or climax).

Fertilization

After the sperm have been deposited in the vagina at the opening of the cervix, they move forward, partly due to a whiplike motion of their flagella, and partly due to muscular movements of the uterus. The sperm pass from the vagina, through the uterus, and into the Fallopian tubes, where union with the ovum may take place. Sperm must be present in the female genital tract for about 4 to 6 h before the ovum can be fertilized. This amount of time is required for an enzyme (hyaluronidase) contained in the acrosomes of sperm to dissolve part of the membrane that protects the ovum.

It appears that many thousands of sperm must be present to produce enough of this enzyme to dissolve the ovum's protective membrane. Only one sperm, however, will penetrate and achieve fertilization. Immediately after this first sperm has penetrated, a chemical membrane barrier is instantly formed, which prevents the entrance of any other sperm present. (See Figure 11.11.) After entry, the tail of the sperm is lost and the 23 chromosomes which are present in the sperm unite with 23 chromosomes present in the ovum. This results in the normal number of 23 *pairs* of chromosomes being established in this, the first cell of a new individual.

Figure 11.11
Sperm surround a portion of the ovum in the Fallopian tube. The large cells are follicular cells, magnified 2000×.

CHECKPOINT

1. *Where do sperm usually meet with an egg or ovum?*
2. *How long must sperm be present in the female genital tract before the ovum can be fertilized?*
3. *Describe what happens when large numbers of sperm surround the ovum at the time of fertilization.*
4. *Why is it important that only one sperm fertilize the egg?*

11.5 Pregnancy and Early Development

The first cell formed by the union of a sperm and ovum is known as the **zygote**. The zygote possesses a full diploid set of chromosomes, half from each parent. Fertilization usually takes place in the first third of the Fallopian tube and the zygote is then moved slowly down the tube toward the uterus, a journey that usually takes two or three days. Immediately after fertilization, before it has even reached the uterus, rapid changes begin to take place in the zygote. The cell divides repeatedly (mitosis), doubling the number of cells present with each division, until a hollow ball of cells is formed, known as a **blastocyst** (blas-toe-sist). (See Figure 11.12.)

As the fertilized ovum passes through the Fallopian tube, a membrane, known as the **chorion** (koe-ry-on), forms around the mass of dividing cells. The membrane is covered with tiny villi secreting enzymes which help to form a path into the tissues of the endometrium so that the blastocyst can nestle close to the maternal blood supply. The blastocyst gradually sinks between the soft cell tissues of the thickened endometrial wall of the uterus. When this process, called **implantation**, has occurred, a successful pregnancy has been established.

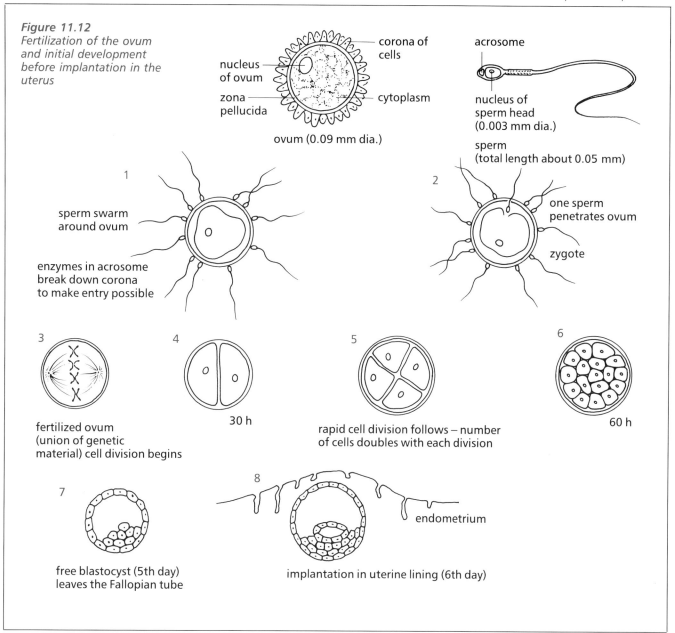

Figure 11.12 Fertilization of the ovum and initial development before implantation in the uterus

The implanted blastocyst continues to divide rapidly, producing more and more cells. Eventually two major groups of cells are produced. One group will eventually become the embryo (the developing baby). The second develops into several structures and tissues that support the growth of the embryo while it is in the uterus.

Structures Within the Uterus

The **amnion** is a thin sac filled with a watery fluid. The embryo develops inside this sac, supported and well-protected from bumps by the fluid. (See Figure 11.13.) The **placenta** develops in close contact with the wall of the uterus. It contains

Figure 11.13
Uterus containing the early embryo and showing the amnion, yolk sac, placenta, and uterine wall

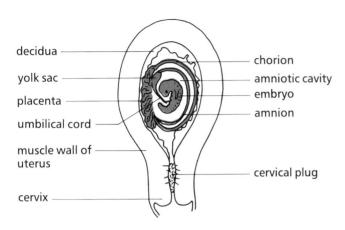

blood vessels both from the mother's circulatory system and from the embryo's blood system. Since there is no direct connection between these two systems, the blood does not mix. In Chapter 5, the problems that can arise when two different blood types are mixed are discussed. Commonly, a developing baby has a blood type different from that of its mother; for this reason, the systems are kept separate. Nutrients, oxygen, and other substances in the mother's blood are transferred to the embryo across the placenta in which capillaries of both systems come close to one another. (See Figure 11.14.) Wastes and carbon dioxide resulting from the cell activities of the embryo diffuse out of its blood vessels and pass into the mother's circulatory system for disposal. The placenta then acts as both a barrier and a bridge. It is the exchange site for all materials entering or leaving the embryo.

The nutrients in the mother's blood are swiftly passed on to the embryo through the placenta. Unfortunately, this system also works for substances such as alcohol and nicotine. The placenta does not have the ability to selectively prevent passage of some substances carried in the mother's blood. Such viruses as German measles or AIDS, and such drugs as thalidomide can pass through the capillary walls and cause deformities in the offspring.

A newborn baby also receives a set of temporary immunities from its mother. If the mother is exposed to certain diseases and becomes immune to them, that immunity, through the placenta, is passed on to the developing baby. The antibodies remain in the blood of the baby for six months or so after birth, which gives it time to become strong enough to withstand many infections. In time, it will build up its own set of protective immunities. A baby has no functional immune system of its own until approximately six months of age. Mother's milk is believed to be an important factor in developing the immune system.

The placenta does a great deal more than just pass nutrients and wastes between the mother and developing baby. When the human body contains foreign proteins, whether it is a wood sliver, bacteria, or even a transplanted organ, it tries to get rid of it. White blood cells in particular perform this function. The baby inside the mother can also be viewed as "foreign protein". If it

Figure 11.14
The placenta forms a barrier between the fetus and the mother, preventing blood cells and large proteins from crossing from one system to the other. Why is this important?

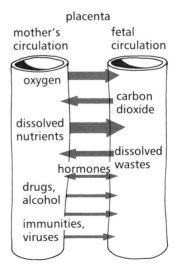

were not for the placenta, the mother's immunological protective system would try to get rid of it.

The **yolk sac** functions only temporarily in human development, although it plays a major role in animals that hatch from eggs. In humans, it produces blood cells for the embryonic circulatory system before the true blood-producing tissues are available.

The last of the structures associated with the embryo is the **umbilical cord**, which connects the developing baby to the placenta. It carries one vein and two arteries. Arteries are defined as vessels that carry blood *away* from the heart and veins carry blood *to* the heart. In adults, blood travelling to the heart through veins is deoxygenated. In the developing baby, the *vein* in the umbilical cord brings oxygen from the mother. Once this oxygen has been used in the developing baby's system, the blood is carried away from its heart in arteries and transported to the placenta where it picks up more oxygen and discharges its burden of carbon dioxide. The primary reason for this difference in the type of vessel carrying oxygen is, of course, the fact that the lungs of the fetus are not working or filled with air; oxygen is obtained through the placenta using the lungs and bloodstream of the mother.

Development of the Embryo and Fetus

When looking at a group of teenagers or adults, it is hard to realize that they were once only 3.5-kg babies. Between birth and about 20 years of age, body mass increases more than 20 times and height about 3.5 times. Even this rate of change, however, is extremely small compared to the changes that take place during the first two months after conception. In the first eight weeks, the overall length of the embryo increases 240 times and mass increases over 1 million times as the single cell develops into a miniature baby.

The nutrient supplies, vital to this amazing growth, flow through the placenta to reach the embryo; thus, the mother's diet during pregnancy has direct and important effects upon the developing baby.

The group of cells which form the embryo undergo such vast and complex changes that only a brief summary is possible here. After dividing many times and forming the supporting tissues already mentioned, the blastocyst soon separates into three layers – the **ectoderm**, **mesoderm**, and **endoderm**. These layers then differentiate (become different types of cells with different functions) into a variety of tissues. (See Table 11.2.)

Table 11.2 Some of the Major Tissues and Organs That Develop from the Three Germinal Layers of the Ectoderm, Mesoderm, and Endoderm

Germ Layer	Tissues and Organs (By Differentiation)
Ectoderm	Nervous system, epidermis, parts of the eye, salivary glands, pituitary gland, adrenal medulla, skin, hair, and nails
Mesoderm	Connective tissue, bone, muscles, cartilage, blood, blood vessels, lymphatics, spleen, adrenal cortex, parts of the reproductive organs
Endoderm	Epithelium of the digestive tract, linings of the lungs and respiratory passages, liver, pancreas, thyroid, parathyroid, and thymus glands

By the end of the first month of pregnancy, the embryo is little more than 3 mm in length, yet its heart has been pumping blood since the eighteenth day after conception. It has the beginnings of eyes, a spinal cord and nerves, lungs, stomach, intestines, liver, and kidneys.

In just eight weeks, the embryo begins to form the first bone cells. This change marks such a major transition that the term embryo is dropped and the word **fetus** is adopted to describe the developing baby. The total period of time that the embryo and fetus are in the uterus is known as **gestation**. In humans, this is 280 days from the beginning of the last menstrual period. (See Figure 11.15.)

Figure 11.15
An ultrasound image of a human fetus

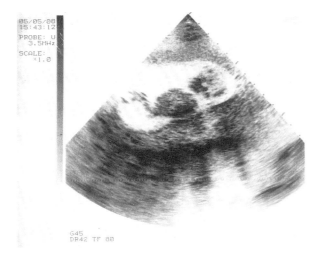

CHECKPOINT

1. *Define implantation.*
2. *Explain the function of the amnion.*
3. *Give three functions of the placenta.*
4. *Why is an umbilical cord necessary?*
5. *What are the three germ layers that form in the early development of the embryo?*
6. *What developmental changes mark the transition from an embryo to a fetus?*

FOCUS ON YOU

Identical twins develop from a zygote which splits into two at a very early stage after fertilization has occurred. Each part develops into a separate individual, yet each individual possesses the same genetic information. Fraternal twins occur when two ovum are released together rather than one. As they are separate cells, they are fertilized by different sperm. Such twins have no greater chance of sharing genetic information than any other siblings.

11.6 Birth and Lactation

When the fetus is finally ready to make its entrance into the outside world, a number of important changes take place in the body of the mother. (See Figure 11.16.) Estrogen, which stimulates the changes in the lining of the uterus during the menstrual cycle, also causes the muscles of the uterus to contract. Progesterone prevents uterine contractions from taking place. During pregnancy, these hormones are kept balanced. Near the time of birth, the level of progesterone drops below a threshold level and the muscles in the walls of the uterus begin to contract signalling the onset of labour.

Figure 11.16
Section through the abdomen. The fetus is shown in position with the head down ready to enter the birth canal at the start of labour.

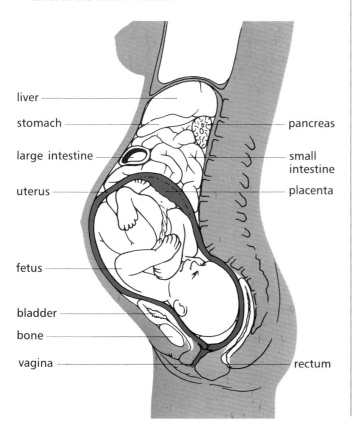

The activity that actually initiates the birth process is still not clearly understood, but hormones produced by the fetus greatly influence the levels of progesterone in the mother. Oxytocin from the pituitary gland aids in stimulating the contractions, which occur in waves, rather like the peristaltic waves in the digestive tract. The contractions are irregular at first, true labour not beginning until the contractions are regularly spaced. As the contractions increase in strength, the amnion which surrounds the fetus breaks and the amniotic fluid is discharged through the vagina. This is sometimes called breaking water.

The flow of amniotic fluid is preceded by the expelling of a mucous plug which has been present in the cervix during pregnancy. (See Figure 11.13.) The passing of this plug, which may be accompanied by some flecks of blood and the waters of the amnion, is usually painless but is soon followed by the onset of true labour.

The muscles of the cervix and the vagina gradually relax and increase the size of the birth canal opening. (Dilation increases the opening of the cervix from about 3 mm to 10 cm.) The ligaments which hold the pubic bones together also relax slightly to allow easier passage of the baby. The hormones prostoglandin and oxytocin both play a role in the relaxation of muscles and the dilation of the cervix before birth.

This first phase of labour may last for about 16 h in the case of the birth of a first baby, but later births usually occur more rapidly. Strong contractions of the uterus eventually force the fetus down toward the cervix and vagina.

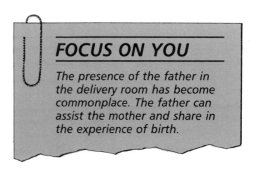

FOCUS ON YOU

The presence of the father in the delivery room has become commonplace. The father can assist the mother and share in the experience of birth.

Delivery

Each contraction moves the fetus further down and forces the head into the opening of the cervix. The second stage of labour involves the expulsion of the fetus from the uterus.

The delivery of the baby is followed a short time later by expulsion of the placenta (afterbirth). The contractions which separate the placenta from the wall of the uterus force it out of the body and also constrict any blood vessels that may be torn to prevent loss of blood. (See Figure 11.17.)

The baby is born still attached to the umbilical cord and the placenta. The umbilical cord must now be tied and cut on the side away from the baby's abdomen. A jellylike substance present within the cord expands when exposed to air sealing off the arteries and veins, so that there is no loss of blood.

For the baby, delivery is a very disturbing experience. Up until this moment, it has been in darkness. Now it is suddenly exposed to light. Its warm, underwater world is suddenly exchanged for a dry one, which is many degrees

Figure 11.17
The delivery

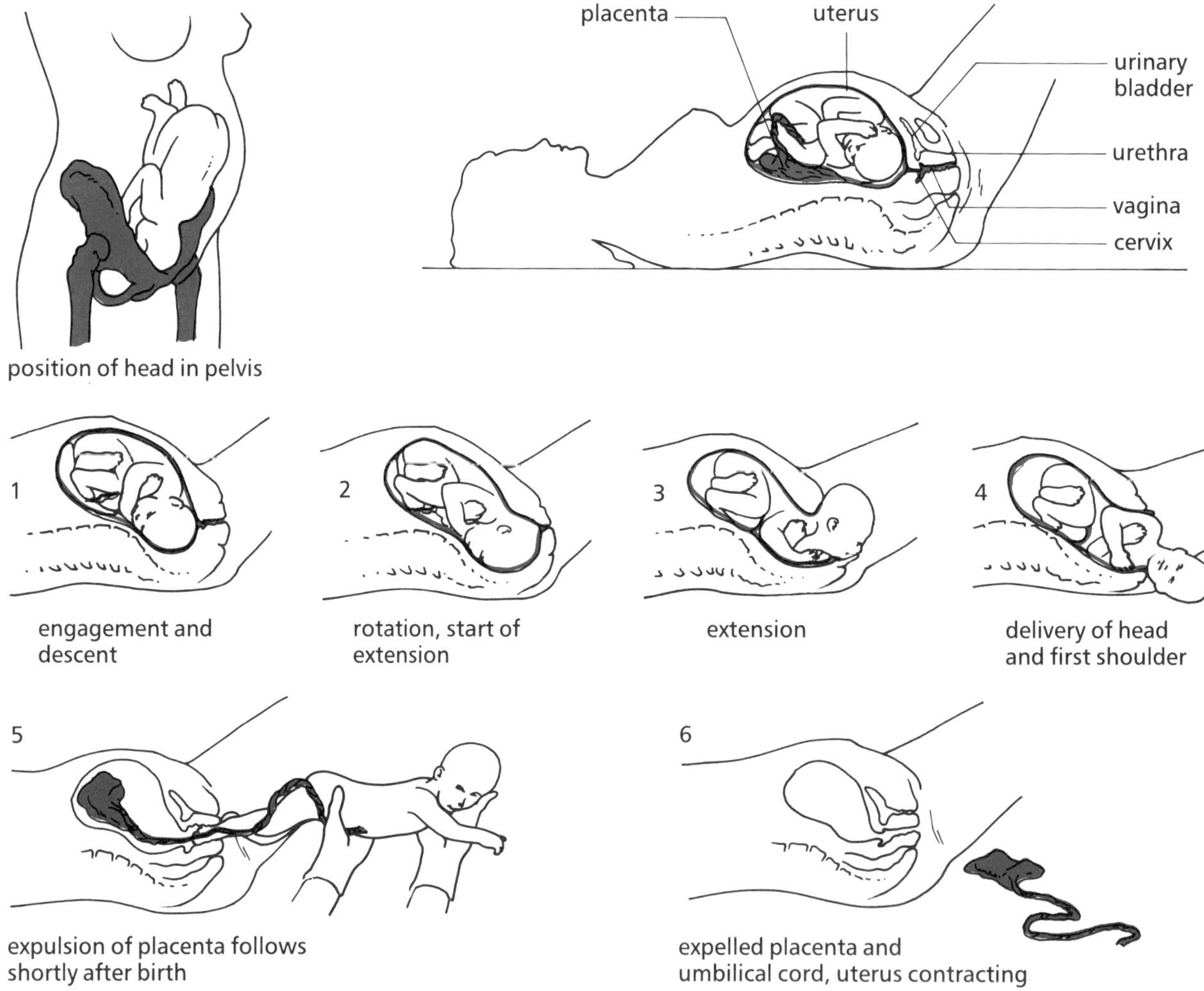

position of head in pelvis

1. engagement and descent
2. rotation, start of extension
3. extension
4. delivery of head and first shoulder
5. expulsion of placenta follows shortly after birth
6. expelled placenta and umbilical cord, uterus contracting

lower in temperature. Its heart must now send blood through its lungs to pick up its own supply of oxygen. These lungs are wet from long immersion in the amniotic fluid. Sometimes, if the baby does not start to breathe readily, the doctor attending the delivery will give the baby a sharp slap, causing it to cry and expand its lungs. The baby must now set its own systems in action to rid its body of wastes.

Caesarean Section
Sometimes there are reasons why a vaginal delivery is not advisable or not possible. Perhaps the mother has a very small pelvic opening, for example, or a baby's umbilical cord has become pinched and the baby's oxygen supply is in danger. In such cases, a **Caesarean section** is performed. An incision made through the abdomen and into the uterus allows the baby to be lifted out rather than have it pass through the birth canal. If it is desired, the mother may be anesthetized below the waist only so that she will be conscious during the birth.

The Mammary Glands: Lactation

The mammary glands, or breasts, are located on the surface of the pectoral muscles between the second and sixth ribs. Their shape and size depend on the presence of various amounts of glandular and adipose tissue. The mammary glands are designed to provide nourishment for babies by secreting milk, in a process known as **lactation**. The size of the breasts does not determine the amount of milk produced.

The mammary glands have a structure similar to that of sweat glands. (See Figure 11.18.) Each is composed of 15 to 20 lobes made up of glandular tissue and fat. Each lobe is connected to the nipple by a **lactiferous** (lak-ti-fer-us) **duct**. Lying along these ducts are sinuses (small chambers) which store the secreted milk until the baby suckles.

The nipple contains many openings through which the lactiferous ducts empty to the exterior. Around the nipple is a circle of darker pigmented skin called the **areola**. It has a rough,

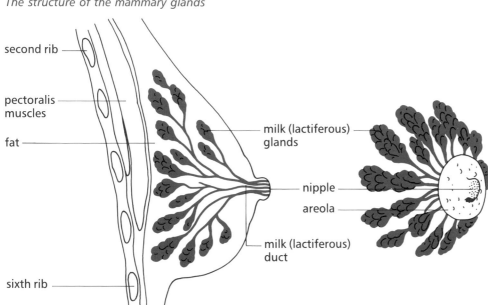

Figure 11.18
The structure of the mammary glands

bumpy surface because of the modified sebaceous glands present. At puberty, the female breasts develop, the ducts become larger, and considerable amounts of fat are deposited. The size of the nipple and areola ring increases and becomes darker in pigmentation. These changes are influenced mainly by the secretion of estrogen and progesterone by the maturing ovaries. Estrogen and prolactin act together during mammary gland development and in lactation. During pregnancy, progesterone adds to the increase in the milk production of the glands.

CHECKPOINT

1. Describe how estrogen and progesterone act together to control the contractions of the uterus which are part of labour.
2. List the major events or stages of a successful birth.
3. What environmental changes does a newborn baby face?
4. What is lactation?

11.7 A Healthy Baby: The Role of the Environment

Sometimes something happens during pregnancy that has little physical effect on the mother, but has a drastic effect on the baby she carries. About one in 16 babies born has a genetic or non-genetic defect. In many instances, it is difficult, or impossible, to determine the cause of a birth defect. The age at which a mother becomes pregnant also contributes to defects. Very young mothers and mothers over 40 years of age are known to bear higher numbers of babies with birth defects than are other mothers. Many of these are due to environmental factors.

- **Viruses**. German measles (**rubella**) in a pregnant woman can cause considerable damage to the fetus, although the mother may not suffer seriously from the disease. If a mother contracts German measles in the early months of pregnancy, her baby may be born deaf or with cataracts. The child may also be mentally retarded or have a heart defect. Other viruses are also suspected of causing birth defects. The HIV (AIDS) virus can be passed on to the fetus.

- **Drugs**. A number of medical and non-medical drugs have been found to be harmful to the fetus. For example, the drug **thalidomide**, which was taken by mothers to combat morning sickness and other unpleasant symptoms of pregnancy, resulted in thousands of deformed babies. These babies were born with hands and feet attached directly to the body, while the main portion of the limb was missing. These children are normal in intelligence and other physical development.

    Excessive doses of otherwise beneficial vitamins have been associated with mental retardation. The **hallucinogenic** drug, LSD, and some other non-medical drugs have been shown to damage chromosomes. Alcohol, aspirin, and chemicals in tobacco smoke have been shown to adversely affect the newborn. Babies of mothers who are drug addicts are usually born already addicted themselves and must suffer the symptoms of withdrawal.

- **Radiation**. The radiation from X-rays, which enables physicians to make many diagnostic decisions, can, if excessive, be harmful. If they are used during the early months of pregnancy when rapid cell division is taking place, X-rays may cause changes in the chromosomes. Pregnant women should always inform the technician of their condition before an X-ray is taken, as the baby is at risk. Whenever X-ray pictures are taken, non-target areas of the body should be shielded, especially the germ cells of the ovaries or testes.

- **Diet**. An important aspect of prenatal care and the prevention of problems is good diet. The developing baby places additional strain upon the mother's supply of vitamins, minerals, proteins, and energy. The mother must take this into account during pregnancy to maintain her own health and ensure the healthy development of her baby.

Timing and the Effects of Drugs

The brief discussion on birth defects may seem very alarming. It is important to know that not all mothers who have German measles deliver babies with birth defects. Not all the mothers who used thalidomide or had X-rays taken during early pregnancy produced defective offspring. Many of them were quite normal.

There are stages during the development of the embryo when a critical development is taking place. Interruption or interference with a process at a vital stage may make a very significant difference. Both medical and non-medical drugs can have significant and dangerous effects upon the unborn during these critical stages of development. When these critical times are passed, the medication may have no adverse effect. For example, if a defect-causing agent reaches an embryo just as its eyebuds are forming, blindness may result. If this stage of development is complete when the drug is taken, then the likelihood of the agent affecting that specific part may be reduced. When in doubt, do not take a drug until checking its safety with your physician.

11.8 Taking Charge of Your Own Sexuality

Sexuality is about relationships between individuals as well as about the function of reproduction. In no other physical activity is it as important to "take charge of yourself" – to understand and take care. There are two particularly important consequences of sexuality you must consider – pregnancy and sexually transmitted disease.

Planned Pregnancy

The arrival of a new baby can, and should be, a special happy time in the lives of any couple. The event, anticipated and planned, is a time of joy and celebration. Planned pregnancy requires careful thought on the part of both persons involved. Before adopting any technique, seek the advice of a physician, preferably one who knows your medical history. Table 11.3 outlines the most common techniques.

Table 11.3 Planned Pregnancies: The Most Common Techniques

Methods	How Effective It Is*	How It Works	Who Uses It	How It Is Used	Need for Medical Services or Concern	Reasons Why It Might Fail	Possible Side Effects
Oral contraceptives (the Pill)	99%*	Pill taken orally prevents follicle from maturing Inhibits ovulation	♀ (female)	Clear instructions on package Must be regular Best taken same time each day Vitamin supplement required Usually taken from day 5 to day 26 of cycle	Yes, examination and prescription Regular check-ups	Forgetting to take the Pill Not following instructions Hormone content not high enough to suppress ovulation	Irregular bleeding, spotting, nausea, weight gain, breast enlargement More risks for women smoking more than 15 cigarettes per day
Condom (sheath, rubber) (with spermicide)	64 to 97%	Fits over penis Prevents sperm entering the vagina Provides protection against STDs	♂ (male)	Placed on penis before intercourse Condom must be held while penis is withdrawn to prevent it slipping off	No	Not put on before any contact with the vagina Not sufficient care in removal Use of petroleum lubricants destroy sheath	None
Diaphragm with jelly or creams	80 to 98%	Prevents sperm from entering uterus Jelly kills sperm	♀	Placed over cervix in vagina Put in place 6 h before intercourse, removed 8 h later	Yes, examination and prescription Fitting instruction Size will change if gain or loss of more than 5 kg	Not correctly placed Spermicide not used Not left in place for 8 h after intercourse Requires advance planning	None
Vaginal sponge	85%	Spermicide	♀	Fits over cervix	None	Not enough spermicide	Unknown

(*As a percentage, the effectiveness of a method indicates the average result for 100 people per year. For example, 99% means of 100 people using a method for one year, one pregnancy will result.)

Table 11.3 Planned Pregnancies: The Most Common Techniques (continued)

Methods	How Effective It Is*	How It Works	Who Uses It	How It Is Used	Need for Medical Services or Concern	Reasons Why It Might Fail	Possible Side Effects
Chemical foams (jellies, creams, foam tablets, and suppositories)	Fair, better than nothing Foams better than the other products listed Should be used with diaphragms, caps, and condoms	Kills sperm but some may escape Blocks entry to uterus Provides protection for 1 h	♀	Place at the cervix in the vagina with special applicator Insert not more than 1 h before intercourse Foams effective immediately; others wait 10 min before intercourse	No	May not completely block the cervix Left too long, has lost its effectiveness Some require waiting period Douching before 8 h is up	May cause irritation May cause allergic reaction
Withdrawal	Not effective	Penis ejaculates outside the vagina and away from it	♂	Withdrawal of penis just before ejaculation	No	Leaking of small amounts of semen at early stages of intercourse Withdrawing too late	None Risk of sexually transmitted disease
Rhythm method (the "safe" period)	53 to 99%	Not having intercourse during part of cycle when ovum is present in uterus	♀ ♂	No sexual intercourse after the ovum leaves the ovary	Medical help to determine cycle and the so-called "safe" period (this usually takes 3 months) Without this, the method is not effective	Irregular periods, especially for teenagers, so dates are never exact Variations in cycles Variation in the time sperm and ovum stay alive	None
Douche	Not effective	Sperm is washed out of vagina after intercourse	♀	Vagina flushed out after intercourse	No	Douche may force sperm into uterus or fail to reach sperm Sperm present in folds of vagina would be especially hard to reach	None

(*As a percentage, the effectiveness of a method indicates the average result for 100 people per year. For example, 99% means of 100 people using a method for one year, one pregnancy will result.)

Chapter 11 / Reproduction

Table 11.3 Planned Pregnancies: The Most Common Techniques (continued)

Methods	How Effective It Is*	How It Works	Who Uses It	How It Is Used	Need for Medical Services or Concern	Reasons Why It Might Fail	Possible Side Effects
I.U.D. (coil, loop)	94 to 99% (maybe less)	Prevents implantation Irritates lining, increases number of white cells	♀	Placed in uterus by doctor, remains in place Copper coils must be replaced every 2 years Lippe's loop can be left in indefinitely Frequent checks required to ensure presence	Yes, examination and insertion Annual check-ups Not suitable for women who have not had children	Some women do not retain the device	Irregular bleeding, spotting, cramps Discomfort in some cases Sterility, PID link, decreased fertility
Sterilization Tubal Ligation	Virtually 100%	Part of tube from ovary to uterus removed No follicle to fertilize	♀ Fallopian tube cut / Fallopian tube tied	Done in hospital under anaesthetic	Yes For mature people finished with producing a family		
Vasectomy	Virtually 100%	Part of the tube from testis is removed Semen but no sperm present	♂ vas deferens cut and tied	Minor surgery in doctor's office	Yes For mature people finished with producing a family	Two sperm counts must be made before considered safe	Temporary soreness or discomfort

(*As a percentage, the effectiveness of a method indicates the average result for 100 people per year. For example, 99% means of 100 people using a method for one year, one pregnancy will result.)

LIFE · SIGNS

SEXUALLY TRANSMITTED DISEASE

Linda was unhappy and worried. She badly needed to talk to someone. She usually shared her personal concerns with her best friend Maria, but this time it wasn't easy to talk to anyone. Linda kept wondering if Maria would keep a confidence this time.

Eventually, Linda called up her friend and suggested that they get together. A request to meet or visit in each other's homes was a common event. The girls went up to Linda's room and put on some records. For some time, they talked of classes, boys, and new records, but Linda couldn't seem to keep her mind on the topic and Maria, becoming impatient, asked her what was bothering her. It took some coaxing to get Linda started, but eventually she blurted out her problem. "I've got VD," she said. Maria was stunned and didn't know what to say at first, and then questions tumbled out. "When?", "How do you know?", "Who was it?" Linda was upset, but gradually she told her friend what had happened.

First, there had been a phone call from a health nurse who said that she had been given Linda's name as a contact. Linda's name had been produced by someone currently being treated for gonorrhea.

Linda had immediately asked the nurse who had given her name, but the nurse would not identify the person being treated. She said that Linda wouldn't like it if she gave her name on the phone to someone else. Linda reluctantly agreed. She had then asked the nurse if she had contacted her mother. The nurse explained that no one else knew, because it was entirely a matter between the two of them. The nurse asked Linda the name of her physician and said that she wanted her to see the physician as soon as possible. She told Linda that she would be telephoning the physician, explaining why Linda was coming in. The nurse also said that she would phone the physician again later to see if Linda had kept the appointment.

As Linda talked with her friend, she gradually became more calm, and in response to Maria's questions, told her about the appointment with the doctor. The doctor first asked Linda how long it was since the contact had taken place, and whether she had been sexually intimate with more than one person. The doctor had examined Linda and, using a sterile swab, had taken a sample from her vagina. The sample was sent to a laboratory where a smear could be examined under a microscope and the swab used to start a test culture.

Linda received another call from her doctor a week later. She was asked to make another appointment. When Linda went the second time, the doctor explained that the test was positive and that she did have gonorrhea.

Linda protested. She had no sores, no pain, no discharge, nothing. She didn't want to believe the results.

The doctor also explained to Linda that, if venereal diseases (or STD – sexually transmitted diseases) went unchecked and untreated, there was a serious risk of a female becoming sterile and unable to have children later. If she had sexual contacts with other persons, she would very probably transmit the disease to them. The doctor said that if everyone with the disease were treated, it would be possible to wipe it out completely. The doctor asked Linda if she was allergic to any antibiotics and then gave her some medication (ampicillin) and some water. She was asked to take it while in the office. She was told that the one treatment would probably clear up the problem, but she was directed to come back in a week's time for a second check, to be sure that she was completely cured.

Maria wanted to know how the nurse knew to call her. Linda explained that, in order to try to stop the spread of the disease, each person was asked to either speak to anyone with whom they had been sexually active or to give the doctor their names and the health nurse would call them, completely confidentially. Linda said that in her case, she was sure that they had kept their promise and that no one else knew.

Sexually Transmitted Diseases (STDs)

In intimate moments each year, some 15 million citizens of North America are enrolled in a new "social club": the "STD Society". The members did not ask to join and were unaware that they had been enrolled until after their initiation. Now they share the dubious privilege of hosting at least one of the more than 28 different viruses, bacteria, fungi, and parasites that are transmitted by sexual contact. The membership is not a happy one. The members worry about their symptoms. Some are sick, some feel lonely and isolated, and can no longer enjoy the companionship they initially enjoyed. Some are in hospital and each year, some members will die. The only people who cannot join this club are those who abstain from sexual contact, have only one sexual partner, and those that understand how these diseases are spread and take the proper precautions to prevent infection. (See Table 11.4.)

There is no immunity from sexually transmitted diseases. You cannot get a vaccination to protect you, which means that you can get some of these diseases more than once. Some STDs, such as syphilis and gonorrhea, can be treated with antibiotics. Others, such as herpes and AIDS, have no known cure as yet. Each STD is caused by a different organism, so it is possible to have more than one of these infections at a time. Some infections produce few symptoms. Some symptoms disappear for long periods, leading a person to believe she or he is cured, only to have the disease break out later in a more advanced stage. The good news is STDs can be prevented if proper precautions are taken.

Ways to Protect Yourself from STDs

- Abstain from sexual activity. If you are sexually active, make sure that both partners have no other sexual contacts.
- Do not indulge in any activity in which semen or blood enters the mouth, vagina, or anus.
- If you are sexually active, use a condom and use it properly. Follow the directions on the package and do not use petroleum-based lubricants, which can cause the condom to fail.
- It is wise to refrain from deep kissing. Sores, cuts, or openings along the gum line, or tiny scratches you are not aware of, can allow a virus to enter the bloodstream.
- Drug users are at special risk and must never share a needle or syringe.
- Avoid drug and alcohol in intimate situations. People are less cautious under the influence of these substances.

Many infected persons show no symptoms at all, but they can pass on the infections. It is not enough to believe your partner is free of STDs. Unless you know someone very well, you should not involve yourself with them in a sexual way. The only absolutely safe way is abstinence.

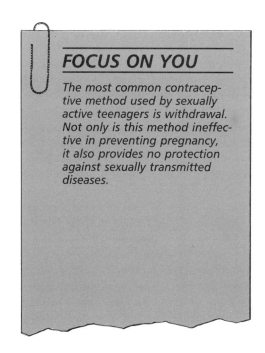

FOCUS ON YOU

The most common contraceptive method used by sexually active teenagers is withdrawal. Not only is this method ineffective in preventing pregnancy, it also provides no protection against sexually transmitted diseases.

Table 11.4 Some Sexually Transmitted Diseases

Disease	Appearance of First Symptoms	Common Symptoms	Method of Transmission	Possible Complications
Gonorrhea (called dose, clap, drop) Cause: a bacterium	2 to 10 days (up to 30 days)	Discharge, white or yellow, from genitals or anus. Painful urination or defecation. Pharyngeal infections usually have no symptoms. Women: Lower abdominal pain, especially after period. Men: May have no symptoms	Direct contact of infected mucous membrane with the cervix, urethra, anus, throat, or eyes	Sterility, arthritis, endocarditis, perihepatitis, meningitis, blindness. Women: Pelvic inflammatory disease. Men: Urethral stricture, erection problems. Newborn: Blindness
Herpes Simplex II (also called Herpes) Cause: a virus	Varies greatly	Clusters of tender, painful blisters in the genital area. Painful urination. Swollen glands and fever	Direct contact with blisters or open sores	Women: May be linked with cervical cancer. Newborns: Severe central nervous system damage or death if infected during birth
Syphilis (also called syph, pox, bad blood) Cause: a spirochete	10 to 90 days (average 3 weeks)	Stage 1: Chancre (painless pimple, blister, or sore) where germs entered body, e.g., genitals, anus, lips, breast, etc. Stage 2: Rash or mucous patches (most are highly infectious). Spotty hair loss, sore throat, swollen glands. Symptoms may re-occur for up to 2 years	Direct contact with infectious sores, rashes, or mucous patches	Adult: Brain damage, insanity, paralysis, heart disease, death. Fetus and Newborn: Damage to skin, bones, eyes, teeth, and liver
Non-Specific Urethritis Chlamydia (called NGU, NSU) Cause: a bacterium, chlamydia	1 to 3 weeks	Small amount white, yellow or clear discharge from genitals, often only noticed in the morning. Women: Often no symptoms. Men: Slight discomfort upon urination	Direct contact with infected area	Women: Pelvic inflammatory disease, cystitis, infertility. Newborns: Pneumonia and conjunctivitis

Table 11.4 Some Sexually Transmitted Diseases (continued)

Disease	Appearance of First Symptoms	Common Symptoms	Method of Transmission	Possible Complications
Trichomoniasis (called trich, TV vaginitis) Cause: a protozoan	1 to 4 weeks	Women: Heavy, frothy discharge, intense itching, burning, and redness of genitals Men: May have slight, clear discharge from genitals and itching after urination (Usually no symptoms)	Direct contact with infected area	Women: Gland infection
Monilial Vaginitis (called monilasis, vaginal thrush, yeast, candidiasis) Cause: a fungus	Variable	Women: Thick, cheesy discharge and intense itching of genitals, also skin irritation Men: Usually no symptoms	Organism often present in the mouth, vagina, and rectum without symptoms Active infection may follow antibiotic therapy or direct contact with infectious person	Women: Secondary infections by other bacteria Newborns: Mouth and throat infections
Venereal Warts (called genital warts, condylomata accuminata) Cause: a virus	1 to 3 months	Local irritation and itching, wet growths usually on the genitals, anus, or throat	Direct contact with warts	Highly infectious, can spread enough to block vaginal, rectal, or throat openings

Acquired Immune Deficiency Syndrome (AIDS)

AIDS stands for **Acquired Immune Deficiency Syndrome**. This disease, in its most serious form, is an infection caused by a virus. The virus, first isolated in 1983, is now officially known as **HIV (Human Immuno-deficiency Virus)**.

AIDS is not a disease itself; rather, it produces a situation in the body which prevents the proper combat and destruction of pathogens or other diseases that try to enter the human body.

The HIV virus seeks out special white blood cells called helper T cells. The virus takes over the DNA of the T cells, forcing it to make copies of the virus. Once as many copies as can be produced from the cell material available have been made, the old cell coat breaks open and the many "new" viruses are released. The process then starts again.

T cells co-ordinate and control our immune system. With the loss of the commanding T cells, the immune system collapses, like an army without leaders; thus, the immune system cannot effectively fight any infection that enters the body. The body is now vulnerable to any disease. It is the assaults of other disease organisms which generally cause the death of the patient, not the HIV virus itself.

The HIV enters the body under disguise enclosed in a T cell from an infected person. The T cell may be in transferred blood or in semen. On arrival in a new victim, the immune system of the new host recognizes the imported T cell, but cannot detect the virus hidden within it. Before action can be taken, the T cells of the new host are infected in turn.

How is the HIV Transmitted?
The medium in which the virus is passed is a body fluid such as blood or semen (saliva is still suspect). People with AIDS do *not* pose a threat to others *except* through the sharing of used needles, blood transfer, or sexual practices. Insects do not spread the virus. Telephones, cutlery, swimming pools, or toilets do not transmit the virus. No case has yet been reported of a family member getting the infection from another member that has the virus, with the exception of babies born to infected mothers.

Symptoms
The early symptoms of AIDS are not specific but may include swollen glands, persistent sweats, fatigue, weight loss for no obvious reason, and diarrhea. These symptoms are common to other illnesses also.

Is There Any Treatment for AIDS?
So far, no treatment has been found which can repair the body's immune system following AIDS. Often the illnesses that a person contracts because of low immunity can be treated, but the constant weakening of the body makes this process more and more difficult. AIDS is still considered a fatal disease. With so many researchers working on possible vaccines or drugs for this disease, it is possible that a cure will soon be found.

There is a great deal not known about AIDS and a lot of information that is, as yet, unconfirmed. Up-to-date information can be obtained from local health offices.

CHECKPOINT

1. (a) Name three methods of family planning.
   (b) Explain how these methods work.

2. What are STDs?

3. Describe three different sexually transmitted diseases other than AIDS.

4. AIDS is not really a disease. People with AIDS do not die from the disease, yet it is considered a fatal disorder. Explain how this can be.

BIOLOGY AT WORK

OBSTETRICIAN
PEDIATRICIAN
OBSTETRICS NURSE
HOME HEALTH CARE WORKER

HEALTH CONCERNS

| herpes virus |
| AIDS |
| chlamydia |
| genital warts |
| crib death |
| PID |
| endometriosis |

Chapter 11 / Reproduction

CHAPTER FOCUS

Now that you have completed this chapter, you should be able to do the following:

1. Describe the structure and function of the parts of the male and female reproductive systems.
2. List the organs of origin and the functions of the following hormones: follicle stimulating hormone, luteinizing hormone, progesterone, estrogen, and testosterone.
3. Explain the function and phases of the female menstrual cycle.
4. Outline the major phases in pregnancy and delivery.
5. Explain the functions of the placenta and how certain drugs may affect pregnant women (tobacco, alcohol, non-medical drugs, antibiotics and thalidomide, etc.).
6. Describe the development of the human embryo and fetus.
7. Describe several natural and artificial methods of fertilization prevention and explain how each functions.
8. Outline the biological basis for the transmission of sexually transmitted diseases, such as AIDS, gonorrhea, syphilis, vaginitis, and chlamydia, and how the transmission can be prevented.

SOME WORDS TO KNOW

Match each of the descriptions given in the left-hand column with a word shown in the right-hand column. DO NOT WRITE IN THIS BOOK.

1. Hormone which initiates the development of a new ovum
2. Hormone produced by the corpus luteum
3. Organ in which the fetus develops
4. The inner lining of the uterus
5. Hormone produced by the follicular cells in the ovary
6. Major male hormone
7. Sperm cells are produced in these tubules
8. Thin sac filled with a watery fluid in which the embryo develops
9. Small temporary structure that provides the developing embryo with blood cells
10. Name given to baby during the first few weeks it is in the uterus

Select any two of the unmatched words and, for each, give a proper definition or explanation in your own words.

A amnion
B embryo
C puberty
D testosterone
E progesterone
F estrogen
G FSH
H LH
I urethra
J yolk sac
K seminiferous
L fetus
M placenta
N uterus
O ovary
P prostate
Q testes
R endometrium

SOME QUESTIONS TO ANSWER

1. List the ducts and organs (in order) through which a sperm passes after developing in the seminiferous tubules, until it leaves the penis.
2. What are the functions of the male accessory glands?
3. Briefly list the sequence of events that takes place during the delivery of a baby.
4. Make a table as follows:

Hormone	Organ Where Produced	Organ That Hormone Affects	General Effect of the Hormone

   SAMPLE ONLY

   List the following hormones and then complete the table: progesterone, estrogen, FSH, LH.
5. What is meant by the following terms?
   (a) implantation
   (b) fertilization
   (c) ovulation.
6. Describe the structure and function of the endometrium.
7. Briefly describe the three phases of the menstrual cycle.
8. What shocks does a baby experience as it enters the world?
9. Make a list of the functions of the placenta and give a brief description of each.
10. Describe three planned pregnancy techniques and state which is the most effective and which is the least effective.
11. List some effective ways to reduce the transmission of STDs.
12. (a) Can a person have more than one kind of STD at a time?
    (b) Can a person have a particular STD more than once?
    (c) Can a person always tell when she or he has an STD?
    Give brief, simple explanations for each of your answers.
13. (a) Why is AIDS such a problem?
    (b) How does it differ from other STDs?
14. (a) List three environmental factors that can cause birth defects.
    (b) Describe one factor in detail.

SOME THINGS TO FIND OUT

1. Select one of the following topics and prepare a written or oral report:
   care of the newborn
   advantages of breast-feeding over bottle-feeding
   hospital volunteers
   the premature baby
   abortion laws
   problems of communication about human reproduction.
2. How does the environment in which a child grows up affect the child's development? As there are several aspects to this question, you may wish to limit your research to just one aspect of development.
3. What is involved in good prenatal care?
4. People who are experiencing physical difficulties in having a family have several options, ranging from medical techniques such as *in vitro* fertilization to adoption. Research one of these options and write a short story based on your research.

MAMMAL DISSECTION

The study of anatomy and physiology is not complete without the opportunity to examine the structures of the body first-hand. Many of the activities in this book have allowed you to observe parts of the body and how they function by looking at the outside of the body. A dissection provides you with the opportunity to see the body's structures, both internal and external. It is also possible to see how the many different structures are arranged with respect to one another.

KEY IDEAS

- *The internal and external structures of the fetal pig are similar to those of a human.*
- *The physical structure of a body part is closely related to its function.*
- *The parts of the body are interrelated; some parts have more than one function.*

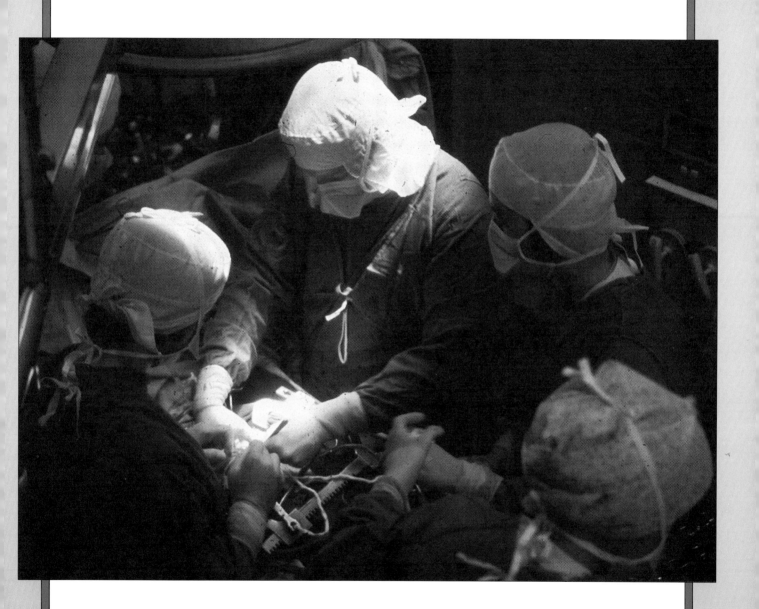

12.1 An Introduction to Dissection

The term **dissect** means "to separate into parts and examine". Before beginning your dissection, take a moment to become familiar with the tools you will use. They are the same basic tools used by surgeons. Adopting a surgical, scientific approach in their use will mean a dissection that will show you all the structures you are looking for, intact and in their place.

Directional Terms

Before you begin to dissect, it is important to learn some of the terms and reference points which are used to locate parts of the body. For example, if you were asked to place your specimen on its back, and then locate its right side, which side would this be? To avoid confusion, the body's parts are always described using the specimen as a frame of reference. The right side is always the specimen's right side, not yours.

You will encounter certain *directional terms* during this dissection. Table 12.1 summarizes the most important of these. Use Figure 12.1 to help you become familiar with these directions.

Many structures of the body are named using the name of the bone closest to it. Muscles, blood vessels, and nerves often use this naming system. Where would you expect to find the radial artery? As you begin to dissect, refer to the names and locations of the major bones of the body, shown in Figure 2.6 on page 45.

Care of Equipment and Specimen

- If your specimen is to be used again, wrap it in a moist paper towel and place it in a plastic bag to be given to your teacher for storage. Be sure your name is on a tag attached to the bag or written clearly in pencil on a piece of paper inside the bag.

Table 12.1 Important Directional Terms

Term	Meaning
anterior (or cranial)	toward the head
posterior (or caudal)	toward the hind end
dorsal (or superior)	the upper and/or back surface
ventral (or inferior)	the lower and/or belly surface
lateral	the sides
medial	close to the midline
superficial	close to the surface
proximal	close to
distal	farther from
pectoral	the shoulder area
pelvic	the hip area
thoracic	within the chest and ribs
abdominal	lower part of body, below ribs

- Dispose of any waste material in the container provided by your teacher, not in a sink.
- Clean and dry your dissection tools carefully. They will remain sharp and rust-free only if stored in this condition. Report dull or damaged equipment immediately. Dull blades will slip and can cause injury.

Materials Required

For your dissection, you will need the following:

- eye/face protection
- disposable gloves
- preserved fetal pig (rinsed in water overnight before first use)
- paper towelling
- dissection tray
- dissection tools (scalpel, blunt-tipped scissors, forceps, blunt probe, pins)
- string
- hand lens or dissecting microscope
- plastic bag for specimen storage

Figure 12.1
Becoming familiar with these terms will help you perform your dissection.

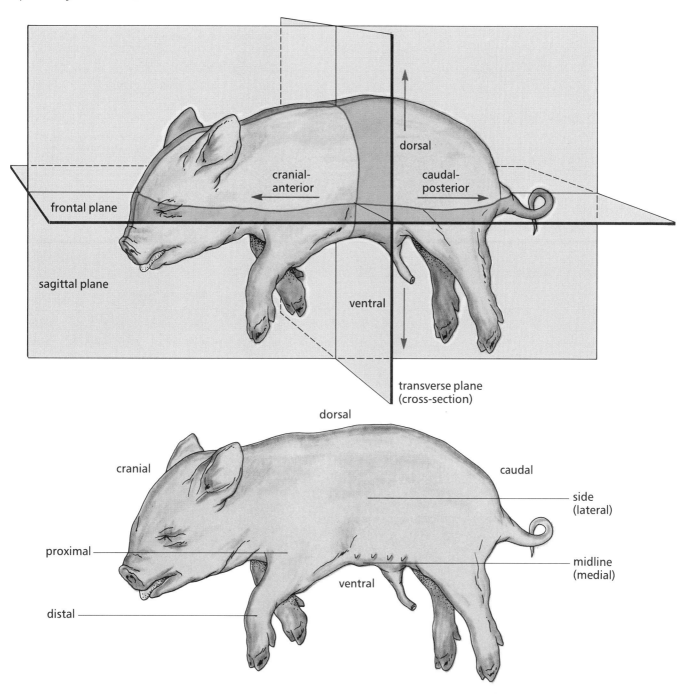

Safety Considerations

A dissection must be performed carefully and patiently. Be sure to follow the instructions in the text, and also those that your teacher gives you. Each time you dissect you should do the following:

- Wear eye/face protection and disposable gloves.

- Wear a laboratory apron, if one is available, to protect your clothing.

- Gently rinse your specimen under running water to wash away excess preservative.

- Work in an area with adequate ventilation.

- Position your specimen so that it is not directly beneath your face and nose.

- Use the dissection tools carefully. Use light pressure when cutting, and cut gradually through layers of tissue.

- Always thoroughly wash your hands and lower arms with soap after completing your dissection work.

CHECKPOINT

1. Explain the difference between the following pairs of terms:
   (a) medial, lateral
   (b) superior, inferior
   (c) anterior, posterior
   (d) dorsal, ventral
   (e) cranial, caudal
   (f) proximal, distal.

2. Sketch a diagram of the human body, and show the difference between a sagittal section and a transverse section.

12.2 Examining the External Features of the Fetal Pig

The fetal (unborn) pig is a mammal, and so are you. Its internal body structures are very similar to yours. This makes it a good choice for a dissection specimen.

Fetal pigs are collected at meat processing plants. They are preserved, and their blood vessels are injected with coloured latex. Arteries are coloured red, and veins are blue (there is one exception which your teacher will point out to you), to make them easier to locate.

Keep in mind when you are dissecting that your fetal pig specimen may have some immature organs and tissues, and also that some of the structures may be slightly different from that of the human body.

Procedure

CAUTION!
You must wear eye/face protection and disposable gloves for this exercise.

1. Obtain your specimen and gently rinse it under a stream of tap water. Use some paper towelling to gently blot it dry, and place the specimen in the dissection tray.

2. Locate the remains of the **umbilical cord** on the ventral surface. *What is the purpose of the umbilical cord?*

3. The **anus**, the opening of the large intestine, is located under the tail.

4. *Determine the sex of your specimen.* A male will have a **scrotal sac**, a pair of swellings between the hind legs, and only one opening under the tail (the anus). (See Figure 12.2.)

Figure 12.2
The male fetal pig

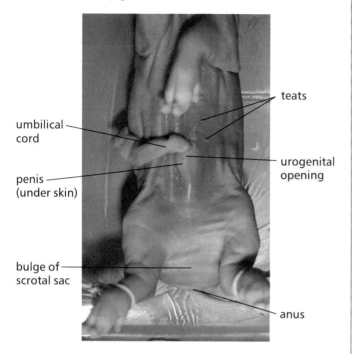

Figure 12.3
The female fetal pig

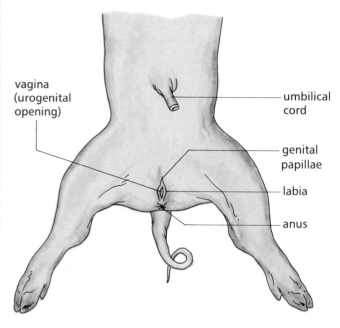

The **urogenital opening** will be found posterior to the umbilical cord. The **penis** is located under the skin. It is a muscular tube that lies behind, or caudal to, the urogenital opening.

Females have two openings under the tail. The one closest to the tail is the anus, and the other is the **vaginal opening**. (See Figure 12.3.)

5. On the abdomen, in both sexes, there are two rows of **teats** (mammary papillae). (See Figure 12.4.) *How many nipples are in each row?*

6. Examine the tissue just underneath the chin. *What do you notice?*

7. Find the **pinna**, or external ear. *How is the structure of the pinna related to its function?*

8. The **external nostrils** or **nares** are located dorsal to the mouth. *What function do these structures have?*

9. Locate the eyes, with their upper and lower lids. Use a scalpel to remove the eyelids. Start the incision at a corner, and cut around the eye. Use the forceps to lift away the eyelids. Find the **nictitating membrane**, a white, crescent-shaped structure in the medial corner of the eye. *What is its function? Do humans have a nictitating membrane?*

10. Use the hand lens to examine the skin of the pig. *Describe its texture, the presence of hair, and any patterns which may be present.* Your teacher may ask you to remove a small section of skin and observe it under the dissecting microscope. *How does it compare to your skin?*

11. *Examine the feet of the pig. How are they different from the human hand and foot?* The pig walks on "tip-toe", using only two **digits** or toes, the third and fourth, in the form of a hoof. The first digit is missing, and the second and fifth digits are much smaller.

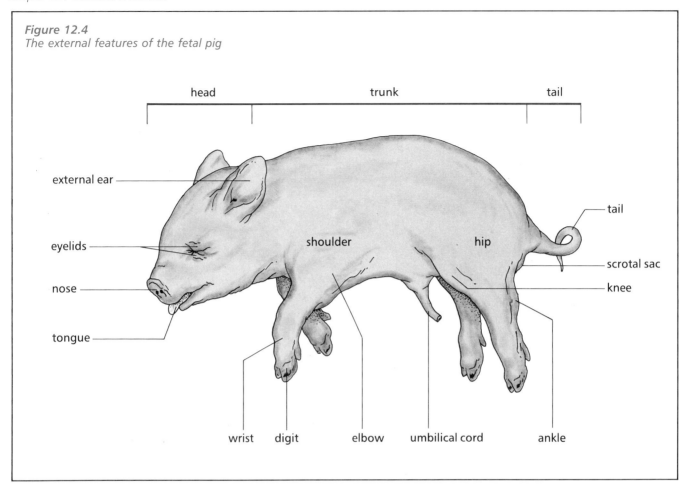

Figure 12.4
The external features of the fetal pig

12. Find the wrist and elbow joints of the front legs. The wrist is just above the hoof. The elbow is located close to the point where the leg meets the body, on the posterior side of the leg.

13. Find the ankle and knee joints of the rear legs. The ankle protrudes from the leg just above the digits. The knee is located close to the point where the leg meets the body. *Will the knee be found on the posterior or anterior side of the leg?*

14. Follow your teacher's instructions for the storage or disposal of your specimen. Be sure to wash your hands thoroughly with soap before leaving the laboratory.

CHECKPOINT

1. *Explain how you can distinguish between a male and a female fetal pig.*

2. *(a) Why is the fetal pig a good choice for a specimen to dissect in order to understand human anatomy?*
   (b) What must you keep in mind concerning the fetal pig as you examine it?

12.3 Dissecting the Digestive System

Procedure

CAUTION!
You must wear eye/face protection and disposable gloves for this exercise.

1. Obtain your specimen and gently rinse it under a stream of tap water. Use some paper towelling to gently blot it dry, and place the specimen in the dissection tray.

2. Use a scalpel to remove a triangular section of skin and muscle from one side of the head and neck. Begin at the corner of the mouth, and continue back to a point behind the ear. (See Figure 12.5.)

3. Locate the **parotid salivary gland**. It lies just below the base of the ear, and is triangularly shaped. Another salivary gland, the **submaxillary gland**, is ventral to the parotid. The **sublingual salivary gland** is anterior to the submaxillary gland. *What is the purpose of the salivary glands?*

4. The **parotid duct** carries the saliva into the mouth. It can be found leading from the base of the gland, under the edge of a rounded muscle into the mouth. (See Figure 12.5.)

5. Use a scalpel to cut through both sides of the mouth. As you make the incision, open the mouth so that you can see where you should continue to cut. The mouth must open wide enough to allow you to see the back of the **oral cavity**. (See Figure 12.6.)

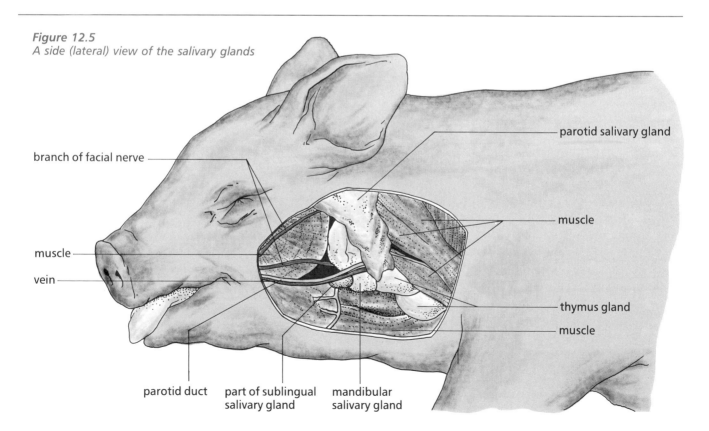

Figure 12.5
A side (lateral) view of the salivary glands

Figure 12.6
The oral cavity of the fetal pig

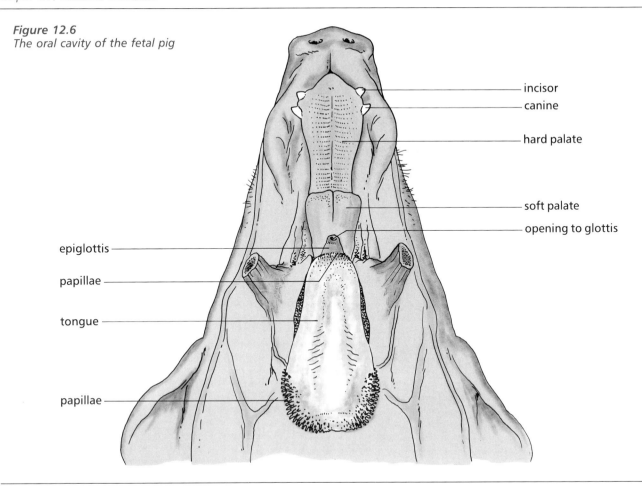

6. Identify the following structures of the mouth and throat: hard palate, soft palate, teeth, tongue, taste papillae, nasopharynx, pharynx, esophagus, glottis, epiglottis. *Sketch the diagram of the mouth shown in Figure 12.6, and label the structures indicated.*
   (a) The **hard palate** is the roof of the mouth. It has a series of ridges.
   (b) The **soft palate** is the continuation of the roof of the mouth. The **pharynx** is the general area at the back of the mouth that we sometimes refer to as the opening of the throat. It includes the region just above the larynx, where it divides into two different passageways, the trachea and the esophagus.
   (c) The full set of **teeth** may not have erupted (come through the gums) in your specimen. If they have, describe the shape and location of the teeth. *How is the shape and location of each tooth related to its function? How do they compare to human teeth?* If teeth are not visible, cut through the gums to see if they are present.
   (d) Use a hand lens or the dissecting microscope to examine a small piece of the **tongue**. *Describe its shape and texture. Where is it attached in the mouth? Where are the* **taste papillae**, *which contain the taste buds, located? How does the tongue of a fetal pig compare with a human tongue?*

(e) Use a scalpel to cut the soft palate along its midline. The space above it is the **nasopharynx**. *What is its function?*

(f) Pull the jaws apart until the opening of the **trachea**, called the **glottis**, and the opening of the **esophagus** can be seen (make additional incisions if necessary). *Which tube is dorsal, and which is ventral?* The **epiglottis** is the small piece of cartilage located near the glottis. *What is its function?*

7. Tie a piece of string around one of the front legs. Pull the string underneath the dissection tray. Gently stretch the legs apart, and tie the string to the other front leg to keep the legs stretched apart. Do the same for the rear legs.

8. Study Figure 12.7, which shows the order of the incisions to be made. Use a sharp scalpel to make the first incision from the little tuft of hair under the chin to the umbilical cord. Remember to use light pressure, and cut gradually through the layers of tissue. Continue to cut through the muscle wall of the abdomen until a small opening appears. Place the *blunt* end of the scissors through this opening, pointing toward the lower end of the abdomen. Lift up as you cut, so that the internal organs are not disturbed. Cut through the rib cage in the same way.

9. Next, use a scalpel to make an incision at the base of the neck to the midline incision, on both sides of the neck. *Do not cut too deeply.* You should not cut through any blood vessels.

10. Proceed with the incisions by cutting around the umbilical cord. Be careful not to cut too deeply. The incisions should continue on either side of the cord straight back to either side of the anus.

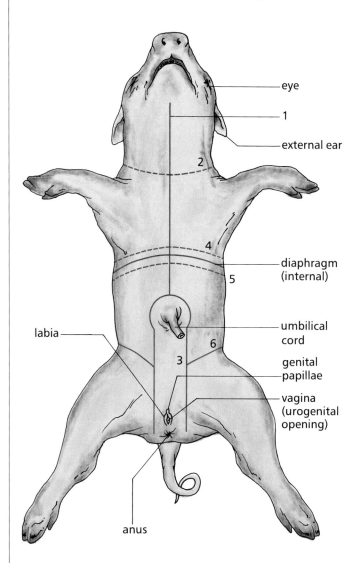

Figure 12.7
Incisions for the dissection of the fetal pig

11. The next incisions will allow you to open the body wall as if it were a set of double doors. Use the scissors to make an incision across the chest at about the point where the front legs meet the body. Make a similar incision close to each of the back legs. (See Figure 12.8.)

Chapter 12 / Mammal Dissection

Figure 12.8
The dissected fetal pig

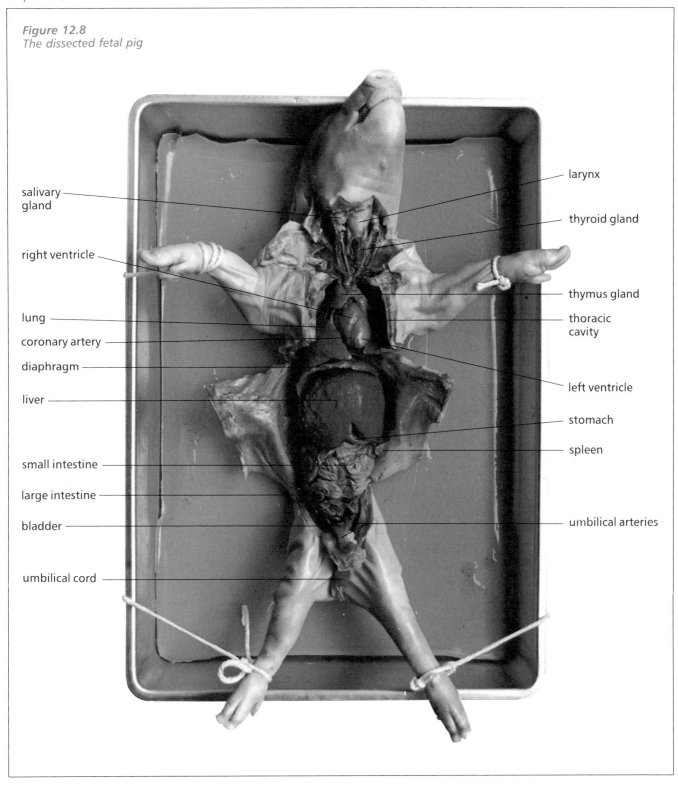

12. Gently lift one side of the body wall of the chest. A band of muscular tissue, the **diaphragm**, is attached to it. Use the scalpel to carefully separate the diaphragm from the body wall as you open the body cavity. Do the same on the other side of the chest. Use dissection pins to hold these flaps out of the way.

13. Lift the umbilical cord and cut the umbilical vein which runs anterior to the liver. Fold back the flap of tissue between the legs. Gently rinse the body cavity under running water. Let the cavity drain before proceeding.

14. Examine the lining of the body cavity on the ventral surface. The thin, shiny membrane is the **peritoneum**. *Describe the pattern and direction of the muscles which are visible in the body wall. Determine the number of muscle layers. Describe the direction of the muscles in each muscle layer. Compare a section in the chest with one in the abdomen.*

15. *Sketch a diagram of the internal organs which can now be seen in the thoracic and abdominal cavities. Label your diagram, using Figure 12.9 as a guide.*

16. The **liver** is the large, brown organ at the anterior end of the abdominal cavity. It has four lobes. Lift up the liver and find the small, greenish-white sac embedded in the base of the liver. This is the **gall bladder**. Follow the **cystic duct** as it leaves the gall bladder, until it joins with the **hepatic duct** from the liver, and becomes the **common bile duct**. Cut the common bile duct close to where it connects to the small intestine.

17. The **esophagus** leads from the pharynx to the stomach. Look between the top of the liver and the diaphragm to find the esophagus as it makes its way to the stomach.

18. Beneath the liver is the **stomach**. *Describe its shape, and list three of its functions.* The **duodenum** is the first part of the **small intestine**, leaving the stomach. The **pancreas**, a somewhat lumpy looking, whitish, leaf-shaped organ, is situated under the stomach. It is attached to the edge of the duodenum. (Do not confuse it with the **spleen** which is a smooth, reddish-brown gland in this same area). *What are the functions of the pancreas?* Find the point of attachment of the pancreas close to the duodenum. Cut this connection, and remove part of the pancreas. Observe it with the dissecting microscope or hand lens, and *make a sketch.*

19. Follow the small intestine until it joins the **colon** or **large intestine**. At this junction, the large intestine continues off to one side, but there is a "dead end" pouch on the other side. This is the **caecum**. *What structure can be found at the end of the caecum in humans?*

20. *How is the shape and location of the fetal pig's colon different from the human colon? Why?*

21. Follow the colon caudally until you find a straight section close to the abdominal wall. This is the **rectum**, the last part of the long digestive tract. It leads to the **anus**, which opens to the outside.

22. Use a scalpel to cut completely through the colon just above the rectum. Cut through the esophagus where it meets the stomach. Carefully remove the stomach, small intestine, and colon. Use the scalpel to free these organs from any other tissues from where they are attached to the back wall of the abdominal cavity. Be careful not to damage or remove any other organs which are in the same area.

23. Remove the stomach from the intestines. Cut the stomach open and observe the stomach lining with the aid of a dissecting microscope or hand lens. The folds in the lining are called **rugae**. *Describe the lining of the stomach.* You may notice a greenish liquid in the stomach. *What is it?*

Chapter 12 / Mammal Dissection

Figure 12.9
The internal anatomy of the fetal pig

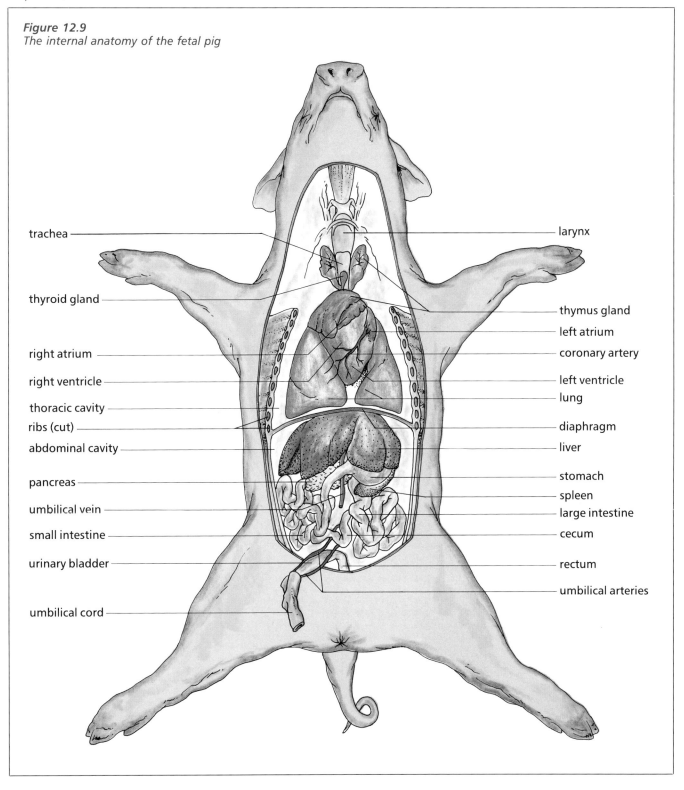

24. Examine the thin layer of tissue, called the **mesentery**, which holds the small intestines together. Use a hand lens or dissecting microscope to examine the mesentery. *Sketch the pattern of blood vessels which are visible.*

25. Remove a 1 to 2 cm section of the small intestine. Cut the tube lengthwise, and open it so it will lay flat. Place it under water in a Petri dish, and observe the lining using the dissecting microscope. *Describe the lining of the small intestine.* Repeat this procedure with a piece of the colon. *How is the colon different?*

26. Follow your teacher's instructions for the storage or disposal of your specimen. Be sure to wash your hands thoroughly with soap before leaving the laboratory.

CHECKPOINT

1. *Sketch and label a diagram of the main organs of the digestive system of the fetal pig. Sketch or trace the outline of the pig from Figure 12.9 to begin your diagram.*

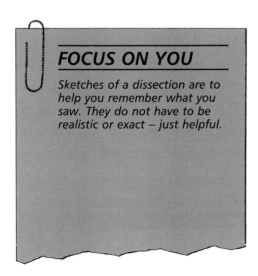

FOCUS ON YOU

Sketches of a dissection are to help you remember what you saw. They do not have to be realistic or exact — just helpful.

12.4 Dissecting the Excretory System

Procedure

CAUTION!
You must wear eye/face protection and disposable gloves for this exercise.

1. Obtain your specimen and gently rinse it under a stream of tap water. Use some paper towelling to gently blot it dry, and place the specimen in the dissection tray.

2. If necessary, prepare the specimen by tying back the legs and pinning the body wall flaps down as described in Section 12.3 (steps 7 and 12).

3. Locate the **kidneys**, in the lower region of the abdominal cavity, resting against the dorsal body wall. (See Figure 12.10.) They are covered with the **peritoneum**, a membrane which lines the inside of the body cavity. Use the scalpel and forceps to remove this membrane. Be careful not to cut too deeply. *Describe the shape of the kidney.*

4. A pair of blood vessels will be found on the medial side of each kidney. These are the **renal artery** and the **renal vein**. *How will you know which one is the artery and which one is the vein?*

5. The **ureter** leaves the kidney in the same area, turns to the posterior, and leads to the **urinary bladder**. The bladder is located between the umbilical arteries, which were folded back out of the way with the umbilical cord. *What is the function of the ureter?*

Figure 12.10
The excretory system of the fetal pig

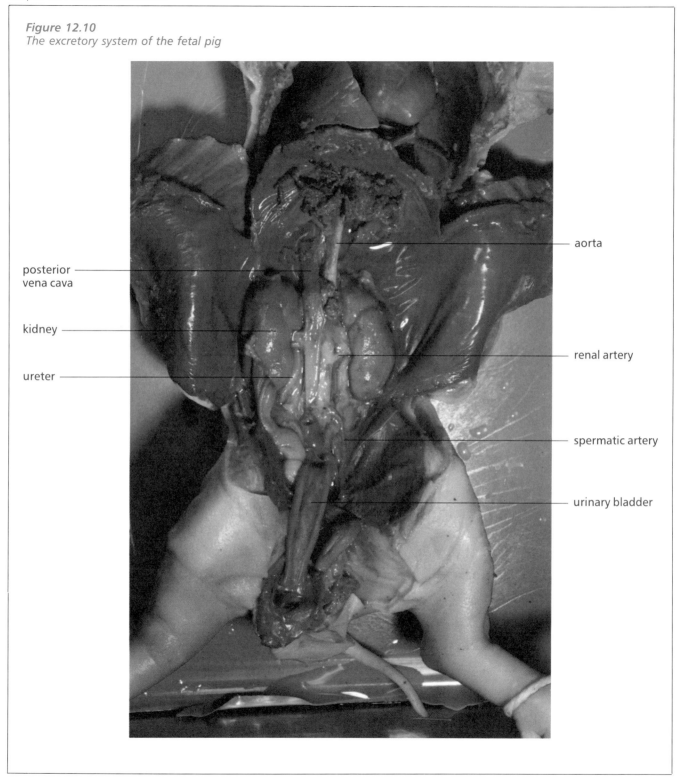

6. Carefully examine the bladder and find the point of attachment of the two ureters. Once you have found these, look at the caudal end of the bladder, and locate the **urethra**. The urethra conducts urine from the bladder to the outside of the body wall.

7. Look on the anterior end of the kidney, toward the medial side, for the **adrenal glands**. These glands are part of the endocrine system. *What hormones do they secrete?*

8. Cut the ureter at a point near the caudal end of the kidney. Lift one of the kidneys, and carefully cut away any tissue surrounding it. Remove the kidney from the body.

9. Use the scalpel to cut the kidney into dorsal and ventral halves. (See Figure 12.11.) Begin the incision near the point where the blood vessels and ureter enter the kidney.

Figure 12.11
Sagittal section of a kidney

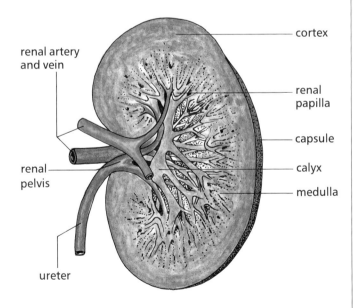

10. Examine the internal structure of the kidney. The ureter becomes the **renal pelvis** as it enters the kidney. Notice the difference in the texture of the outer portion of the kidney. This is the **cortex**. Lying inside this is an area called the **medulla**. *Where would you expect to find the microscopic nephrons?*

11. Follow your teacher's instructions for the storage or disposal of your specimen. Be sure to wash your hands thoroughly with soap before leaving the laboratory.

CHECKPOINT

1. Sketch and label a diagram of the main organs of the excretory system of the fetal pig. Sketch or trace the outline of the pig from Figure 12.9 to begin your diagram.

12.5 Dissecting the Respiratory System

Procedure

 CAUTION!
You must wear eye/face protection and disposable gloves for this exercise.

1. Obtain your specimen and gently rinse it under a stream of tap water. Use some paper towelling to gently blot it dry, and place the specimen in the dissection tray.

2. If necessary, prepare the specimen by tying back the legs, and pinning the body wall flaps down as described in Section 12.3 (steps 7 and 12).

3. Open the mouth as wide as possible. Find the following structures of the mouth and throat: nasopharynx, pharynx, esophagus, glottis, epiglottis. If you need to review, look back to Section 12.3, and see Figure 12.6.

4. Carefully cut through the tissue of the neck and locate the **trachea** and the **larynx**. The trachea looks like a corrugated hose, with alternating bands of cartilage and soft tissue. *What is the purpose of the bands of cartilage?*

5. The larynx is an enlarged bulge at the cranial end of the trachea. It contains the vocal cords. Use a scalpel to cut open the larynx, and examine the vocal cords. *What do they look like?* Insert a blunt probe into the larynx and push it anteriorly through to the glottis (look from the mouth to see it).

6. Examine the outside of the larynx. At its base is a oval, reddish gland, the **thyroid gland**. *What is the function of the thyroid gland?* The **thymus glands** are located on either side of the larynx, and join together below the thyroid gland. They look something like the letter Y. The thymus glands are part of the lymphatic system. They form an important part of the immune system which protects the body from disease.

7. Follow the trachea as far as possible. It leads to the **lungs**. The **heart** lies in between the lungs. The **bronchi** can be seen if you look dorsal to the heart. Cut the trachea. Using a syringe inserted into the trachea, try to inflate the lungs.

8. Notice that the lungs are divided into lobes. *How many lobes can you find in each lung?* Find the thin membrane which covers the lungs. If it is not visible, lift the lungs and look on the dorsal surface. This is the **pleural membrane**. *What is its function?*

9. The **diaphragm** is a smooth sheet of muscle lying beneath the lungs. You have already cut part of it away from the ventral body wall. Notice that the esophagus passes through it. *What is the function of the diaphragm?*

10. Follow your teacher's instructions for the storage or disposal of your specimen. Be sure to wash your hands thoroughly with soap before leaving the laboratory.

CHECKPOINT

1. Sketch and label a diagram of the main organs of the respiratory system of the fetal pig. Sketch or trace the outline of the pig from Figure 12.9 to begin your diagram.

12.6 Dissecting the Circulatory System

Procedure

CAUTION!
You must wear eye/face protection and disposable gloves for this exercise.

1. Obtain your specimen and gently rinse it under a stream of tap water. Use some paper towelling to gently blot it dry, and place the specimen in the dissection tray.

2. If necessary, prepare the specimen by tying back the legs, and pinning the body wall flaps down as described in Section 12.3 (steps 7 and 12).

3. Carefully cut away the tissue of the neck and chest anterior to the heart to expose the major blood vessels. (See Figure 12.12.) The vessels will be visible since the arteries have been injected with red latex, and the veins with blue latex.

Chapter 12 / Mammal Dissection

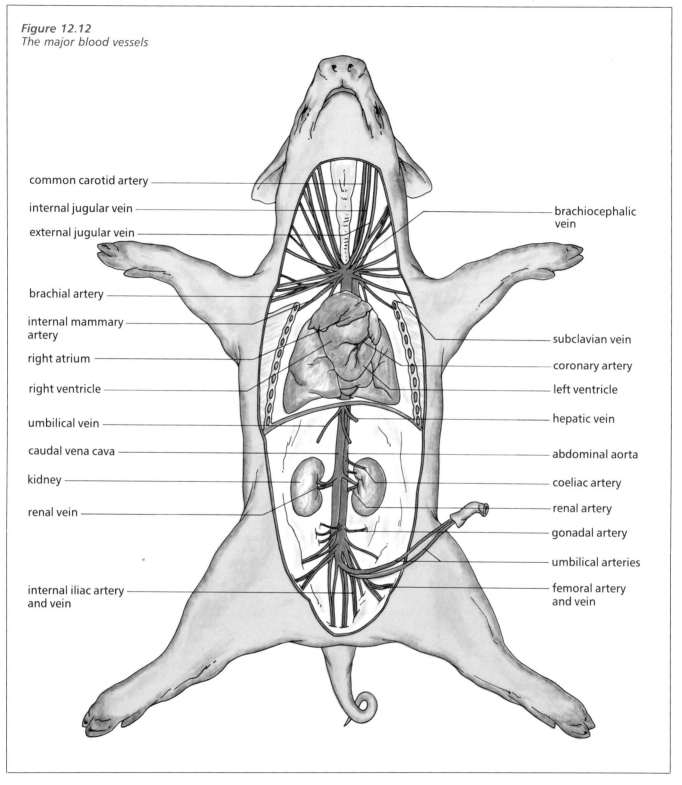

Figure 12.12
The major blood vessels

4. Find the arteries going to the head and neck. These are the **carotid arteries**. The veins returning blood to the heart are the **jugular veins**.

5. Locate the **subclavian vessels** which branch from the carotid arteries and the jugular veins to supply the tissues of the front legs.

6. The major vessels going to and coming from the lungs are the **pulmonary arteries** and **pulmonary veins**. In this instance only, veins will be coloured red, and arteries coloured blue. *Why are the colours reversed for these blood vessels?*

7. Examine the surface of the heart. Carefully remove the thin membrane which surrounds it. This is the **pericardium**. If you cannot find it, look at the dorsal surface of the heart. On the surface of the heart are the **coronary arteries**, which supply the heart muscle with oxygen and nutrients. *What happens if the coronary arteries are blocked?*

8. Lift the heart and find the **aorta**, the main blood vessel leaving the heart. It starts anterior to the heart, and curves to run beneath or dorsal to the heart. This curve is called the **aortic arch**. Below the arch, it is called the **descending aorta**. (See Figure 12.13.)

9. Find the **renal arteries** which branch from the descending aorta, and supply blood to the kidneys.

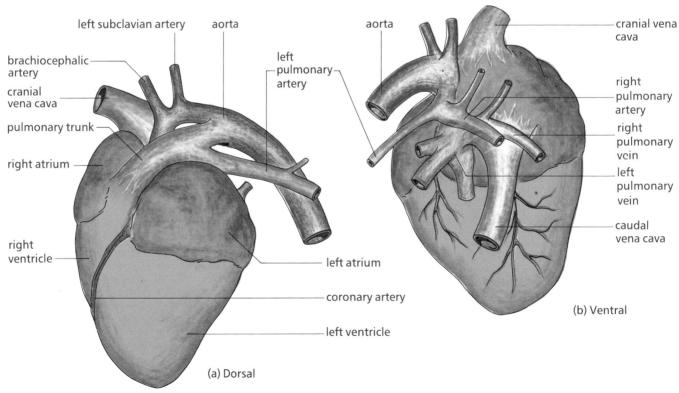

Figure 12.13
Fetal pig heart

10. Find the **renal veins**, which leave the kidneys and join the **posterior vena cava** as it returns to the heart.

11. Locate the **celiac artery** which leaves the descending aorta anterior to the left renal artery. It carries blood to the stomach, spleen, and liver. The **spleen** is the long, thin, reddish-brown organ located on the left side, just below the liver (actually below where the stomach was before it was removed). The veins which collect blood from the digestive tract and the spleen, and take it to the liver, make up the **hepatic portal system**. The **hepatic veins** return the blood in the liver to the vena cava.

12. Follow the descending aorta in a caudal or posterior direction until you see the arteries branching off to the rear legs. These are the **external iliac arteries**. Follow one artery until it branches. The artery is now called the **femoral artery**. *From where does it derive its name?* The large veins leading from the rear legs are the **iliac veins**. The **umbilical arteries** can be found in this same area. The **umbilical vein** which leads to the liver was severed when you folded back the umbilical cord.

13. Lift the heart, and cut the blood vessels leading to it. Use the scalpel and cut through the wall on each side of the heart, from top to bottom, so that you can look inside. Locate the four chambers of the heart, and distinguish between the atria and the ventricles. *Compare the thickness of the muscle walls in each chamber. Why is there a difference?* Look for the openings between the chambers. *What controls the flow of blood between the chambers?*

14. Follow your teacher's instructions for the storage or disposal of your specimen. Be sure to wash your hands thoroughly with soap before leaving the laboratory.

CHECKPOINT

1. Sketch and label a diagram of the main organs of the circulatory system of the fetal pig. Sketch or trace the outline of the pig from Figure 12.9 to begin your diagram.

12.7 Dissecting the Reproductive System

Procedure

CAUTION!
You must wear eye/face protection and disposable gloves for this exercise.

1. Obtain your specimen and gently rinse it under a stream of tap water. Use some paper towelling to gently blot it dry, and place the specimen in the dissection tray.

2. If necessary, prepare the specimen by tying back the legs, and pinning the body wall flaps down as described in Section 12.3 (steps 7 and 12).

3. Determine the sex of your specimen (see Section 12.2, step 4). If it is a male, omit steps 4 to 7, and begin at step 8. If it is a female, continue with step 4. After you have completed your own dissection, be sure to observe a fetal pig of the opposite sex.

4. If your specimen is a female, separate the rear legs of the pig, and cut through the skin, body wall, muscles, and cartilage along the midline of the pelvic region. Continue until you find three tubes. (See Figure 12.14.) The tubes should be, in order from ventral to dorsal, the **urethra**, the **vagina**, and the **rectum**. *What is the function of each tube?*

Figure 12.14
The reproductive organs of the female fetal pig

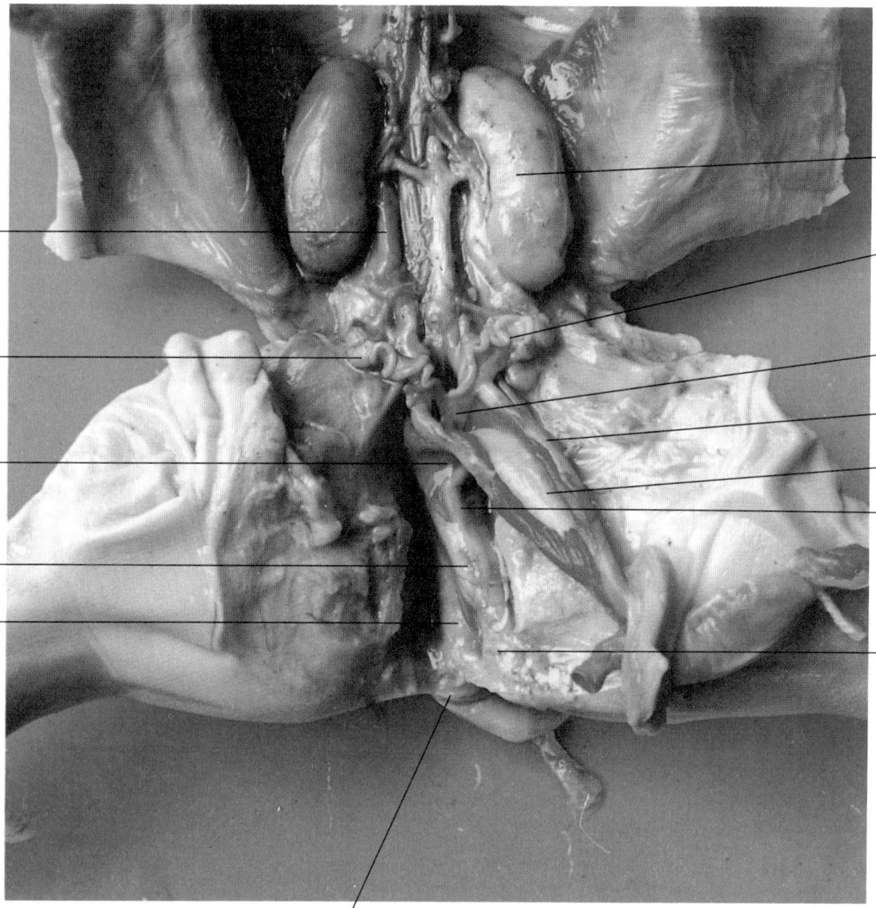

Labels: ureter, ovary, cervix area, vagina, rectum, kidney, oviduct, uterus, umbilical artery, bladder, urethra, urogenital sinus, genitalia region

5. Find the **ovaries**, one on either side, along the dorsal body wall, caudal to the kidney (or to the cavity left when the kidney was removed).

6. Follow the **oviduct** or **Fallopian tube** to the base of the umbilical arteries. The oviducts lead to the **uterus**, which leads to the tube identified in step 4 as the vagina. A slight constriction in the tube marks the division between the uterus and the vagina. This is the **cervix**. The section of this tube which is closest to the outside of the body is the **urogenital sinus**. In such a young animal, the vagina and the urethra share this opening. An adult female will have two separate openings.

7. Examine the urogenital opening located just under the tail. The **genitalia**, or external reproductive structures, are located around this opening.

8. If your specimen is a male fetal pig, locate the **scrotum** between the rear legs. (See Figure 12.15 on the next page.) Use the scalpel to make a cut in one of the scrotal sacs along its midline. The **testes** are two small oval structures. In some specimens, the testes will be within the abdominal cavity. In more mature fetuses, the testes will have descended into the scrotal sac. The testes are found inside a thin pouch. Carefully cut away this membrane to examine the testis and the **epididymis**, which lies along one side of the testis.

9. Find the **spermatic cord** which contains the spermatic artery, vein, and nerve, and the sperm-carrying duct. It leads from each testis through the opening between the abdominal cavity and the scrotal sac.

10. Place the flap of tissue containing part of the umbilical cord back in its proper position. Find the **urogenital opening** which is posterior to the umbilical cord. The **penis** is located under the skin. It is a muscular tube that lies directly behind the urogenital opening. It extends back toward the anus. The **urethra** leads through the penis.

11. Follow your teacher's instructions for the storage or disposal of your specimen. Be sure to wash your hands thoroughly with soap before leaving the laboratory.

CHECKPOINT

1. *Sketch and label a diagram of the main organs of the reproductive system of a male and a female fetal pig. Sketch or trace the outline of the pig from Figure 12.9 to begin your diagrams.*

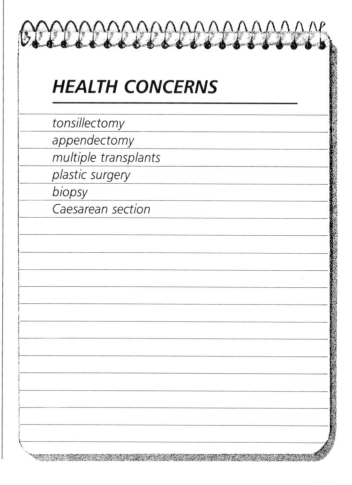

BIOLOGY AT WORK

PATHOLOGIST
SURGEON
ANATOMIST
MEAT INSPECTOR
BUTCHER
FIELD NATURALIST

HEALTH CONCERNS

tonsillectomy
appendectomy
multiple transplants
plastic surgery
biopsy
Caesarean section

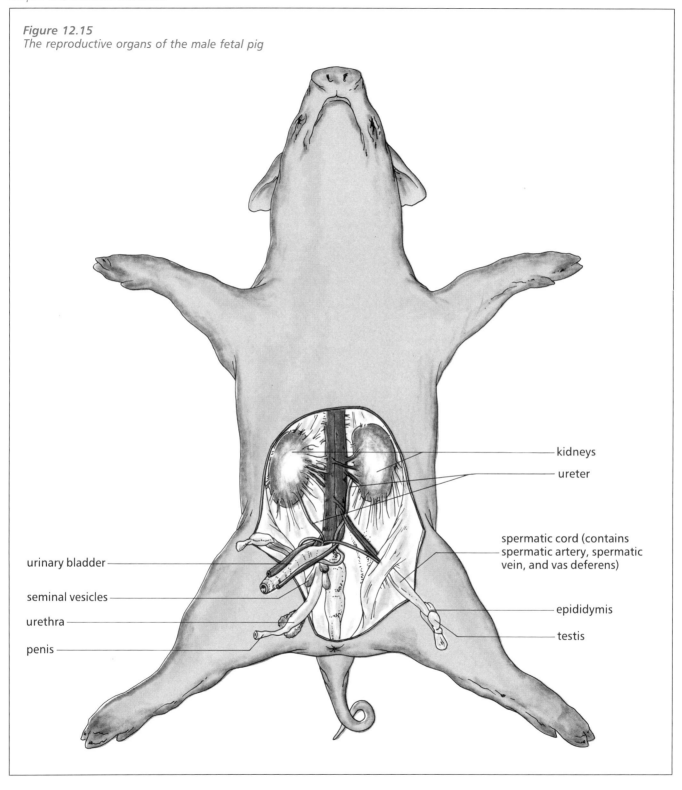

Figure 12.15
The reproductive organs of the male fetal pig

CHAPTER FOCUS

Now that you have completed the dissection in this chapter, you should be able to do the following:

1. Describe the external features of the fetal pig.
2. Draw diagrams to show the major organ systems of the fetal pig.
3. Describe the physical location of the major organ systems with respect to each other and with respect to their function.
4. Compare the anatomy of the fetal pig to that of the human.
5. Perform the techniques of dissection and observation.

SOME THINGS TO FIND OUT

1. Based upon your knowledge from this dissection, describe the steps that a surgeon might follow to repair a defective valve in the heart.
2. Investigate the surgical procedures required for a kidney transplant.
3. Lasers, remote sensing devices, and ultrasonics are replacing the surgeon's scalpel in many procedures. Prepare a report giving an example of how this type of technology is being used. Why is it so important? What are the disadvantages, if any?

13

A QUESTION OF SURVIVAL: MANAGING HUMAN WASTE

Within your lifetime, the availability of drinkable water may become the most serious long-range threat to human survival on this planet – but not because water is scarce. The problem is contamination of water supplies by improperly managed human waste. Pesticides and toxic compounds from soil have been found in water deep underground. Thousands of lakes have suffered biological damage as the result of acid rain from airborne pollution. Each day, wastes are pumped directly into the earth's rivers, lakes, and oceans.

These problems are now being faced locally and worldwide. As each person, city, and country begins to take the responsibility of waste management seriously, we come a step closer to ensuring our own survival.

KEY IDEAS

- Biodegradable waste can be broken down by decomposers.
- The sources of water pollution are litter, enriching substances, and toxic waste.
- Water must be purified at treatment plants before and after use.
- There are three main kinds of solid waste – domestic, industrial, and hazardous waste.
- Sanitary landfills, incineration, and composting can be used to dispose of solid waste.
- Combustion, vaporization, and mechanical action produce air pollution.

Chapter 13 / Managing Human Waste

13.1 What is Waste?

Waste is what we do not use. It is what we discard, materials we want to get rid of before they accumulate into amounts that interfere with our activities and way of life.

When people lived in small communities, they could adequately dispose of their waste by burying it, burning it, using earth outhouses, or allowing it to enter waterways. Micro-organisms decomposed much of the waste, which plants then used for growth. When wastes were released into the waterways, the water diluted the waste. If the quantity of waste was small, and the volume of water was large, this method caused few problems.

Sometimes, however, diseases were unknowingly spread through the water supply. In the early 1800s, the city of London, England used the Thames River as its source of water as well as for disposal of untreated human waste. A physician, John Snow, studied the incidence of cholera in the city. Cholera causes vomiting, cramps, and diarrhea, and is often fatal. Dr. Snow noticed that people who drank water taken from the lower Thames, downstream from the sewer outlets, had a high incidence of cholera while those who lived on the upper Thames largely escaped the disease. He correctly concluded that the disease-causing organisms originated in the polluted water supply.

As the human population continued to increase, the quantity of waste became greater than natural decomposition could control. In addition, as manufacturing processes produced new substances, wastes were produced that could not easily be decomposed. As these wastes entered the water supply, it became increasingly polluted. Treating water before its use became necessary for health. Before long, however, it became obvious that wastewater also needed to be treated before being returned to the environment and so to the water supply.

CHECKPOINT

1. What is waste?

2. Explain why Dr. Snow's observations about cholera showed a need to be concerned about water purity.

3. Why did manufacturing increase the amount of pollution in the water supply?

Figure 13.1
A waterway polluted with litter

13.2 Sources of Water Pollution

Looking at some waterways is enough to convince anyone not to drink the water there. At other times, water may not appear polluted, but could still be unsafe to drink. The wastes which pollute natural water systems can loosely be grouped into three categories – litter, enriching substances, and toxic substances.

Litter

Litter includes anything dumped into waterways. Some examples are old tires, pop cans, bottles, styrofoam, plastic bags, even old cars and steel drums. (See Figure 13.1.)

Enriching Substances

Enriching substances are rich in nutrients such as nitrates and phosphates. These stimulate plant growth, causing algae in the water to grow rapidly. This is called an **algal bloom**.

It would seem, at first, that the enriching substances are not a big problem. The fertilizers that run off the farmer's fields, and the nitrates and phosphates from biological wastes feed plants, and many animals feed on these plants. But such plant growth is so rapid that it chokes the waterway. (See Figure 13.2.) As the algae die, decomposers such as bacteria and fungi increase. These organisms use up much of the oxygen in the water.

Biochemical Oxygen Demand (BOD) is a measure of how much oxygen is required to break down the organic material present in a particular sample of water. The more organic material, the higher the BOD. If the BOD is continually high, the oxygen in the water decreases. Eventually, oxygen-dependent organisms such as fish and crayfish die from suffocation. (See Figure 13.3.) The death of these creatures again supplies more nitrates and phosphates, and the blooms of algae begin again. This cycle accelerates the process of **eutrophication**, in which nutrient-rich debris gradually accumulates in a lake, encouraging plant growth; eventually, the lake fills in with soil, becoming dry land.

Figure 13.2
Enriching substances cause algae in the water to grow rapidly.

Chapter 13 / Managing Human Waste

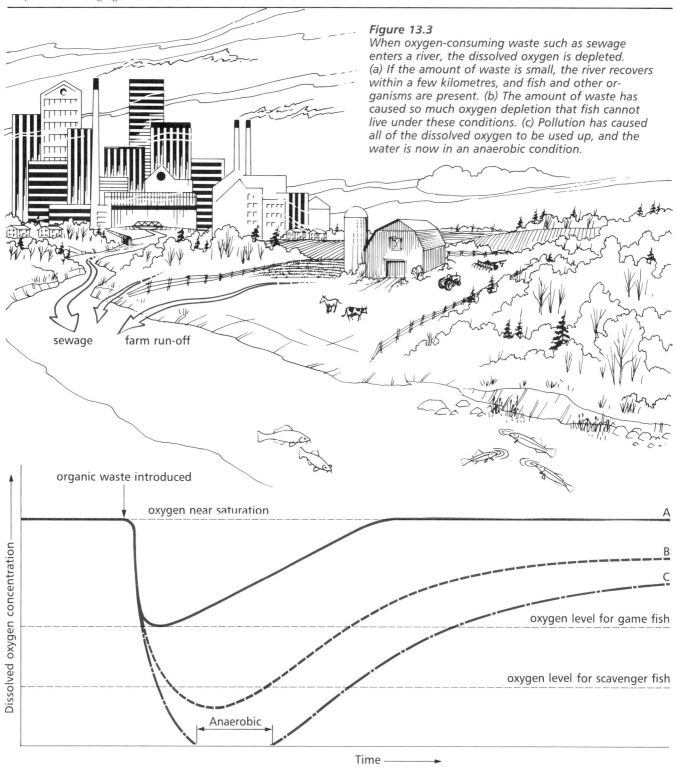

Figure 13.3
When oxygen-consuming waste such as sewage enters a river, the dissolved oxygen is depleted. (a) If the amount of waste is small, the river recovers within a few kilometres, and fish and other organisms are present. (b) The amount of waste has caused so much oxygen depletion that fish cannot live under these conditions. (c) Pollution has caused all of the dissolved oxygen to be used up, and the water is now in an anaerobic condition.

Toxic Substances

Some substances produced by industries such as mining, plating processes, and paint manufacturing are **toxic** (poisonous). These substances make water unfit for use. If present in high enough concentrations, the toxins destroy aquatic plants and animals.

Many of the toxic substances produced by industry are metals. The so-called "heavy metals", which include mercury, lead and cadmium, tend to accumulate in the bodies of organisms. (See Figure 13.4.)

Figure 13.4
Pollutants such as heavy metals become more concentrated in organisms higher in the food chain. The final consumers, in this case humans, may receive a dose which damages their nervous system and may cause death. Where has this been a problem in Canada?

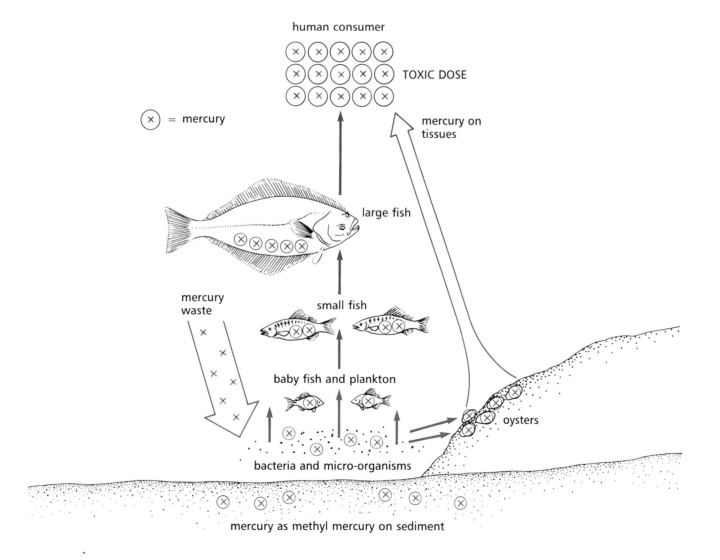

Chapter 13 / Managing Human Waste

Table 13.1 A Summary of Major Water Pollutants

Pollutant	Sources	Examples	Notes
Oxygen-demanding wastes	Animals, oil refineries, food processing, paper making	Sewage, manure, decaying vegetation, some industrial wastes	Decomposed by aerobic bacteria if oxygen present
Inorganic chemicals and minerals	Mining, smelting, oil production	Salts, acids, metal compounds	Washed into rivers, etc., not decomposed, cumulative
Synthetic organic chemicals	Agriculture, industrial sources	Pesticides, herbicides, detergents, industrial wastes	Magnified in the food chain, not usually biodegradable
Fertilizers and nutrients	Manures, fertilizers, farm run-off, detergents, sewage	Nitrogen and phosphorus	Increased plant growth, clogs rivers, eutrophication, algal blooms
Pathogens	Human and animal wastes, meat processing plants	Infectious organisms, viral and bacterial diseases such as typhoid fever	Contaminated groundwater enters food processing
Erosion and sedimentation	Run-off, building and agricultural development, mining, natural erosion	Soil, mineral leaching, sand and dust, storm drains	Flooding, poor soil conservation, natural run-off, contamination by oils and salts from roads

FOCUS ON YOU

Not only the waste we allow into our water today should concern us. Residents of the rural community of Elmira were shocked to learn their wells had been contaminated by chemicals long buried beneath a local industry. Plans to supply safe water to the town include a multi-million dollar pipeline from the Grand River, which itself is at risk of contamination through the movement of groundwater.

Activity 13A

Examining Water Samples

Problem

How do water samples from different sources in your area differ?

Materials

five plastic containers (100 to 200 mL), with lids
marking pen/grease pencil
microscope
clean microscope slides and cover slips
eyedropper
distilled water
disposable gloves

Procedure

1. *Prepare a table in your notebook with the following headings:*

Source	Environment	Physical Appearance and Odour of Sample	Description Under Microscope (with Sketch)	
			Water on Slide	After Drying
SAMPLE ONLY				

2. *Make a list of five different places where you could obtain a water sample, including one from a tap. A sample of distilled water will be used as a control. Submit your list to the teacher for approval before proceeding further.*

3. Label each container with a short form or number code so that you will know where the water sample was taken from.

 CAUTION!
Use disposable gloves when handling the water samples. Wash your hands thoroughly when finished.

4. Obtain a water sample from each approved source. You do not need to collect any more than 100 mL. To obtain the sample, remove the lid and invert the plastic container. Place it 10 to 20 cm below the surface of the water (if possible), and then turn it right side up. When the container is full, bring it to the surface, and replace the lid. Keep the sample container closed until you are ready to examine the water. While you are at the collection site, *briefly describe the environment at the place where you obtain the sample,* e.g., presence or absence of plants and animals, roads, drainage pipes, etc.

5. *Describe the appearance and odour, if any, of each sample.*

6. Label a microscope slide with the same code as the sample bottle. Use the eyedropper to transfer some of the water from the sample bottle to the slide. You will only need a few drops. Cover with a clean cover slip.

7. Examine the sample using the microscope. Begin with the low-power objective lens. Remember to adjust the light intensity.

8. *Describe what you see, and make a sketch. Record your observations in the table you have prepared. If you find any living things, indicate which organism seems most plentiful.*

9. After your observations are complete for this sample, remove the cover slip from the slide, and add another few drops of the sample.

10. Allow the slide to dry overnight. When the slide is dry, examine it again for any material which was left behind when the water evaporated. *Record your observations.*

11. Continue your examination for each water sample by repeating steps 4 to 10.

12. Prepare a microscope slide using a few drops of distilled water as your sample. Examine it in the same way.

13. Dispose of each sample as instructed by your teacher. Wash your hands with soap and warm water.

Questions

1. Which sample(s) contained the most living organisms?

2. Which sample(s) contained the most suspended non-living materials?

3. Compare the residue left on the slide for tap water and distilled water. Is there a difference? Why or why not?

4. Explain how the environment of the collection site may contribute to the materials (or lack thereof) found in the water sample.

CHECKPOINT

1. Name the three categories of wastes which pollute natural water systems, and give an example of each.

2. What is biochemical oxygen demand?

3. Explain how an excessive supply of nutrients can cause damage to a water ecosystem.

4. Explain how humans can be poisoned by heavy metals even if they do not drink the water from polluted waterways.

13.3 Testing Water Quality

Aquatic Organisms

Certain organisms that are present in rivers and lakes can be used as indicators of the condition of the waterway. For example, many organisms, such as the stonefly nymph, will only be present if the oxygen content of the water is high, while others, such as tubiflex worms and midge larva, are present in waters that are low in oxygen. "Coarse" fish, such as carp, will be found in low oxygen situations whereas game fish, such as trout, will not. Species of blue-green algae can also tolerate low oxygen levels.

Fecal Coliforms Test

Organisms which cause intestinal infections are excreted in large numbers in fecal material, and this material may contaminate water. Unfortunately, these pathogenic bacteria are difficult and time-consuming to identify. A non-pathogenic bacterium, the fecal coliform, *Escherichia coli*, is found in the human intestinal tract, and is not normally found in soil or water. This bacterium is easily identified and counted in a water sample. For these reasons, it is used as an indicator of the amount of pollution from human waste in water.

In a **fecal coliforms test**, a sample of water is taken, and passed through a filter capable of

Figure 13.5
Fecal coliforms. Each colony grew from a bacterium present in the water sample being tested. What does their presence indicate?

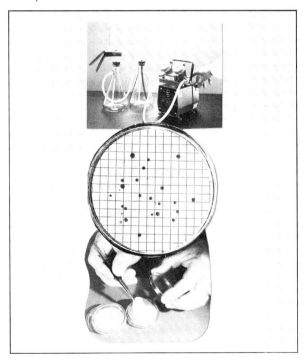

catching any bacteria present. The filter is then placed in a nutrient solution which encourages the growth of *E. coli* bacteria. (See Figure 13.5.) The colonies which develop thus provide a count of the number of fecal coliform cells which were in the sample of water being tested.

The results of fecal coliform tests are often published in the newspapers to provide an indicator of the cleanliness of local beaches. It must be pointed out that water may be free of coliform bacteria, but still be highly dangerous because it could contain pathogens from other sources, or toxic wastes.

Clarity and Colour

The clarity or transparency of water can also be an indicator of its condition. Suspended organic particles, minerals, and plant and animal organisms will cause the water to be somewhat cloudy, or **turbid**. A white plate, with two red or black pie-shaped sections, about 25 cm in diameter, called a **Secchi disc**, is lowered into the water. The line used to lower the disc is marked at constant intervals. The depth at which the disc disappears from view is an indication of the amount of suspended material in the water. Tropical waters can produce readings of 50 m or more. Northern lakes may give readings in excess of 25 m. At what depth would the Secchi disc disappear in the river or lake nearest your home?

The **Forel-Ule colour scale** is used to determine the colour of water. Each colour is an indicator of the dissolved and suspended materials in the water.

Additional Tests

Chemical analyses can be done if desired to give a more accurate picture of the quality of the water. The amount of dissolved oxygen and carbon dioxide, acidity (pH), hardness, ability to absorb acids, and the presence of certain elements and compounds can be measured.

Activity 13B

Analyzing Water Samples

Certain chemicals which are dissolved in water samples can be identified by using other chemicals as indicators. If a reaction occurs when the indicator is added to the water sample (usually shown by a colour change), the chemical is present. Commercial test kits contain the indicators needed to test for a variety of dissolved chemicals. They are easy to use, and provide a simple, yet effective, method of analyzing water samples. Be sure to follow the safety instructions provided with the kit.

After obtaining your teacher's permission, gather a number of different water samples, including tap water, and samples from the surface and near the bottom of a small lake, pond, stream, or river. (How will you take a bottom sample? What safety measures must be taken?) Your teacher may wish to include the sources used in Activity 13A.

CAUTION!
Use disposable gloves when handling water samples.

Follow the instructions in the test kit to test for pH, total alkalinity, hardness, and dissolved chemicals such as oxygen and carbon dioxide. *Record the results of each test in your notebook.*

What do the results of each test indicate about the quality of the water sample? Explain your conclusions.

CHECKPOINT

1. What simple indicators can be used as a measure of water quality?

2. Why is *Escherichia coli* used as an indicator of the presence of human waste in water?

3. (a) What is a Secchi disc?
   (b) How is it used?

4. What chemical tests can be done to determine water quality?

13.4 Water Purification: Water Fit to Drink

Cities and towns in Canada take their water supply from local rivers, large lakes, or wells. This water contains a variety of other substances, from living organisms to debris and possible chemical contaminants. **Water purification plants** must therefore treat water using some or all of the following methods to ensure it is safe to drink. (See Figure 13.6.)

Treatment Methods

1. Vertical Screens – keep out large organisms and debris.

2. Chlorine – kills micro-organisms, including pathogens.

3. Chemical Coagulants – bind fine particles together into clumps called floc so they settle out more quickly.

4. Mixing Tanks – in which chemical coagulants are mixed with water.

5. Settling Tanks – in which suspended material, including floc, settles to the bottom for removal.

Figure 13.6
A water purification plant. Note the stages in the treatment.

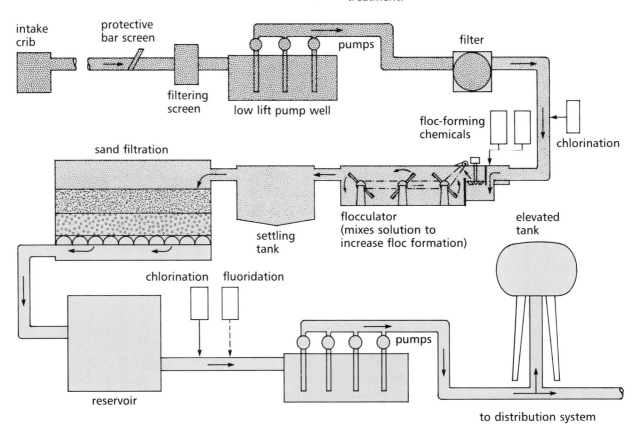

6. Filtration Through Sand and Gravel – to remove very fine, suspended solids.

The amount of water needed by a community varies during the day and during the year. Water purification plants store water in *reservoirs* against times of increased demand. Before the water leaves the reservoir, the level of chlorine is checked and more is added if necessary. In many areas, hydrofluosilicic acid (fluoride) is added at this time to improve dental health.

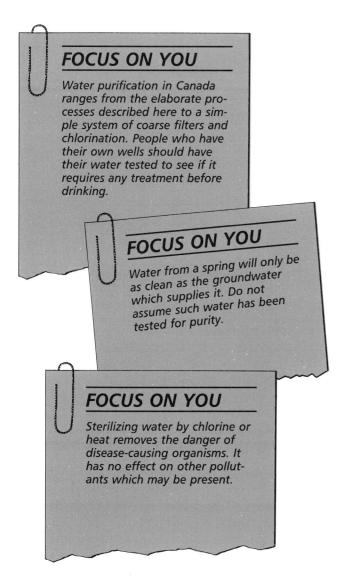

FOCUS ON YOU

Water purification in Canada ranges from the elaborate processes described here to a simple system of coarse filters and chlorination. People who have their own wells should have their water tested to see if it requires any treatment before drinking.

FOCUS ON YOU

Water from a spring will only be as clean as the groundwater which supplies it. Do not assume such water has been tested for purity.

FOCUS ON YOU

Sterilizing water by chlorine or heat removes the danger of disease-causing organisms. It has no effect on other pollutants which may be present.

Activity 13C

Water Purification

How do floculation and filtration help purify water?

Materials

sterilized soil
aluminum potassium sulphate (alum)
250-mL graduated cylinder
calcium hydroxide (limewater) solution (saturated solution)
sodium phosphate solution, dilute
250-mL Erlenmeyer flask
large plastic container (or flower pot)
gravel, medium size
crushed stone
sand, coarse
sand, fine (or flint shot)
disposable gloves
eye protection
(Your teacher may provide you with additional water samples.)

Procedure

1. (a) Add approximately 100 mL of the sodium phosphate solution to a 250-mL Erlenmeyer flask.

    CAUTION!
   Wear disposable gloves and eye protection when mixing or adding chemical reagents.

   (b) Slowly add the calcium hydroxide solution to the flask until a change takes place. *Record your observations.*

2. (a) Fill the graduated cylinder about two-thirds full with tap water.
   (b) Add a small amount of soil (approximately 50 g) to the cylinder, and shake well. Allow the mixture to settle. Describe the water in the cylinder after the soil has settled for a few minutes.

    CAUTION! Calcium hydroxide can irritate skin and sodium phosphate is moderately toxic. Avoid spilling these substances on your skin.

   (c) Add a small amount of alum (5 to 10 g) to the cylinder, and shake the mixture again. Allow the mixture to settle. *Record your observations.*

3. (a) Make a small hole about the size of your finger in the bottom of the plastic container.
   (b) Add enough gravel to the container to cover the bottom.
   (c) Cover the gravel with a 10-cm layer of crushed stone. Add the coarse sand to completely cover the crushed stone.
   (d) Make a 10-cm layer of fine sand on the top.
   (e) After mixing up the solution in the cylinder from step 2, slowly pour it into the top of the pail, and collect the water which escapes from the bottom. *Describe the appearance of the water before and after it passes through the filtering layers.*

Questions

1. (a) What kinds of materials are removed from water by flocculation and filtration?
   (b) What materials are not removed by these processes?
2. Predict what will happen to the filter in step 3 after it has been used several times. Suggest two or more ways in which this problem can be corrected.

If you have travelled to other countries on holidays, you know that many countries do not enjoy safe drinking water supplies. Canadians are fortunate. Not only do we have the most abundant freshwater supplies of any nation, but we also have some of the purest water sources and a very high standard of water quality control. Our good fortune requires a responsible attitude to see that our waters are not contaminated by careless or irresponsible disposal of wastes.

CHECKPOINT

1. Why is chlorine added to water?
2. (a) What is floc?
   (b) Why are floc-forming chemicals added to water?
3. List the main steps in the water purification process, and briefly explain what is done at each step.

13.5 Wastewater Treatment: Return to the Environment

It has been estimated that the average amount of water used per person in one day is about 340 L; however, we only take into our bodies about 2.1 L of it. Much of the water that we use carries away waste substances, such as sewage.

The Composition of Wastewater

Domestic sewage is the waste from toilets, sinks, laundry tubs, etc. Together with any commercial or industrial wastes, which are legally or illegally washed into the sewer system, this mixture of **wastewater** arrives at a treatment plant. These plants are primarily designed to treat sewage.

Toxic substances from industrial wastes may pass through such systems without being effectively removed, as well as, in some cases, destroying the bacteriological balance of the system, so that it fails to treat the normal domestic sewage.

Wastewater contains solids, liquids, and some gases. Among the solids are paper, rags, garbage, feces, sand, and silt. The remaining solids are either *suspended solids*, which are large enough to be seen and filtered out, or *dissolved solids*, which cannot be filtered out.

Septic Systems: Treating Small Amounts

The wastewater produced from a single dwelling may be adequately treated by a **septic system**. (See Figure 13.7.) The word septic means to decay or rot. In such a system, solid wastes are decomposed within an underground tank. Liquid wastes trickle out into the soil through a series of pipes called a leaching or absorption field, with soil micro-organisms acting to break down the organic material. There must be enough area for this decomposition to occur before the water reaches ground or surface water. The soil beneath a septic system must therefore be at least 150 cm deep. In addition, septic systems cannot be placed close to each other, near wells used for drinking water, or near open water.

Processing Large Quantities of Wastewater

There are two main processes in municipal wastewater treatment – **primary treatment**, which removes settleable solids, and **secondary treatment**, which breaks down organic matter by aerobic bacterial action. In Canada, the majority of plants combine these two processes into an **activated sludge system**. (See Figure 13.8.)

Primary Treatment

Primary treatment removes the small settling solids and reduces the suspended solids by 40 to 60 percent. It requires a series of steps:

1. **Screens** block the entry of large, hard, solid objects.

2. **Comminutators** shred and grind solids that have passed through the screens into smaller particles.

3. **Settling tanks** or **clarifiers** are large tanks that are usually aerated. The bubbles of air produce a tumbling action which keeps organic matter nearer the top and allows heavy particles of grit and sand to fall to the bottom. Floc-forming chemicals may also be added. The grit is continuously removed from the bottom by scrapers or other techniques, and, if free of toxic substances, will be taken to landfill sites for disposal. Floating solids and scum are skimmed off during this process.

Figure 13.7
A septic tank system

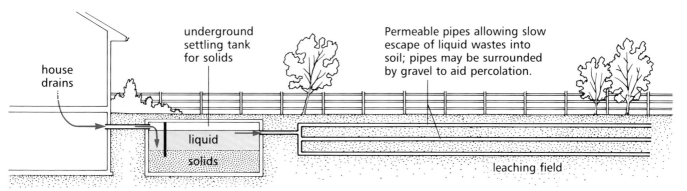

Chapter 13 / Managing Human Waste

Figure 13.8
Wastewater treatment by conventional activated sludge system

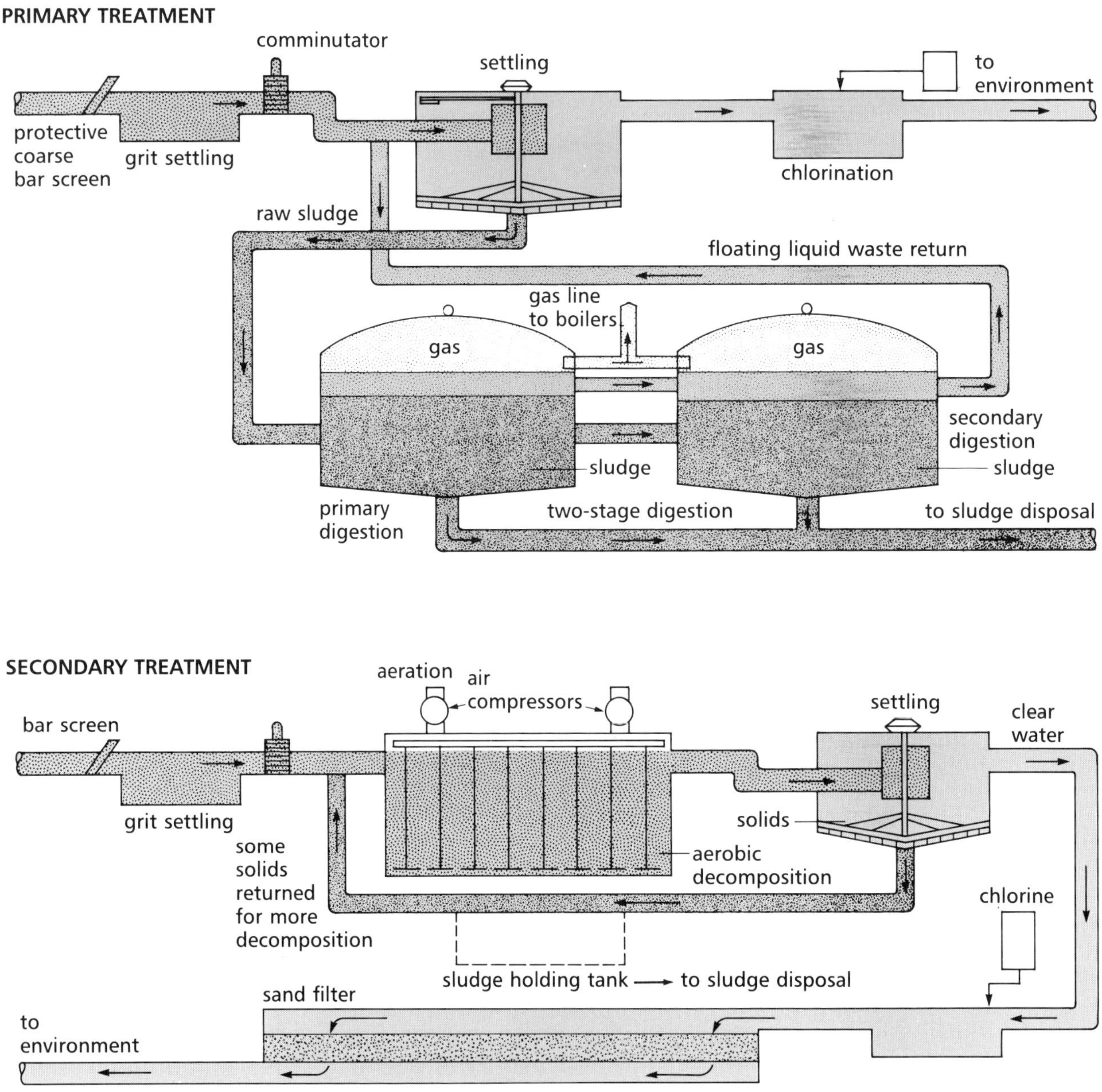

The resulting liquid still carries 60 to 70 percent of its original organic material. Primary treatment alone does not greatly improve the water quality; if it is foul, poisonous, and smelly, it remains so. In some treatment plants, it is released directly into the environment, such as a local river, where aquatic decomposers must finish the process.

Secondary Treatment
Most natural ecosystems cannot cope with the amount of organic material remaining in wastewater after primary treatment; thus, secondary treatment consists of an "artificial" ecosystem, in which micro-organisms in **aeration tanks** break down the organic matter in the presence of oxygen. (See Figure 13.9.)

Other tiny organisms are also present, acting as scavengers and feeding on some of the free bacteria present. These organisms are separated from the treated wastes, settled out in the final clarifier, and used again. It is vital that no toxic substances enter these tanks. If organisms in the tanks are killed, the treatment process would be ineffective.

The liquid that flows from the secondary treatment plants still contains nitrogen and phosphate chemicals. Many other dissolved substances will also be present.

Figure 13.9
Aeration by bubbling or spraying provides the oxygen required by the bacteria. What would occur if oxygen were not constantly added to the water?

Sludge

During both stages of wastewater treatment, suspended and dissolved materials are collected. These materials are known as **sludge** – **raw sludge** coming from primary treatment, and **digested sludge** from secondary treatment. Sludge is currently disposed of in one of three ways, depending upon the contaminants it contains – by incineration, being spread on soil as fertilizer (for crops not directly intended for human consumption), or in landfill sites.

Tertiary Treatment

Tertiary treatment consists of chemical processes designed to remove specific contaminants, such as phosphates, remaining in the water following secondary treatment. The type of tertiary treatment will depend on what the water usually contains, which will vary from one city to another. The water leaving a tertiary plant is usually pure enough to drink, but the process is too expensive to be widely used.

Water Quality Standards

Water is monitored constantly before, during, and after its discharge from wastewater treatment plants to see that it meets specific quality standards. (See Figure 13.10.) Records and monthly reports must be submitted to the appropriate government agency. Each report contains an analysis of factors such as hardness, turbidity, temperature, calcium, phosphates, dissolved oxygen, and more than 20 heavy metals. Coliform counts are expected to be zero.

Being able to flush waste down the drain is not something to take for granted. If more wastewater enters a plant than it can hold for treatment, as may occur after a torrential rainstorm, or if an excessive amount of contamination destroys the aerobic bacteria, there may be no alternative but to chlorinate then discharge wastewater directly into the environment – the same environment from which our drinking water must come.

CHECKPOINT

1. What is the purpose of the tank in the septic tank system?

2. Why is the spacing of the absorption fields of a septic tank system of concern?

3. What is the main purpose of each of the two main processes in wastewater treatment?

4. Briefly explain how suspended solids are removed from wastewater in the primary treatment stage.

5. Why is the wastewater aerated in the secondary treatment stage?

6. Describe the method used to dispose of sludge.

7. Why is a tertiary treatment process only used in some cities, and not others?

Figure 13.10
Discharges into natural waterways are constantly tested.

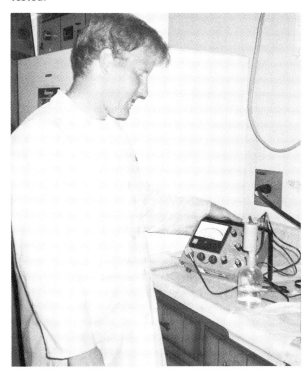

13.6 Solid Waste

Each Canadian discards nearly 2 kg of domestic garbage every day. (See Figure 13.11.) This amounts to about 20 million tonnes per year. If all that waste were placed in garbage bags and lined up, it would stretch across Canada some 300 times.

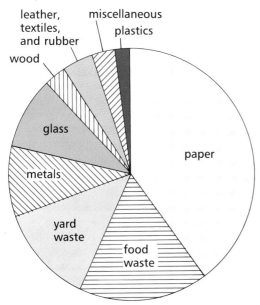

Figure 13.11
The composition of urban solid waste

- **Domestic solid waste** is waste generated by households and, in larger communities, its transport and disposal is the municipality's responsibility. It may include paper, glass, cans, food substances, dust, packaging material, leaves, brush, and garden refuse. Domestic solid waste also often includes garbage collected from stores and small businesses, consisting of large amounts of cardboard and paper.

- **Industrial solid waste** includes a wide variety of materials. In this category falls construction debris, wood, brick, mortar, concrete, and steel, as well as machine scrap or mine tailings. Food industries produce large amounts of organic wastes. Non-hazardous industrial solid waste may be accepted by some landfill sites, but it is usually the responsibility of the producer company to transport these products.

- **Hazardous wastes** are substances that are toxic or potentially harmful to humans and the environment. They are not just produced by industry. Each household discards over 30 L of hazardous waste each year. Old medicines, paints, thinners, oven cleaners, herbicides, rat poisons, oils, and antifreeze are some of these substances. The containers from these products often hold a considerable quantity of the substance. When broken and crushed, and placed in landfill sites, their contents may be released, endangering groundwater.

Some municipalities are now collecting household hazardous waste at special depots, or sponsor hazardous waste collection days. The average annual cost of collecting these wastes can range from $50 to $150 per household, and normally only two percent of all households participate, even though such substances should not be discarded with normal garbage.

Once collected, the hazardous wastes are separated into different chemical groups. Usable substances such as paint or oil are forwarded to recycling companies. The remainder is sent to waste disposal companies for treatment.

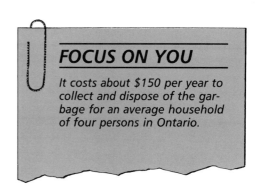

FOCUS ON YOU

It costs about $150 per year to collect and dispose of the garbage for an average household of four persons in Ontario.

Activity 13D

Composition of Urban Solid Waste

Problem

What kinds of things make up most of our garbage?

Procedure

1. *Prepare a table in your notebook as shown at the bottom of this page.*

2. Prepare a box with a plastic bag liner for each type of solid waste. It is easiest to sort the waste as it is discarded. Ask the members of your family to co-operate with you in this activity for a one-week period. As items are discarded, *record the number of items,* and place them in the proper container.

3. At the end of one week, compare the relative masses and volumes of each of the containers *and record them in your table. Rank the containers in descending order by volume and by mass.* (A bathroom scale may be used to estimate the mass of each container.)

4. *Total the estimated masses and calculate the proportion of each type of waste.*

Questions

1. (a) What items might not have been put into the garbage if reusable products had been used?
   (b) What proportion of the entire amount is this?

2. (a) What things could be removed from the waste for recycling or composting?
   (b) What proportion of the entire amount is this?

3. Use the results obtained from Questions 1 and 2 to calculate the mass of materials that could be saved by a city of 1 million people.

Type of Waste	Number of Items	Estimated Mass	Percent of Total Mass	Rank	
				Mass	Volume
Paper					
Food waste					
Yard waste					
Metals					
Glass					
Wood					
Plastics					
Textiles, rubber					
Other					
Total			100%		

Dumps and Landfill Sites

In the past, the way to dispose of solid waste in a small community was to transport it to some isolated area, not too far from town, but away from people's homes, and just dump it. (See Figure 13.12.) Now, this solution is no longer acceptable. Such dumps attract disease-carrying scavengers as well as potentially poisoning local wildlife, and can be breeding sites for flies and pathogenic organisms. They can also catch fire. Precipitation falling on the material in the dump dissolves chemicals in the waste. This contaminated water, called **leachate**, can run off and pollute nearby rivers or lakes. It may also seep into the groundwater.

Most of the problems with open dumps are eliminated, or reduced, by using a **sanitary landfill** process. (See Figure 13.13.) Much of our solid waste is **biodegradable**, that is, it can be decomposed or broken down and made harmless by natural processes. Within each cell of the landfill, biodegradable substances are decomposed by **aerobic** organisms until the oxygen supply is depleted. The process is then continued more slowly by **anaerobic** organisms. During the process of anaerobic decomposition, methane gas is produced which, in some sites, can be collected as a source of energy.

Figure 13.12
What problems are associated with this method of waste disposal?

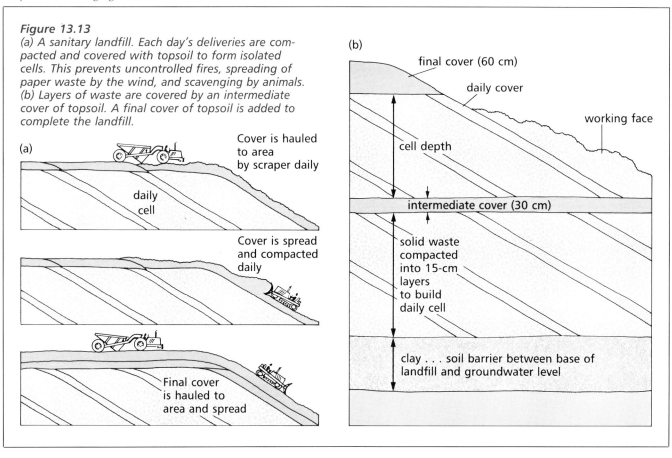

Figure 13.13
(a) A sanitary landfill. Each day's deliveries are compacted and covered with topsoil to form isolated cells. This prevents uncontrolled fires, spreading of paper waste by the wind, and scavenging by animals.
(b) Layers of waste are covered by an intermediate cover of topsoil. A final cover of topsoil is added to complete the landfill.

Other materials are **non-biodegradable**, such as certain plastics, glass, and metals. These materials resist being broken down by heat, light, and decomposer organisms, and remain for a very long time in the environment.

Landfill site selection is important, as any toxic leachate from its contents must be kept away from groundwater by an impervious base such as heavy clay or rock. (See Figure 13.14.)

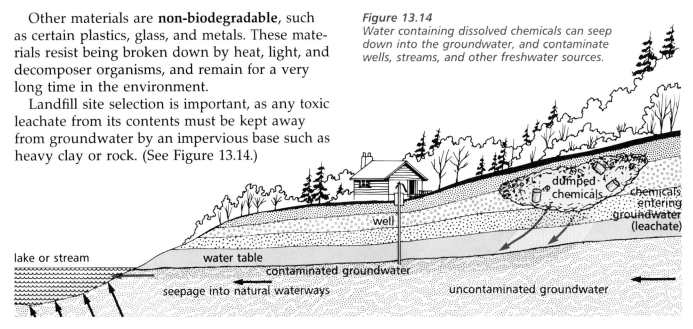

Figure 13.14
Water containing dissolved chemicals can seep down into the groundwater, and contaminate wells, streams, and other freshwater sources.

Sometimes liners of plastic, concrete, or other waterproof materials are used to line the bottom of the landfill site. In either case, in modern sites, a leachate collection system is usually installed to carry away contaminated water for treatment.

When a landfill site is full, the land is unsuitable for heavy buildings. Old landfills often become valuable recreation properties, from golf courses to ski hills to parkland.

Incineration

For many large cities, the increasing cost of transporting solid waste to landfill sites, plus the growing cost of land within a reasonable distance, makes **incineration**, or the burning of waste, an economical alternative.

Modern city incinerators accept garbage through a hopper. The waste passes over various sorting grates and some metals are recovered by magnets. The remaining material then passes through a rotating **kiln** to be burned. About 85 to 90 percent of the volume of waste is burned to ash. The remaining waste and ash must still be disposed of in landfill sites.

While all incinerators produce some air pollution, including poisonous gases from burning plastics, modern designs include devices, such as **scrubbers** and **electrostatic precipitators**, which collect some gases and small particles, to reduce this problem.

One advantage of incineration is that it provides a source of heat energy that can be used to heat buildings or generate electricity. A **solid waste reduction unit (SWARU)** in the Hamilton-Wentworth region of Ontario burns about 400 t of solid waste each day, reducing the volume of the waste by 95 percent. (See Figure 13.15.) The energy produced is used to generate 14 million kilowatt hours of electricity annually.

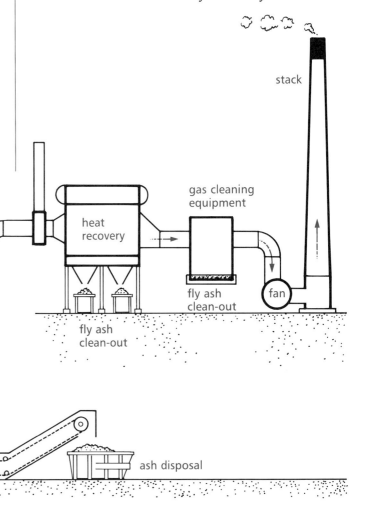

Figure 13.15
A simplified incinerator

Activity 13E

Assessing a Waste Disposal Site Proposal

Your community has determined that its current waste disposal site will be filled within three years. A community meeting is scheduled to discuss the selection of a new site.

Your teacher will divide you into small groups, and each group will present one of the following points of view at the "community meeting":

- site location committee
- municipal government
- city engineer
- local politician (trying to satisfy everyone)
- environmental activist
- consultant (hired by site location committee)
- irate neighbours (two) who live near a potential site
- irrational, negative loudmouth who is against everything
- calm, rational, interested citizen
- local media (to report on meeting)

Each group will present its viewpoint, based on logical arguments. Students will be expected to present a case, answer questions, defend their proposals, and point out weaknesses in other proposals.

Use actual information from your area as the basis for your arguments, where possible. For the location selected, find out about soil types, distance from collection points, existing roads, traffic patterns, land expropriation required, costs, proximity to homes, waterways, parks, quantity and type of waste to be disposed of, size of site required, water table and leachate problems, and the predicted number of years of operation. Include any additional relevant information your group decides is useful.

The media report may be prepared in any format. (Videotapes should be edited to a maximum 10-min presentation.) Each group will have the opportunity to evaluate the media coverage of its viewpoint.

CHECKPOINT

1. List the three types of solid waste and give an example of each.

2. What is leachate?

3. (a) Why is it illegal to dispose of household hazardous wastes in landfills?
   (b) What other means of disposal are available?

4. Define the following terms, and give an example of each: biodegradable, non-biodegradable.

5. (a) What can landfill sites be used for once they are filled and closed?
   (b) What can they not be used for? Why?

6. What are the advantages of solid waste disposal by incineration?

FOCUS ON YOU

The boom has already begun for biologically safe substitutes for hazardous substances used in the home. So called "green products" may cost up to 50% more, but consumers seem willing to pay the difference. Companies are springing up around the world to satisfy the environmental shopper.

13.7 Alternatives to Disposal

One solution to the problem of waste disposal is to reduce the amount of waste that is produced. This requires a change in our thinking and also in the way we live and shop. With some effort on the part of consumers, solid waste could be reduced by 75 percent.

The Four R's of Waste Reduction

- **Reuse**. Repair, share, borrow, rent, or buy a previously owned product; avoid disposable products (for example the estimated number of disposable diapers used by the average baby from birth to two years is 7000); look for needed items at garage sales and second-hand shops; use soft rags for cleaning and packaging rather than paper towels.

- **Reduce**. Buy only what you need; buy in packages which minimize wastes, since packaging normally makes up 38 percent (by mass) of all garbage; reduce the amount of non-essential paper products you buy.

- **Recycle**. Up to 80 percent of household waste is recyclable; paper, glass, metal, and some plastics can be recycled to reduce the amount of garbage by 28 m³ per household each year.

- **Recovery**. Composting of grass clippings and kitchen leftovers recovers nutrients; recovery systems now reclaim the silver from photo processing, chromium acid from chrome-plating waste, and waste oil; in 1988, 3500 Canadian companies converted 340 000 t of wastes into useful products; regional waste exchanges match waste-producing companies with other companies who can put the waste to good use.

Recycling Programs

The four main recyclable components are paper, glass, cans, and plastic. These must be separated before proceeding, and this is the reason for the blue boxes. (See Figure 13.16.) Householders separate these materials from other non-recyclable garbage. Biodegradable (compostable) materials, such as kitchen and garden wastes, have recently been added to the list of materials recycled by municipalities. Table 13.2 provides a summary of some of the factors to be considered in this program.

Figure 13.16
The blue box has become a common sight as more and more communities encourage recycling. This material was separated from one week's waste of a family of four.

FOCUS ON YOU

Recycled paper reduces the energy required to produce new paper by 50 percent. It also saves 17 trees per household per year.

BIOTECH

Composting to Reduce Waste

Organic materials, when separated from garbage, can be used for **composting**. Waste from food preparation and meals, leaves, lawn trimmings, and garden trash decompose very quickly into simpler compounds. If these compounds are returned to the soil, they act as fertilizer and enrich the soil. Some municipalities collect compostable materials. In other areas, composting is done on an individual basis.

To prepare a site for backyard composting, a small section of the yard can be enclosed with wire netting to form a loose boxlike structure. A large garbage can, barrel, or wooden box with the bottom removed and holes punched in the sides can also be used. The height of the enclosure should range between 0.3 and 1.5 m for best results.

Begin with a thin layer of commercial fertilizer, and then add dry and wet layers of materials in thicknesses of between 8 to 15 cm. Cover each layer with a sprinkling of soil. The pile of materials should be turned or mixed every two or three weeks, to ensure that the pile is aerated. It must be kept moist, but not soggy.

Micro-organisms decompose the waste to a dry, odourless material called **compost**. The compost will be ready to use when it no longer produces heat after each turning. When it is added to the garden, compost lightens the soil and improves its water retention ability. In addition, it provides a natural supply of soil nutrients.

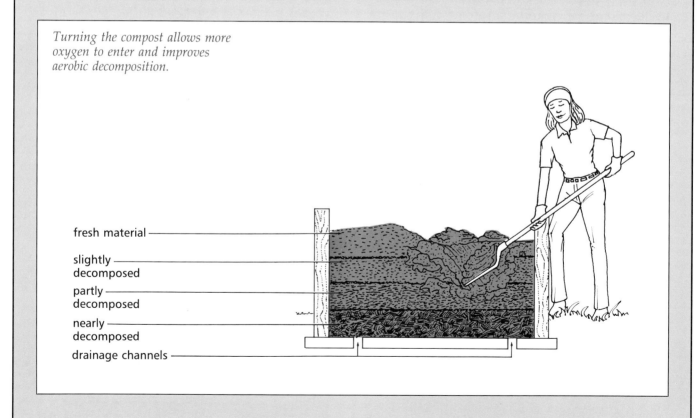

Turning the compost allows more oxygen to enter and improves aerobic decomposition.

Table 13.2 The Four Main Recyclable Components

	Newspapers	Glass	Cans	Plastic
Types of Buyers	Papermills, cellulose insulation and building product manufacturers, dealers and brokers	Glass container manufacturers	Scrap dealers and de-tinning plants	Manufacturers of recycled packaging
Contamination	Newspapers must be free of non-paper items and glossy magazines	Glass must be sorted by colour; no ceramics or stones; some buyers insist on no caps or neck rings	Excessive food or residue contaminants may cause load rejection	Styrofoam, "biodegradable" or brittle materials cannot be used
Preparation	May be baled or loose according to plant facilities; string and baling ties are no problem	Crushing depends on company; if crushed, no mixed glass; bulk decreases costs	Need not be baled but crushing or compacting allowed; must flow freely into plant equipment	Shredding
Prices	Fluctuate according to demand	Fairly stable prices; vary with contamination; better prices for regular large loads	Stable, but a limited market	Increasing market as construction material and for packaging

CHECKPOINT

1. List the four R's of waste reduction.

2. Explain how a typical secondary school student could reduce the amount of waste that is produced
   (a) at home
   (b) at school.

3. What are the four main components of recycling?

13.8 Industrial Hazardous Waste

Industrial hazardous wastes include any substance which is ignitable, corrosive, reactive, toxic, radioactive, infectious, phytotoxic (interferes with photosynthesis), or mutagenic. These substances may only be present in very small amounts or concentrations. The concentration at which each of these substances present a danger to people, plants, animals, and the environment is known as its **threshold level**.

Regulations have been made to control the disposal of these substances. Each industrial producer of hazardous waste must be registered with the government. Special storage and disposal facilities must be used. Companies which transport or dispose of these wastes must also be registered, and licensed by the government. (See Figure 13.17.) These records are kept on a computerized data base for monitoring purposes.

Figure 13.17
Companies which transport hazardous wastes must be registered and licensed by the government.

Disposing of Hazardous Waste

Hazardous waste disposal is managed in a number of different ways. One legal requirement is that hazardous wastes must be separated from other wastes for more efficient disposal. If possible, the waste produced should be minimized, or alternatively, non-toxic materials should be used. The toxic waste from one plant may be used productively by some other industry (*waste exchange*). Wastes may be reprocessed to recover energy and to reuse or recycle recoverable materials. If these options are not available, hazardous wastes must be disposed of in a manner that conforms to strict regulations to maintain a safe environment.

The *Canadian Environmental Protection Act* requires that new substances be assessed before they are used. Anyone has the right to ask that a substance be assessed, and to ask for information about a substance. A minimum of two people are required to request that the government investigate possible cases of pollution. Polluters can be assessed fines of up to $1 million a day, and possible jail sentences.

Activity 13F

Waste Management Technology

Disposing of waste poses many problems, some of which may be confronting your own community. Select one of the types of waste problems (for example, domestic solid waste, wastewater, or hazardous waste), and investigate how it is being dealt with in your community. Attempt to find out what kind of technology is being used to process the waste, and how effective it is. *Prepare a brief written report summarizing your findings.*

CHECKPOINT

1. *What substances are considered to be hazardous wastes?*

2. *Why is it important that hazardous waste be monitored?*

3. *List five ways that the disposal of hazardous wastes can be managed.*

13.9 Airborne Wastes

We usually refer to airborne wastes as **air pollution**. The concentration of pollutants in an air sample varies from one location to another, and from one day to another, but no place on earth is free of air pollution. One new government building in Ottawa could not be used after its completion because the materials used in its construction produced hazardous fumes.

The Pollution Index

The **pollution index** is an alert system used in most industrial areas of the country. The index is based on the levels of sulphur dioxide and suspended particles in the air. The index forms an ascending scale starting at zero, with any reading below 32 considered acceptable, that is, has no measurably adverse effects on human health. At an index reading of 58, persons with respiratory disorders may be at risk. At 100, prolonged exposure could have mild effects on healthy people, and serious effects on persons with heart or respiratory diseases.

A value of 32 is considered an *advisory level*. At 50, a *first alert* is declared, and at 75, a *second alert*. A level of 100 is considered the *air pollution episode threshold level*. The prevailing weather conditions are also considered. In conditions of low winds and aggravating temperatures, industries may be advised to prepare for possible slowdowns in their operations. At 50 or 75, they can be ordered to shut down all operations which produce major air pollution emissions. At 100, all businesses not essential to the health and safety of the public can be ordered to close down. Fortunately, if the early steps have been followed, the reading of 100 is rarely reached.

Basically, air pollution is caused by three processes – combustion, vaporization, and mechanical action.

Combustion

Combustion involves the burning of materials, such as fuels or solid waste. The complete combustion of hydrocarbons (compounds of carbon and hydrogens) will produce carbon dioxide and water vapour. Other materials which are present during combustion may be released as pollutants.

Carbon Monoxide
Carbon monoxide is a colourless, odourless, and tasteless gas produced by incomplete combustion. When it is inhaled, it quickly associates with hemoglobin in the blood. This blocks the vital transport bond sites of oxygen, causing anoxia, a condition in which the tissues have an insufficient supply of oxygen. It is therefore extremely dangerous to leave an automobile running in an enclosed or poorly ventilated area, such as a garage. Gas heaters and furnaces must be properly vented to the outside of a building. Even cigarette smoking in a closed car can build up levels of carbon monoxide high enough to cause disastrous effects on a person's ability to drive. Keeping car engines and furnaces properly tuned and ventilated can reduce the carbon monoxide emissions.

Sulphur Dioxide
Fuels such as coal, which contain sulphur, produce *sulphur dioxide* as a combustion product. Mining, smelting operations, and metal producing mills release more than 1000 t of sulphur dioxide per day into our atmosphere. This pollutant is a colourless gas, with a pungent, suffocating odour. It irritates the eyes and respiratory system in concentrations as low as five parts per million (ppm). For years, the usual way to get rid of this gas was to use tall chimney stacks, so the gas did not reach the neighbouring cities. (See Figure 13.18.) As the cities expanded, the chimneys were built higher and higher. This reduced the concentration of the gas in the local

area, but spread the pollution to previously unaffected areas.

Emissions of this gas are now greatly reduced by washing and scrubbing processes before the gas is released from the stacks. Such processes produce sulphuric acid which can be sold at a profit.

Figure 13.18
What is the function of a smokestack and what problem does it cause?

Nitrogen Dioxide
The high temperatures of engines and furnaces cause the nitrogen and oxygen in the air to react and form *nitrogen oxide* compounds. Nitric oxide combines with oxygen in the presence of sunlight to form nitrogen dioxide, a major contributor to photochemical smog, which irritates the eyes and respiratory pathways. Emission control devices in automobile exhaust systems reduce the emission of these nitrogen compounds.

Lead
Lead accumulates in the body tissues, affecting the nervous system, and can cause brain damage in high concentrations. The use of a lead compound in gasoline as an "anti-knock" has released lead into the atmosphere. Leaded gasolines have been replaced by unleaded fuels, since a program was announced by the government to end production of leaded fuels in 1990.

Vaporization

Some organic compounds are easily vaporized. These volatile substances are primarily gasoline or organic solvents, such as benzene, used in industry. The unburned hydrocarbons play a role in the formation of photochemical smog, and they may react to form other compounds which irritate the eyes and lungs.

Mechanical Action

The mechanical action of drilling, grinding, sanding, and sweeping causes dust and other particulate matter to enter the atmosphere. Workers using such tools must wear protective masks at all times to prevent these substances from entering the lungs and causing health problems.

Asbestos fibres, for example, cause irritation and swelling in the lungs, clog airways, and produce shortness of breath. This, in turn, may cause more serious disorders. (See Figure 13.19.)

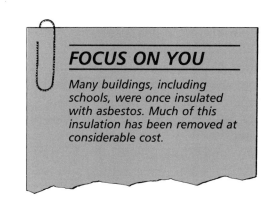

FOCUS ON YOU

Many buildings, including schools, were once insulated with asbestos. Much of this insulation has been removed at considerable cost.

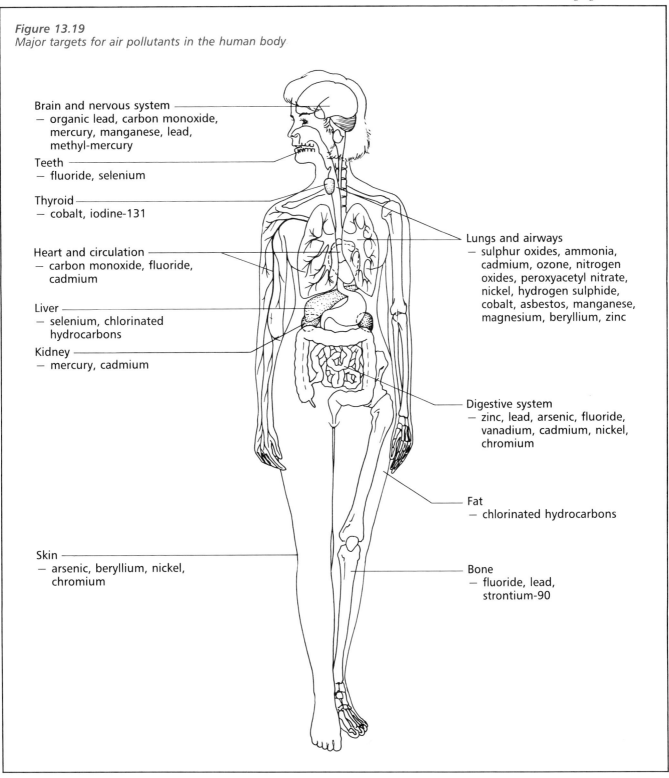

Figure 13.19
Major targets for air pollutants in the human body

Particles of different types are found in the atmosphere. Those particles with diameters greater than 10 μm tend to fall out of the atmosphere very quickly and are not considered to be extremely harmful. Medium-sized particles (between 1 and 10 μm), for instance, coal dust and fly ash, remain in the air much longer. Of greatest concern are fine particles (less than 1 μm). Particles of this size are suspended for long periods of time, and are small enough to penetrate the natural defenses of the lungs. (See Figure 13.20.)

Figure 13.20
A micrometre is one-millionth of a metre. It is represented by the symbol "μm". To give you some idea of its size, imagine that you could slice the tiny part of a ruler that lies between two of the millimetre markings into a thousand pieces. (A millimetre is one-thousandth of a metre.) One of the pieces would be 1 μm in width.

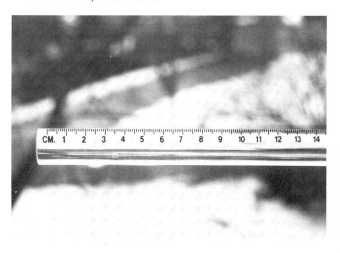

Smog

Originally, the word **smog** meant a combination of smoke and fog. It now applies to a haze produced in the atmosphere by pollution.

- **Photochemical smog** tends to occur in cities which have sunny, warm, dry climates. It contains a mixture of nitrogen oxides, carbon monoxide, ozone, and other chemicals called PANs (peroxyacyl nitrates), produced by the action of sunlight on automobile and industrial exhausts. (See Figure 13.21.)

- **Industrial smog** tends to occur in cities which have long, cold, wet winters, and depend heavily on coal and oil for heating, manufacturing, and producing electricity. It contains mainly particulate matter and sulphur dioxide.

Activity 13G

What Goes Up, Must Come Down

Problem

What kinds of particles are in the air we breathe?

Materials

three microscope slides
three cover slips
petroleum jelly
microscope
plastic ruler with millimetre divisions
marking pen/grease pencil
spatula

Procedure

1. Prepare a table in your notebook with the following headings:

Location	Description of Particles	Range of Particle Sizes
	SAMPLE ONLY	

Figure 13.21
Photochemical smog is produced by the action of the sun on automobile and industrial exhausts.

2. Select three different locations (both indoors and outdoors), and label one microscope slide for each location.

3. Cover one side of each microscope slide with a fine layer of petroleum jelly. Use your finger to apply a thin layer, and then remove any excess with the edge of a spatula. The layer of petroleum jelly should be even and as smooth as possible.

4. Place the slides in the locations you have selected, and allow them to remain there for 24 h.

5. Determine the width of the field of view for the microscope you will use to examine the slides. This will allow you to estimate the size of the particles that are found on the slide.

 To do this, follow these steps:
 (a) Place the clear plastic ruler on the stage.
 (b) Using the low-power objective, focus on the ruler and determine the field of view based on the number of millimetre divisions you can see.
 (c) To determine the field of view for the other objective lens, divide the magnification of the objective lens by the magnification of the low-power objective lens, then divide the field of view obtained in (b) by this number.

6. Place a cover slip on an area of the slide that appears to have particles on it. Observe the particles using the low-, medium-, and high-power objective lenses. *Describe the particles you find. Are any of the particles coloured? Do any of them have regular shapes? What might they be? Estimate the sizes of the particles you find.*

Questions

1. What might have produced the particles that you collected from each location?
2. What kind of activities are conducted near your collection site that may have increased the number of particles found in the air?

Reducing Air Pollution

It is not always possible to completely prevent air pollution. Some concessions have to be made, whether for economic or technical reasons. The pollution index helps monitor and control air pollution. There are laws to enforce pollution reduction, with heavy fines for corporations and individuals who do not comply. Convicted individuals can also face imprisonment.

The governments of some provinces have introduced legislation to provide financial help to companies where the installation of *abatement equipment*, such as scrubbers, can reduce pollution. Loans, grants, sales tax exemptions, and capital cost programs are also used to encourage companies to consider abatement programs.

CHECKPOINT

1. What three processes are the main causes of air pollution?
2. What is the pollution index and why is it used?
3. Why is carbon monoxide difficult to detect?
4. What are the sources of most of the sulphur dioxide and nitrogen dioxide emissions?
5. How can you protect yourself from breathing harmful particulate matter?
6. What is the difference between photochemical smog and industrial smog?
7. What devices help to reduce emissions which cause air pollution?
8. What steps has the government taken in an attempt to reduce air pollution?

13.10 Acid Precipitation

Most of the sulphur dioxide and nitrogen dioxide which is released into the atmosphere combines very quickly with water vapour and oxygen to produce acids. The chemical word equations for these reactions are shown here:

- During combustion:
  sulphur + oxygen → sulphur dioxide

  In the atmosphere:
  sulphur dioxide + water + oxygen → sulphuric acid

- During combustion:
  nitrogen + oxygen → nitrogen dioxide

  In the atmosphere:
  nitrogen dioxide + water → nitric acid

Droplets of these acids dissolve in rain or snow, and fall out of the atmosphere. This is **acid precipitation**, more commonly known as **acid rain**. The acidity of such precipitation ranges between pH 3.0 and 5.0. The pH of rain that has not been contaminated with pollutants is 5.6. (See Figure 13.22.)

The Effects of Acid Precipitation

The effects of acid precipitation continue to be studied, but some disturbing results are already well-known. The thawing of acid snow in the spring can quickly and dramatically lower the pH of lakes and rivers. Under such conditions, frog eggs do not hatch, and fish fry die soon after emerging from their eggs. As the acidity of the water increases, fish suffer from skeletal deformities.

BIOTECH

pH and Acid Rain

The levels of acidity of water solutions can be expressed using pH. Pure water has a pH of 7.0. Solutions with a pH of less than 7.0 are considered to be acidic; those above 7.0 are basic. The lower the pH value, the greater the acidity. A solution with a pH of 4.0 is more acidic than one at 5.0. Each whole-number change in pH represents a tenfold increase in acidity. For example, a solution with a pH of 4.0 is ten times more acidic than a solution at 5.0. It will be 100 times more acidic than a solution with a pH of 6.0, and 1000 times more acidic than pure water (pH 7.0). The pH of acid rain can be as low as 3.0, which is 300 times more acidic than normal rainfall.

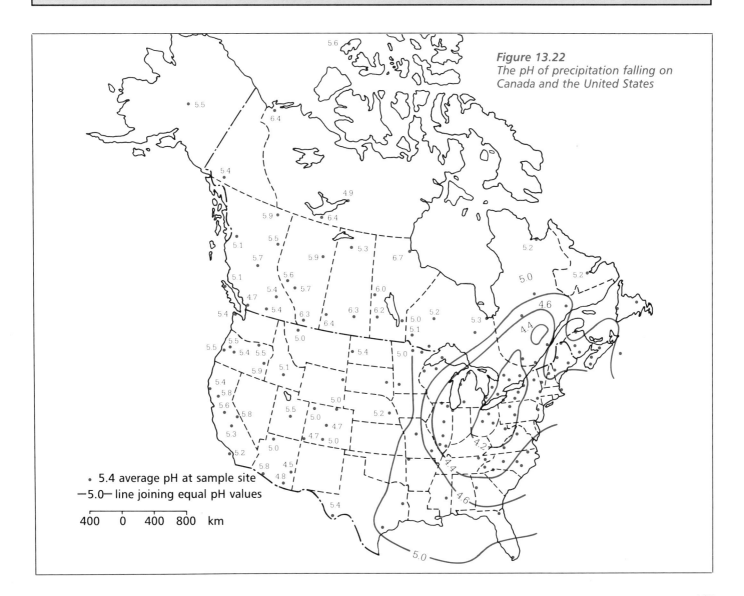

Figure 13.22
The pH of precipitation falling on Canada and the United States

In areas such as the Canadian Shield, the acids leach toxic substances such as aluminum salts into the waterways. High levels of aluminum have a detrimental effect on the development and functioning of fish gills. Ecologists estimate that more than 350 000 lakes in Eastern Canada (from Ontario's western border and south of James Bay) are sensitive to the increase in acidity. Already there are 148 000 lakes with a pH of less than 6, and with damaged aquatic life. There are 14 000 lakes with a pH of 4.7 or less, which are completely dead. If the current trends continue, 48 000 Canadian lakes will soon not be able to support life. (See Figure 13.23.)

High acidity also has a detrimental effect on the micro-organisms in soil which are responsible for breaking down organic matter. Without their contributions, carbon and nitrogen cannot be recycled through the food web. Seed germination is inhibited, and the waxy covering on leaves is damaged, opening the way to damage from diseases and pests.

Scrubbers installed in the smokestacks of coal burning plants have significantly decreased sulphur emissions, but do not remove most of the nitrogen oxides. Nitrogen oxides form nitric acid in the atmosphere. Most plants in North America do not yet have this equipment. In addition, the fuel efficient automobile engines developed after the energy crisis of the mid-1970s produce more nitrogen oxides in their exhaust. This problem will require long-term solutions and even when solved, its long-term effects will remain for some years to come. (See Figure 13.24.)

Figure 13.23
The addition of lime can reduce the acidity of a particular lake, but it is only a temporary solution.

Figure 13.24
This structure was eroded by acid precipitation.

Chapter 13 / Managing Human Waste

FOCUS ON YOU

When a fire was deliberately set in the mammoth tire dump near Hagersville, it was estimated that one litre of oil was released to the ground per tire burned. Along with the toxic substances released into the air, this made the Hagersville tire fire potentially one of North America's worst environmental disasters. Extinguishing the blaze took 17 days and cost over 1.5 million dollars. Cleaning the contaminated soil and water will take far longer and will be far more costly.

Activity 13H

Producing Acid Rain

Problem

How does the burning of sulphur contribute to acid rain?

Materials

deflagrating spoon
250-mL beaker
tap water
oxygen
two gas bottles
glass plate
Bunsen burner
bromothymol blue indicator
sulphur
face/eye protection
apron/lab coat
fume hood

Procedure

 CAUTION!
This will be a demonstration by the teacher. Eye/face protection and an apron or lab coat must be worn.

1. Fill a gas bottle with oxygen by the downward displacement of water. Leave about 1 cm of water in the bottle. Cover the mouth of the bottle with a glass plate.

 CAUTION!
This activity must be done in a fume hood.

2. Place the gas bottle in the fume hood. Place a 250-mL beaker, about half full of water, in the fume hood.

3. Obtain a very small quantity of sulphur (about as much as would sit on your smallest fingernail) in a clean deflagrating spoon.

417

4. Hold the spoon in the edge of the Bunsen burner flame, and ignite the sulphur.

5. Quickly insert the spoon into the oxygen bottle.

6. Allow the sulphur to burn for a few seconds, then remove the spoon and plunge it QUICKLY into the water to extinguish the flame. Cover the gas bottle with the glass plate. *Describe what you see in the gas bottle.*

7. Hold the glass plate on top of the gas bottle and shake the bottle vigorously.

8. Add 2 or 3 drops of bromothymol blue indicator and *describe the colour produced.*

9. Add a small amount of calcium carbonate to the bottle, and *describe any change in the colour produced.*

10. To another clean gas bottle, add enough tap water to cover the bottom of the bottle to a depth of about 1 cm. Add 2 or 3 drops of bromothymol blue indicator. Shake the bottle from side to side to mix the water and the indicator. *Describe the colour produced.*

Questions

1. What colour is bromothymol blue indicator in
   (a) acid?
   (b) base?
   (c) tap water?

2. What substance is produced when sulphur is burned?

3. What happened to this product after shaking the bottle?

4. Describe the colour of the indicator before and after the calcium carbonate was added.

Activity 13I

How Does Acid Rain Affect the Environment?

Problem

What effect does acid rain have on plants, metals, and calcium carbonate?

Materials

distilled water, pH 7
acidic water, pH 3-4
two spray mist bottles
two green plants
iron filings
magnesium turnings
marble chips (calcium carbonate)
four test tubes
test tube rack
two 250-mL beakers
millimetre ruler
apron
goggles

Procedure

• Effects on Plants

1. Select two green plants that are approximately equal in size and amount of foliage. Label one "pH 7" and the other "Acid Rain". Place them in an area where they will receive an adequate amount of light, at room temperature.

2. Prepare two spray bottles, one labelled "pH 7", and the other labelled "Acid Rain".

CAUTION!
Wear an apron and goggles when spraying your plants.

3. Water each of the plants as necessary, using distilled water for the plant labelled "pH 7", and acidic water for the plant labelled "Acid Rain". Each time that the plants are watered, use the appropriate spray bottle to mist the leaves as well. Clean any spilled liquid from the area. Wash your hands with soap and water.

4. Observe the appearance and growth of the plants daily, for an extended period of time. *Record any differences you notice.*

- Effect on Metals

1. Fill two test tubes about half full with distilled water, and label them "pH 7". Place the tubes in the test tube rack.

2. Fill two test tubes about half full with acidic water, and label them "Acid Rain". Place the tubes in the test tube rack.

3. Place a small amount of iron filings in one of the test tubes labelled "pH 7", and one of the test tubes labelled "Acid Rain".

4. Add a few small pieces of magnesium turnings to the other two test tubes.

5. Observe the test tubes carefully, and *record your observations.*

6. Allow the tubes to stand overnight and observe them again. *Record your observations.*

- Effect on a Common Building Material (Calcium Carbonate)

1. Fill a 250-mL beaker about half full with acidic water, and label it "Acid Rain".

2. Fill another 250-mL beaker about half full with distilled water, and label it "pH 7".

3. Select two small marble chips, approximately the same size.

4. *Measure and record the size of one of the chips.* Place this chip in the beaker labelled "Acid Rain".

5. Repeat step 4 for the other chip, and place it in the beaker labelled "pH 7".

6. Observe the beakers and *record your observations.*

7. Leave the chips in the beakers overnight. Next day, remove the chips and measure them. *Record your measurements and observations.* Wash your hands with soap and water.

Questions

1. (a) What effect does acid precipitation have on plants?
   (b) How might this affect farmers and the crops they grow?

2. What effect would acid precipitation have on structures such as bridges, or buildings made of marble and limestone?

Activity 13J

Acid Precipitation and You

Is the precipitation in your area acidic? It is relatively easy to check. All you will need is a clean, wide-mouth plastic container and some pH paper (or a pH meter, if one is available). Place the container where the rain or snow can fall directly into it. After some rain or snow is collected (allow the snow to melt first), test the pH of the precipitation. Would you expect the pH of the precipitation to be different at another time of the year? Explain your answer.

CHECKPOINT

1. Write the chemical word equations for the reactions which cause acid precipitation.
2. Describe the effects of acid precipitation on
   (a) plants
   (b) animals
   (c) buildings
   (d) bridges.

13.11 We Are Learning

- In 1989, more than 1800 companies with 28 000 employees provided advice and equipment to protect the environment.
- Phosphates from detergent use have all but disappeared in the Great Lakes since phosphate-free detergents were developed.

- More than $300 million has been spent by government and industry to clean up Hamilton Harbour, once described as one of the most polluted bodies of water in the Great Lakes. More than 50 species of fish have returned to the waters, and water birds are also returning in increasing numbers.

- A Toronto engineering firm has developed a high-tech combustion chamber for the burning of solid waste which has doubled the efficiency of previous models. The system is designed to convert more than two-thirds of the energy in burning garbage into electric power.

- The phasing out of leaded gasolines in the United States has reduced lead emissions into the air by 86 percent over a ten year period. Similar results are expected in Canada.

- The citizens of the Sudbury area have worked to reforest the Sudbury area, once described as a "moonscape". Experiments by Dr. Keith Winterhalder of Laurentian University showed that adding lime to acidified soil could increase the probability of plant growth. The residents of the area have participated in a major liming and planting program. Companies such as Inco Ltd. have contributed to this project by establishing a tree nursery in an abandoned mine shaft.

- The City of Mississauga, Ontario has become the first city in North America to recycle plastic waste. The plastic will be used, among other things, to produce weatherproof construction materials.

BIOLOGY AT WORK

WATER QUALITY TECHNICIAN
TOXICOLOGIST
PUBLIC HEALTH BACTERIOLOGIST
CUSTOMS OFFICER
SOIL SCIENTIST
AQUATIC BIOLOGIST
CIVIL ENGINEER
ENVIRONMENTAL LAWYER

HEALTH CONCERNS

heavy metal poisoning
   (lead, mercury, cadmium)
cholera
typhoid
asbestosis
environmental allergies

Chapter 13 / Managing Human Waste

CHAPTER FOCUS

Now that you have completed this chapter, you should be able to do the following:

1. List the three categories of wastes which pollute natural water systems, and give an example of each.
2. Define Biochemical Oxygen Demand and explain how it is an indicator of water quality.
3. Describe how algal blooms can contribute to the process of eutrophication.
4. Explain how pollutants in water can be concentrated within organisms.
5. Describe several methods of testing water quality.
6. List the main steps in the water purification process and describe the purpose of each step.
7. List the main steps in wastewater treatment and describe the procedures used at each step.
8. Draw and label a diagram of a typical septic tank system.
9. List the three types of solid waste and give an example of each type.
10. Describe the alternatives for disposing of solid waste and give advantages and disadvantages for each method.
11. List and explain the four R's of waste reduction.
12. List five ways that hazardous waste disposal can be managed.
13. List the three main causes of air pollution.
14. Explain the cause of acid precipitation and describe some of its effects on the environment.

SOME WORDS TO KNOW

Match each of the descriptions given in the left-hand column with a word shown in the right-hand column. DO NOT WRITE IN THIS BOOK.

1. Measure of oxygen requirements
2. Indicator of the amount of human waste present
3. Small clumps of particles
4. Sewage and water entering a treatment plant
5. Contaminated water seeping from landfill
6. A process that requires oxygen
7. Cycle of rapid algae growth and decomposition in lakes
8. Collects small particles in smokestack emissions
9. Material that can be broken down by decomposers
10. Material removed from wastewater
11. Solid waste reduction unit
12. Mixture of decomposed organic material
13. Air pollution alert system
14. Cloudy or murky
15. Atmospheric haze

A absorption field
B fecal coliform test
C pollution index
D electrostatic precipitator
E wastewater
F acid precipitation
G biodegradable
H eutrophication
I Biochemical Oxygen Demand
J aerobic
K compost
L sludge
M SWARU
N turbid
O floc
P incineration
Q leachate
R Secchi disc
S smog
T threshold level

Select any three of the unmatched words and, for each, give a proper definition or explanation in your own words.

SOME QUESTIONS TO ANSWER

1. Explain how the location of a house (or a community) may have caused serious illness in England in the 1800s.
2. (a) What causes an algal bloom?
   (b) What problems can result?
3. (a) What can cause water to be turbid?
   (b) Why is this a problem?
4. Why is alum added to water during the purification process?
5. Why are water reservoirs necessary?
6. What tests are performed on water leaving a purification plant?
7. Explain how toxic wastes could interfere with the operation of a wastewater treatment plant.
8. What remains in the wastewater after the primary and secondary treatments?
9. Why should householders with septic tanks not use bleach and other strong chemicals to clean their toilets?
10. Explain with the use of a diagram how waste is disposed of in a sanitary landfill.
11. What things must be considered when selecting a site for a sanitary landfill?
12. Explain the operation of a solid waste incinerator.
13. How has the development of energy efficient cars contributed to the problem of acid precipitation?
14. Give three examples of success stories in the fight against pollution.

SOME THINGS TO FIND OUT

1. When in history was water pollution first recorded as a problem? Where did this problem occur? What was done to solve the problem?
2. Describe the conditions which led to the deaths of hundreds of people in the air pollution disasters of London, England and New York City.
3. Investigate the penalties in your community for illegal dumping of litter into rivers and streams. Survey your community to determine the seriousness of this problem.
4. Locate the source of the drinking water for your community and report on the condition of that source.
5. If hydrofluosilicic acid is added to the water supply in your community, find out when this began and what positive (and negative, if any) effects it has had on the community.
6. How are hazardous wastes disposed of in your community?
7. How do scrubbers and electrostatic precipitators work?

OUR ENVIRONMENT: WHAT WILL WE MAKE OF IT?

When a balloon is fully inflated, it becomes a fragile shell, and can be destroyed by the point of a pin. The surface of the earth is like that balloon. All living things are found in its thin, fragile shell, called the biosphere. In the past, it could withstand the poking, pulling, and stretching that the human race inflicted upon it. But now, it has been stretched to a point close to the limit.

KEY IDEAS

- Ecology is the study of the interaction of living things and the environment.
- Matter and energy move through the ecosystem in the food chain.
- Food production has increased, but continues to lag behind population growth.
- The disposal of untreated wastes damages the environment.
- Decisions by individuals can affect the future of the environment.

14.1 The Human Population

In July of 1987, the population of the earth reached the 5 billion mark, and it continues to grow. As of mid-1989, the world population stood at 5.234 billion and the 6 billion mark is expected to be reached by 1997. In the time it takes to read just a few sentences on this page, 150 people will be born; each and every hour, about 10 000 people are added to the population. At that rate, it would only take three and one-half months to arrive at the entire population of Canada. Totalling up these numbers for a year, we arrive at more than 85 million people.

If you were to simply say hello to each of these people, it would take you more than two and one-half years, working at it 24 h a day. And when you had finished, you would have to greet the over 200 million more who were born during that time!

Most of the world's population growth has occurred during the last 10 000 years. Although population records and census figures were not available until recent times, estimates suggest that in the year A.D.1, the population was close to 300 million. By A.D.1800, the population had exploded to 1 billion, and it doubled within the next 100 years. Population experts at the United Nations have predicted that the world population will reach 6.323 billion by the year 2000 and 8.330 billion by 2020. (See Figure 14.1.)

Figure 14.1
The human population continues to grow.

Activity 14A

Graphical Analysis

Problem

How can a graph be used to make predictions about future trends?

Materials

calculator
graph paper

Procedure

1. *Prepare a table in your notebook with the following headings*: (Note: Your table should have spaces for 35 years.)

Year	Amount
0	$0.01
1	
2	
3	
⋮	
35	

2. Imagine that you were hired to work at an annual salary of one cent. You are guaranteed a raise at the end of every year which will double your salary. Calculate the amount you will be paid each year by multiplying the previous year's salary by 2. *Record the amounts in the appropriate spaces in your table.*

3. When you have completed your calculations, *prepare a graph by placing years on the x-axis (horizontal) and amount on the y-axis (vertical). Plot the data on the graph, and draw the line of best fit.*

4. *Graph the data in Table 14.1 on world population statistics. Place time (year) on the x-axis, and population on the y-axis.*

Table 14.1 World Population

Year (A.D.)	Population (millions)
1	200
500	250
1000	300
1600	500
1850	1 000
1930	2 000
1960	3 000
1975	4 000
1987	5 000
1989	5 234

Questions

1. (a) Describe the shape of the curves for each graph.
   (b) What kind of curve (or growth) is this?

2. What was the world's approximate population in 1950?

3. Extend the curve and determine the projected population for the year 2010.

4. Would you take a job for a salary of one cent per year, to be doubled every year? Why or why not?

Population Changes

Predictions of the growth of the human population are made by examining such things as **birth** and **death rates**, **fertility rate**, and **age distribution**.

Birth and Death Rates

If the birth rate is equal to the death rate, the total number of individuals alive at one time will not change. Birth rates and death rates vary from country to country, and from region to region.

(See Figure 14.2.) At the present time, there are about 2.8 live births for each death throughout the world. **Demographers**, or population specialists, calculate the birth rate (or death rate) using the following formula:

$$\text{birth (or death) rate} = \frac{\text{births (or deaths) per year}}{\text{mid-year population}} \times 1000$$

The birth rate and death rate can be used to calculate the percentage growth rate for a population. It is calculated using this formula:

$$\% \text{ annual growth rate} = \text{birth rate} - \text{death rate}$$

Table 14.2 1989 Birth Rates, Death Rates, and Infant Mortality Rates

Region	Birth Rate	Death Rate	Infant Mortality Rate
World	28	10	75
More Developed Nations	15	9	15
Less Developed Nations	31	10	84
Africa	45	15	113
Asia	28	9	78
Europe	13	10	12
Latin America	29	7	55
North America	16	9	10
Oceania	20	8	26

Note: These rates give the number of births, deaths, or infant deaths per 1000 persons in the population on July 1, 1989.

Source: Population Reference Bureau, 1989 World Population Data Sheet, Washington, D.C.

The world's annual growth reached its highest point in 1965, at two percent, and has declined somewhat since. The United Nations estimates that the present growth rate of 1.8 percent may decline to 1.5 percent by the year 2000. Even with this decrease in the growth rate, the global population will still continue to increase by 95 million persons per year.

Figure 14.2
Estimated birth and death rates, and rates of natural population increases

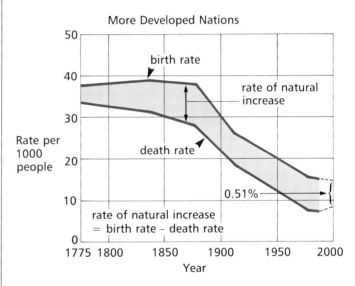

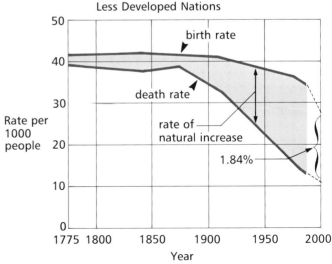

The dramatic increases in the world's population since the last century are the result of a decrease in the death rate. As food supplies, medical care, technology, and sanitation improved, there was a general decline in death rates. The average life span in the world is now 63 years. (See Figure 14.2.)

Fertility Rate

The **fertility rate** is an estimation of the average number of children each woman will have in her childbearing years. The average fertility rate in the world in 1989 was 3.6 children per woman. Less developed countries had a rate of 4.7, while in the more developed countries, the rate was 1.9. (See Table 14.3.)

A fertility rate of 2.1 children per woman is the *replacement rate* in the more developed countries of the world. At this rate, two children replace two parents, and any female children who die without reaching childbearing age. In less developed countries, the replacement rate is 2.7. Why is it higher?

If the fertility rate is greater than the replacement rate, the population will continue to increase.

Age Distribution

The **age distribution** of a population is the percentage of people at each age level. A profile or **age structure diagram** can be constructed based on the percentage of people born within five year intervals. (See Figure 14.3.)

Figure 14.3
Age structure diagrams. How does the age structure of Sweden compare with that of Mexico?

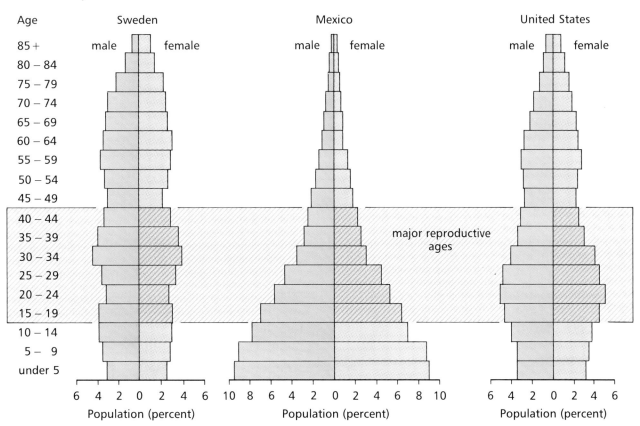

Table 14.3 Average 1989 Total Fertility Rates for Geographical Regions

Region	Average Total Fertility Rate
World	3.6
More Developed Nations	1.9
Less Developed Nations	4.1
Africa	6.3
Asia	3.6
Europe	1.7
Latin America	3.6
North America	1.9
Oceania	2.7

Note: These rates give a projection of the average number of live births per woman for those women who live at least to age 44.

Source: Population Reference Bureau, 1989 World Population Data Sheet, Washington, D.C.

The shape of the profile indicates whether the population will increase, decrease, or stay the same. The world's less developed countries, with rapidly increasing populations, have a triangular-shaped profile. There are a large number of people in the reproductive age group, ages 15 to 44 years, or in younger groups, ages 0 to 15 years.

In countries such as Mexico, with this large proportion of young people, the population will continue to increase as these people have children, even if the fertility rate is low.

A more developed country, such as Sweden, has a profile with sides that are almost vertical. This indicates that in each age bracket, there are similar numbers of individuals. The number of people who die is equal to the number of people who are born, and thus, the population remains constant. What would the profile look like for a population that was declining?

In the more developed countries, the proportion of older people in the population is increasing. The proportion of those 65 years of age and older is three times that of the less developed countries. What special needs and problems will this create?

CHECKPOINT

1. How many people are added to the world's population each hour?

2. What is the birth rate for Canada, assuming its mid-year population is 26.3 million, and 394 500 babies were born during the year?

3. Calculate the annual growth rate for the following countries:
   (a) China, birth rate = 21, death rate = 7
   (b) Hungary, birth rate = 12, death rate = 13
   (c) Zimbabwe, birth rate = 47, death rate = 11.

4. Sketch an age structure diagram for
   (a) a country with a stable population
   (b) a country whose population will increase rapidly in the next few years
   (c) a country which has a decreasing population.

14.2 The Study of Populations and the Environment

Living things have an effect on each other, and on the non-living, physical environment in which they live. In return, the environment affects living things. Together, this interacting system is called an **ecosystem**. The study of this interacting system of organisms and environment is **ecology**. A group of organisms of the same species, living together in the same area, is called a **population**. All of the different populations living together in the same area are a **community**.

Each population plays a role in the functioning of the community. For example, most plants are **producers**, which means that they produce their own food. **Consumers** are those organisms which get their food by eating other organisms. Decomposers feed on non-living organic matter, such as wastes, or dead plants and animals. The role that an organism plays in the community is its **niche**, which may be considered to be its occupation. Its **habitat**, or where it lives, may be considered to be its address.

The organisms of an ecosystem are constantly changing, taking in energy and matter from the environment, and releasing wastes. New organisms appear, grow, reproduce, and die. But, even with all these changes, the overall ecosystem tends to stay the same. Although matter is constantly being taken in by organisms, the decomposers eventually return what has been used to the environment. This recycling of nutrients maintains a steady flow of matter within the system. (See Figure 14.4.)

Figure 14.4
Nitrogen, a vital nutrient, is continuously recycled through the ecosystem.

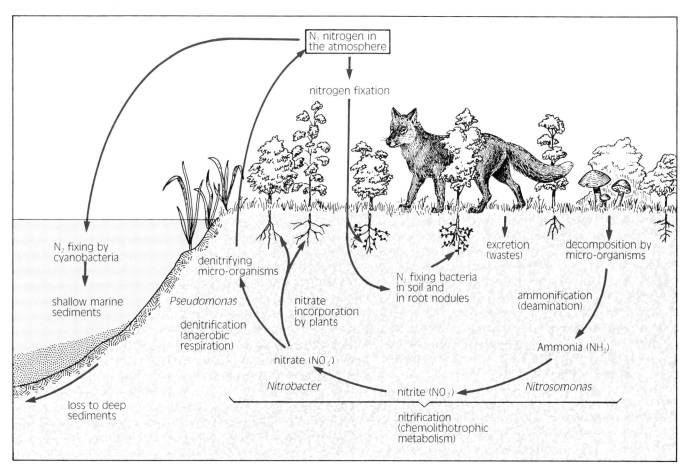

Homeostasis

As long as the system is balanced (the energy in equals the energy used, and matter is recycled), there is no net change. All organisms and ecosystems, even the organs within your body, maintain a steady state through a process of self-regulation called **homeostasis**. If one part of the system is altered, another part (or parts) will adjust to regain the balanced state. To do this, there must be some method of feedback control. For example, while listening to music, you are in control of the volume. If the music is too loud, you turn it down. If you can't hear it, you increase the volume. This is an example of a **negative feedback control system**. If the volume decreases, you move the proper control to increase it.

In an ecosystem, there are similar controls, based on the relationships of the organisms concerned. For example, in the Canadian Arctic, the Arctic hare population may increase if certain conditions, such as the availability of food, are favourable. As the hare population increases, this increases the food supply for its predator, the lynx. The lynx population will then increase, but the hare population will begin to decline. Eventually, the lynx population will also decline as its food supply decreases. This is known as the **predator-prey relationship**. (See Figure 14.5.) Although these populations increase and decrease, the ecosystem as a whole remains in balance.

Years ago, the human population was simply another part of the ecosystem, and had only a limited effect on the other organisms. People lived a day-to-day existence, hunting and gathering food. Today, however, circumstances have changed. The decisions and actions of even one person can have an impact on thousands of organisms, and the effects may be evident for many years. We are becoming aware that there is a high price to pay for the power to disrupt the balance of the ecosystem.

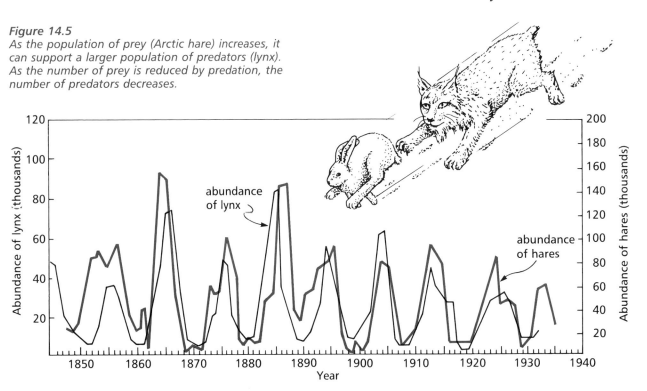

Figure 14.5
As the population of prey (Arctic hare) increases, it can support a larger population of predators (lynx). As the number of prey is reduced by predation, the number of predators decreases.

CHECKPOINT

1. Define or explain the meaning of the following terms: ecology, ecosystem, population, community.

2. What is the difference between an organism's habitat and its niche?

3. What is homeostasis?

14.3 Factors Affecting Population Size

If there is nothing to restrict the growth of the population, that is, there is an abundance of food and ideal conditions for survival, reproduction will increase the population's numbers as quickly as possible. This maximum reproductive rate is called the **biotic potential** of the species. A graph of such a population over time will have a curve that starts out slowly, but eventually sweeps upward to form a J shape. This type of growth is called **exponential growth**, or growth by doubling. (See Figure 14.6.) You graphed exponential growth in Activity 14A.

Figure 14.6
A population which continues to reproduce as quickly as possible shows an exponential growth in its numbers.

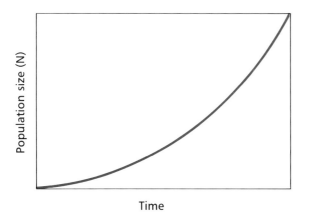

Environmental Resistance

In any ecosystem, there are one or more **limiting factors** which will slow the growth of a population. Factors such as predators, the availability of food, competition, disease, and disasters are referred to as the population's **environmental resistance**. The environmental resistance limits the biotic potential of a population, slowing its growth to match the **carrying capacity** of the particular ecosystem. The carrying capacity is simply the maximum number of individuals of a species the particular ecosystem can support. (See Figure 14.7.)

Figure 14.7
When a population meets environmental resistance, the curve tends to flatten out, and takes on an S shape.

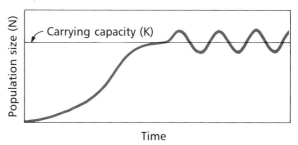

Factors Affecting Populations

Factors which affect a population may be grouped into one of two categories. Those which include living things such as bacteria, fungi, single-celled organisms, plants, animals, and humans are the **biotic** factors. The **abiotic** factors are those non-living things, such as water, oxygen, carbon dioxide, temperature, and minerals.

Human populations are also limited by environmental resistance. In the past, malnutrition, famine, disease, and war were the main pressures. Thomas Malthus first described these factors in 1798, predicting that the world's food supply could not keep pace with the population growth.

Factors Tending to Increase Population Size

Advances in medicine and sanitation have brought many diseases under control, improved life expectancy, and sharply decreased the infant mortality rates. Geneticists have improved the yield of food-producing plants and animals. New fertilizers, herbicides, pesticides, and irrigation practices have increased food production. Technology has improved the harvesting, storage, and transportation of food. All of these factors have increased the earth's carrying capacity for the human organism, but not sufficiently. The population has continued to grow, and a large part of the world's population exists with barely enough food, water, and shelter to live. (See Figure 14.8.)

The human species has the most efficient reproductive system of any large animal species in the world. Humans are not limited to one season for childbearing. Unlike most animals, humans do not have to cope with predators (other than other humans!). They can manipulate the environment to suit their needs for shelter, water, and food. The natural limiting factors that control animal and plant populations are largely being eliminated or certainly minimized.

Figure 14.8
What problems will be faced by the world as its population continues to grow?

The Consequences of Human Population Growth

Population growth contributes to many of the major problems we face today. Too many people results in too much waste, leading to water, air, and soil pollution. Water supplies are depleted, arable land is poorly used, and pressures of every kind are brought upon the environment.

An obvious problem is the provision of enough food and water for all these people. Food is a **renewable resource**, that is, it can be replaced by growing more crops, and raising more animals for food. Other resources, such as minerals, fuels, and land, are **non-renewable**. There are only limited amounts of these resources, and once used, they cannot be replaced. But there are other concerns as well. How will we deal with the demand for sanitation, hospitals, housing, schools, and transportation?

The solution to these problems must be carefully considered, since a solution to one problem may cause another problem. The building of more homes, factories, and highways, for example, to provide more living space may also reduce the amount of available farmland. Because of these limited resources, and the problems of waste, some believe that we have already reached the limit of the earth's carrying capacity.

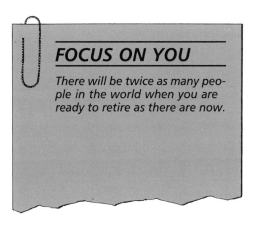

FOCUS ON YOU

There will be twice as many people in the world when you are ready to retire as there are now.

CHECKPOINT

1. *Sketch a graph of a population with exponential growth.*

2. *What is carrying capacity?*

3. *Explain the difference between biotic factors and abiotic factors, and give two examples of each.*

4. *Define environmental resistance and give three examples.*

5. *What is biotic potential?*

6. *Describe five things which have led to an increase in the population.*

7. *What is the difference between a renewable resource and a non-renewable resource? Give one example of each.*

14.4 Food Consumption

What have you had to eat today? Are you careful about what you eat? Some 13 percent of North Americans are dieting and another 34 percent say they are careful about what and how much they eat. If you are anything like the average North American, you consume about 61 kg of red meat, 40 kg of poultry, and 54 kg of potatoes in a year. What about something to drink? Do you drink 42 L of fruit juice, 39 L of milk and 176 L of soft drinks each year (that's over 625 cans!)? The average North American does. What do you eat for dessert? Once again, our average person eats 14 L of ice cream and 9 kg of candy each year.

The average daily food energy intake per person is just over 15 000 kJ. This is above the average kilojoule requirement per day for males and females. (See Chapter 8.) It is not surprising, then, that in Western countries, as many as one in five men, and one in three women are overweight. It must be pointed out that approximately

20 percent of the food prepared in North America is discarded as waste.

People on other continents are faced with different problems. The United Nations has estimated that one out of every ten people on earth is underfed. Other estimates suggest that one in every six people is either too poor to buy food, or does not have enough land to grow the amount of food needed. The developing countries, with about three-quarters of the world's population, account for less than one-half of the world's production of major food crops. While food production has increased faster than the population over the last 20 years in some areas of the world, it has declined or remained the same in many developing countries, especially in the Middle East and Africa. Africa's annual growth rate in food production between 1980 and 1986 was only 2.9 percent. Its population grew by at least three percent in that same time. (See Figure 14.9.)

Compounding the problem of food shortages is the fact that food is not equally distributed to all people in a region or country. Average food consumption is generally lower in urban areas,

Figure 14.9
The growth of food production in different areas of the world compared to the growth of the population

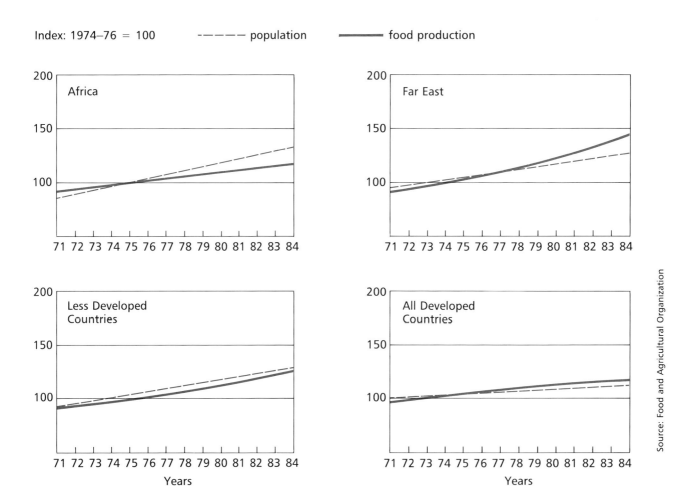

Source: Food and Agricultural Organization

especially in low income areas. The infant mortality rate, which is an indicator of nutrition, among other things, can be three times higher in these areas. Malnutrition and anemia may be twice the national average.

Food may also be unequally distributed within the family unit. Often young children are underfed, with most of the food going to those who are able to work and support the family unit.

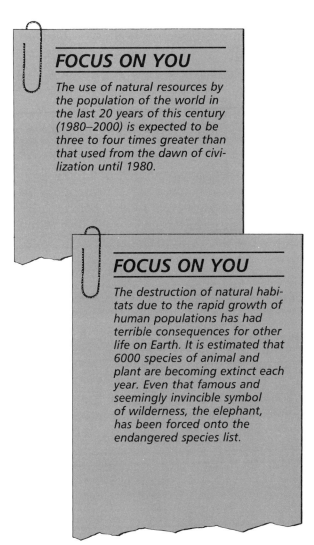

FOCUS ON YOU

The use of natural resources by the population of the world in the last 20 years of this century (1980–2000) is expected to be three to four times greater than that used from the dawn of civilization until 1980.

FOCUS ON YOU

The destruction of natural habitats due to the rapid growth of human populations has had terrible consequences for other life on Earth. It is estimated that 6000 species of animal and plant are becoming extinct each year. Even that famous and seemingly invincible symbol of wilderness, the elephant, has been forced onto the endangered species list.

Activity 14B

Use of Resources

What quantity of resources does the average person use in a lifetime?

Procedure

1. Use the information in Table 14.4 to calculate the amount of each resource the average North American uses in a lifetime of 76 years.

Table 14.4 Resource Consumption

Resource	Average Daily Consumption Per Person
Water	340 L for domestic use 7400 L for domestic, agricultural, and industrial use
Iron and steel	1.8 kg
Petroleum	3.0 L gasoline 71.5 L crude oil
Paper	0.22 kg

2. Now use the information in Table 14.5 to calculate the amount of waste the average North American produces in a lifetime of 76 years.

Table 14.5 Waste Production

Type of Waste	Average Daily Production Per Person
Garbage	2.4 kg domestic garbage 8.6 kg domestic, commercial, and industrial garbage 1 bottle or can
Air pollutants	0.86 kg
Sewage	454 kg domestic, agricultural, and industrial sewage

Questions

1. (a) How would the world change if everyone on earth used the same amount of resources as the typical North American?
   (b) How would the environment be affected if North Americans changed their lifestyles, and used the same amount of resources that are used in less developed countries (about 2.5 percent of what North Americans use)?

2. The people in the less developed countries of the world produce much less waste than North Americans do. Explain why this is so.

3. (a) How many new automobiles and trucks does the average North American purchase in a lifetime if 14.2 million are sold each year? (The population of North America was estimated to be 275 million in mid-1989.)
   (b) What happens to these automobiles once they are no longer on the road?
   (c) Suggest what might be done to make better use of the resources used to make an automobile.

CHECKPOINT

1. How much meat and potatoes does the average North American eat in a year?

2. How much of the prepared food in North America is discarded as waste?

3. (a) What proportion of the world's population is underfed?
   (b) How is it possible for people in areas such as North America to be underfed?

14.5 Increasing the Food Supply

The problem of food shortages is not as easy to solve as it might first appear. A simplistic answer is to increase the amount of food that is grown. The so-called **Green Revolution** of the 1960s introduced new varieties of high-yield wheat and rice to the less developed countries of the world. The grains were bred to produce two to three times the yields of traditional grains. Increases in world grain production did follow in the next decade; however, these new varieties require water and fertilizer which is not always available or affordable.

Reducing Crop Losses

Reducing crop losses from pests and diseases could dramatically increase the world's food supply. Damage to the crop as it is grown, or the destruction of the harvested food product reduces the amount of food for consumption by 45 percent. The use of pesticides and herbicides can increase food production, but it also causes environmental problems as the chemicals accumulate in the food chain. Estimates suggest that 500 000 people become seriously or fatally ill from the effects of these chemicals each year.

The lack of storage and transportation facilities to bring crops to market continues to be a problem. In fact, the grain currently grown in the world could provide an adequate diet for everyone, if it could be distributed to the areas of need.

Increasing Farmland

Increasing the amount of land used for farming could increase food production. The amount of cropland in the world has increased by more than 20 percent since the 1950s, but this increase has not been able to keep pace with the increase in population. Estimates suggest that the present amount of cropland could be doubled if all potential areas were cultivated.

The problems associated with these new lands are numerous. (See Figure 14.10.) Most of this land is currently covered with tropical rainforest, and is susceptible to erosion. In addition, the soils require massive amounts of fertilizer, and contain high levels of undesirable elements, such as aluminum. These lands are some distance from population centres, which adds to cultivation and shipping costs. Also, the effects of such massive changes to the naturally occuring vegetation of the earth are only now becoming understood.

While some countries propose methods of increasing the amount of farmland, the more developed countries of the world have subsidy programs to encourage farmers not to grow crops. The subsidies limit the amount produced, which in turn keeps prices up at a desirable level. If prices decline, the farmers are not able to earn enough money to buy seed, fertilizer, and equipment.

Figure 14.10
As cities expand, valuable farmland is lost forever.

Changes in Diet

In North America, about 75 percent of the grain produced each year is used to feed livestock. Worldwide, this figure is about 33 percent. A change in diet to reduce the amount of meat consumed would shorten the food chain. This would not only reduce the energy spent in food production, but also would make more food available.

Other suggestions include getting more food from the oceans, developing new foods, using fabricated foods such as simulated meat made from soybeans, or harvesting edible bacteria grown on waste. Some have even suggested that the resources used to feed pets be directed to the production of food for humans. (See Figure 14.11.)

Figure 14.11
In North America, the amount of pet food purchased in a year is equivalent to the protein needed to feed a population of 24 million people.

Population Control

Many experts believe that increasing cropland is not the answer, but limiting the population is. Dr. Norman Borlaug, who was awarded the Nobel Prize in 1970 for his work in developing new strains of high-yield grains, said that he would rather have found a birth limiting wheat. He stated that all his discovery did was give the world a little more time. Some countries have instituted family planning programs and financial incentives in an attempt to limit family sizes. In China, for example, regulations were established to limit couples to only one child.

There are many factors to consider when proposing solutions for world hunger and malnutrition. Technical, environmental, logistical, economic, ethical, and social problems all need to be wrestled with. Whose responsibility is it to be concerned? Can individuals do anything that will help? What is your responsibility?

CHECKPOINT

1. What proportion of the world's food supply is lost to pests and diseases each year?

2. What problems prevent the food that is grown from getting to the people who need it?

3. Explain why the conversion of tropical rainforests to cropland may not help the world's food shortage.

4. List at least five ways that more food could be made available throughout the world.

5. Explain how a reduction in the amount of red meat in the diets of North Americans could provide more food.

14.6 Waste Not, Want Not

The large amount of solid waste we produce each day is a reflection of our "throw-away society". We are encouraged to buy new items through a constant barrage of advertising. Many of these items are designed to be used only once, and then discarded. Much of what we choose to discard would be used for some useful purpose in the less developed countries of the world. Some countries find uses for up to 60 percent of the waste they produce.

Since we have been faced with a crisis in waste disposal, many areas have introduced recycling programs. Not only does recycling reduce the amount of waste that must be disposed of, it also saves energy at the manufacturing level. The energy needed to develop virgin materials and products is greatly reduced. Recycling aluminum from old scrap, for example, uses only one-twentieth of the energy needed to extract new aluminum from aluminum ore.

Waste management costs to the public are also reduced. The life of the waste collection equipment is extended, and the capital costs for new landfill sites is reduced. Most importantly, water, land, and air pollution are reduced. Both renewable and non-renewable resources are conserved, and new jobs are created.

Saving Water

The amount of water used, on a per capita basis, in the more developed countries of the world is about three times what is used in the less developed countries. Part of this difference is based on a reduced supply, rather than demand. About 80 countries, with 40 percent of the world's population, are experiencing water shortage problems. (See Figure 14.12.) This situation is expected to get much worse as populations continue to increase, and as the demand for water increases.

Massive projects have been suggested to meet this demand in several areas of the world. In the United States, water diversion projects have been proposed which would divert the water from major rivers, including those which flow from Canada into the Arctic Ocean. The water would be distributed to the areas in Canada, the United States, and Mexico which are experiencing shortages. The Soviet Union has also planned to divert the water from 12 north-flowing rivers, which also empty into the Arctic Ocean. Scientists are concerned that this reduction in the amount of fresh water reaching the Arctic Ocean would increase its salinity. The higher concentration of salt would act like antifreeze, possibly causing polar ice to melt.

Figure 14.12
Water shortage problems affect 40 percent of the world's population.

The Dilemma of Sludge Disposal

The difficulties of human population, waste management, and protecting the environment are well illustrated by the treatment of sewage. If sewage is not treated, human health and the health of the environment suffers. We can clean the water – but what of the remaining sludge? Sludge from wastewater treatment contains high levels of nitrogen and phosphates, which are the basic requirements for plant growth, and thus treated sludge makes an excellent fertilizer; however, if the sludge has been contaminated by toxic wastes or heavy metals, it cannot be used in this way. If it were, the toxins could easily enter the food and water supply with disastrous effects. (See Figure 14.13.) Communities in Japan have suffered from cadmium poisoning because rice fields and drinking water were contaminated with cadmium-rich mine wastes. Incinerators can be used to burn the sludge, but this may introduce toxic materials into the air, and still leaves an ash residue.

Figure 14.13
One "solution" to waste management problems has been to dump sludge and other solid waste into the ocean. Another is to transport such waste to other countries. Why is this of concern?

CHECKPOINT

1. *What are three reasons to recycle waste?*

2. *What proportion of the world is presently experiencing water shortages?*

3. *Describe how sewage sludge is disposed of, and give one disadvantage of each method.*

14.7 Pollution of the Oceans

The seemingly endless ability of the oceans to "take away" our waste is an illusion. The most powerful indications of this arrived on beaches in the summer of 1988. During that summer, flows of hypodermic needles, sutures, catheter bags, and other medical wastes began to litter the East Coast of North America. During that same year, 7000 seals in the North Sea died in an epidemic that some scientists believe was linked to the dumping of industrial waste. More than 50 beluga whales washed up on the shores of the St. Lawrence River in a four year period beginning in 1986. These whales died of bladder cancer, and a failure of the immune system. High levels of organic pollutants were found in their tissues.

The accidental dumping of millions of litres of oil from ships like the Exxon Valdez (1989), the Amoco Cadiz (1978), and the Torrey Canyon (1967) have killed millions of marine creatures. (See Figure 14.14.) These incidents only happen occasionally, but each year over 5 million tonnes of oil and petroleum products are added to the oceans. Most of this comes from the disposal of lubricating and engine oils in urban areas, and the discharges from oil tankers.

Figure 14.14
The accidental dumping of oil into the oceans kills millions of marine creatures.

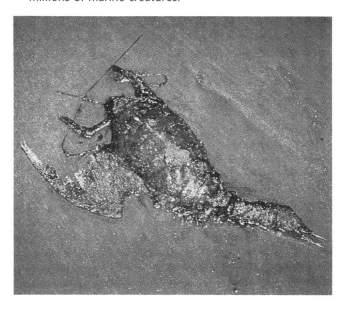

Plastic pollution is a problem that affects every form of marine life. A study of commercial fishing vessels reported that every year over 23 000 t of plastic packaging and 135 000 t of fishing gear were dumped or lost at sea. Merchant vessels discard 690 000 plastic containers into the water each day. Whales and turtles eat these pieces of plastic, and later die of starvation as their digestive systems are blocked. Other animals become tangled in the packaging and fishing nets, and drown, starve, or die from the injuries they receive. It has been estimated that 40 000 seals are killed in this way each year in the Bering Sea.

Some 54 nations agreed to stop the dumping of pollutants, such as oil and radioactive materials, into the oceans as far back as 1975. But, by 1987, there were only 29 countries which had signed an agreement to forbid ships from discarding plastics. Lawsuits launched in some countries have successfully overturned regulations which banned ocean dumping, on the grounds that ocean dumping is less hazardous to humans, and also less costly, than other disposal methods.

CHECKPOINT

1. Describe at least three effects of dumping wastes in the oceans.

2. (a) How much oil and petroleum products are added to the oceans each year?
   (b) Where do these products come from?

14.8 How Much Is A Tree Worth?

If we are to determine the worth of a tree, it is first necessary to recall why trees are valuable. Trees provide firewood, timber for building projects, and wood for paper products. These uses for "dead" wood are only some of the reasons why trees are valuable. Living trees play an important role in the repair and maintenance of the environment. They add oxygen to the atmosphere, and assist in the recycling of carbon and nitrogen. Forests absorb rain and hold it, releasing it slowly into the atmosphere and into the earth. Acting as a sponge, trees reduce erosion and contribute to soil building. They provide habitats for organisms, and absorb some air pollutants. Although only an estimate, it has been calculated that a typical tree that lives for 50 years provides almost $200 000 of ecological benefits, including about $35 000 worth of oxygen.

About one-third of the world's land mass is covered by forests. These forests are presently shrinking by about one percent each year as trees are cut to clear land for farmland, or to provide lumber and firewood. Nowhere is this more evident than in the tropics, where 20 ha of rainforest are cut down every minute of the day. (See Figure 14.15.)

Figure 14.15
The United Nations' Food and Agriculture Organization calculates that between 10 and 11 million hectares of tropical forest are being destroyed each year.

Although these tropical rainforests cover only seven percent of the world's land mass, they contain 40 percent of all plants and animals. Scientists have predicted that if current trends continue, about 20 percent of all plants and animals will disappear with the forests by the year 2020.

Most people are familiar with tropical forest plants such as cacao, coffee, bananas, and citrus fruits. Some people are aware that products such as waxes, dyes, flavourings, latex, and resins are produced by other tropical plants. However, many people are not aware that some of our most valuable medicines come from these forests. We have not even named over 80 percent of the tropical plants found in these rainforests, let alone determined their usefulness.

But the effects of the destruction of the rainforests are more than economic. Erosion follows the cutting of the forests, and agriculture further depletes the topsoil. In addition, scientists are noticing that in areas where large areas of tropical rainforests are cut, there has been a decrease in the amount of rainfall. The island of Haiti, once covered with dense tropical vegetation, has become like the North African desert. Some speculate that the Amazon River region of South America is already experiencing changes leading to the same end.

Ecologists and forestry experts suggest that in order to preserve the tropical forests, governments and logging companies must sponsor reforestation programs and set aside forest preserves. Countries with tropical rainforests must manage the use of their forest resources, or the world will be faced with the consequences.

CHECKPOINT

1. *List five ways in which forests play an important role in the repair and maintenance of the environment.*

2. *Why are forests being cut down?*

3. *How might the destruction of a tropical rainforest affect you?*

4. *What should be done to preserve tropical rainforests?*

FOCUS ON YOU

The word vandalism comes from an ecological disaster. In A.D. 324, the vast farm-rich African provinces of the Roman Empire were invaded by a barbarian tribe called the Vandals. In less than a decade, they destroyed the irrigation system which supplied 250 000 km^2 with water. Without water, the plants died and the desert expanded to cover the area.

14.9 The Atmosphere

Local air pollution problems are highly noticeable. Residents in areas with a high level of air pollution can often detect the presence of tiny particles of soot and grit, foul odours, or photochemical smog. Some of the other air pollutants are not as noticeable when they are released, but they cause serious damage to the environment.

Some air pollutants are chemical compounds such as sulphur dioxide and nitrogen dioxide. (See Chapter 13.) These compounds combine with the water in the atmosphere. The acids produced eventually fall back to earth in some form of precipitation, usually called **acid rain**.

Acid precipitation is responsible for weakening or destroying forests and the organisms in lakes around the world. The forests of Eastern Canada, the New England area of the United States, and the forests of Scandinavia and Central Europe have suffered significant damage. (See Figure 14.16.)

The Greenhouse Effect

The natural end products of the complete combustion of carbon-bearing fuels are water and carbon dioxide. Carbon dioxide is an odourless, tasteless, non-toxic gas. It has a minimal effect on plants and animals, especially since it is also a product of cellular metabolism, and is used by plants for photosynthesis. It may, however, affect all living things because of the **greenhouse effect**.

In the atmosphere, carbon dioxide molecules act like the glass in a greenhouse. The sun's rays penetrate the molecules and strike the earth's surface, warming it. The warm earth radiates infrared rays. These rays are trapped by the carbon dioxide molecules. As a result, the atmosphere stays warmer. Scientists believe that if the average global temperature were to rise by even eight-tenths of a degree per decade, the polar ice

Figure 14.16
Possible consequences of the greenhouse effect

caps would melt. The additional water could cause a 10-m rise in sea level, which would flood many coastal cities. Rainfall patterns would also change, affecting freshwater supplies and the water levels of the Great Lakes.

The soot and other particulate matter in the atmosphere have the effect of reflecting sunlight. This reduces the heating of the earth's surface, and may to some extent counteract the greenhouse effect. In fact, this shielding effect is thought to be responsible for the cooling of the global climate since 1940.

The Ozone Layer

Another air pollution problem, the destruction of the *ozone layer* by **chlorofluorocarbons (CFCs)**, was identified in 1974. CFCs are manufactured for use as propellants in aerosol sprays, coolants in air conditioners and refrigeration units, solvents for computer chips, and as a foaming agent in styrofoam and foam rubber. As these substances enter the atmosphere, they eventually rise to its upper levels. Here the CFC molecule breaks apart to release chlorine. The chlorine atoms break apart the molecules of ozone which form a protective layer around the earth. This layer shields the earth from ultraviolet radiation. (See Figure 14.17.)

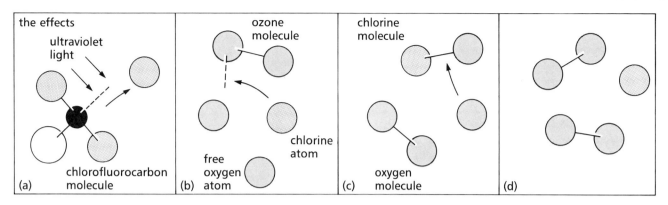

Figure 14.17
Each chlorine atom is capable of destroying 100 000 ozone molecules.

In (a), ultraviolet light in the upper atmosphere breaks off a chlorine atom from a chlorofluorocarbon molecule.

In (b), the chlorine attacks and breaks apart an ozone molecule.

In (c), a molecule of ordinary oxygen and one of chlorine monoxide are formed.

In (d), once a free oxygen atom breaks up the chlorine monoxide, the chlorine is free to continue the process. The lower diagram illustrates how thin patches of the ozone layer will move in the upper atmosphere, distributing the harmful effects of increased intensities of ultraviolet radiation to all nations.

As the ozone layer continues to be destroyed, patches or windows in the layer will get larger and larger. The increase in the intensity of ultraviolet light is expected to increase the incidence of skin cancer, cataracts, damage to UV sensitive crops, and possibly marine organisms which live close to the surface. Scientists have estimated that an 85 percent cut in CFC production would be required to stabilize the chlorine levels in the atmosphere. Canada, the European Economic Community, and 24 other countries have signed an agreement (The Montreal Protocol on Substances that Deplete the Ozone Layer) to freeze the production levels of CFCs beginning in 1989, and to reduce some production levels by as much as 50 percent by 1998.

Setting Environmental Standards

Did you ever wonder how the government determines the standards for the environment, or how acceptable levels of atmospheric pollutants are determined?

Standards are produced by the Ministry of the Environment, in response to a concern brought to the Ministry's attention by an individual, a group, or a corporation. Usually, this occurs when a request is made for the inspection and approval of a process which may release emissions into the atmosphere. At other times, it may be the result of an individual's concern about a possible pollution problem, or through the efforts of a group of concerned citizens.

The research literature and the standards in other areas of the world are constantly reviewed to make sure that decisions about acceptable levels of pollutants are based on current information.

The standards for environmental quality which are now being set reflect a change in the way we view pollution. In the past, the standards for air quality were based on the premise that pollutants would be diluted in the environment. Today, the standards reflect an effort to control the emission of pollution at the source. Also, the effects of pollutants over a person's lifetime are now being considered, rather than the effects of short-term exposure.

Once the standards have been determined, they are published as regulations, and made public. The administrative team of the Ministry conducts constant reviews to see if the present objectives for air quality are being met, and if revision is necessary.

The Ministry of the Environment has limited resources, and this makes it impossible to monitor all emissions all of the time; however, the standards serve as a control, in the same way that the posted speed limit controls the traffic. The end result is a cleaner environment.

Anything to Declare?

The pollutants released into the air and waterways of the world are not restricted by immigration and travel regulations. They cross international boundaries without a passport, and are never refused admission. The new "host" country pays the price of the unwelcome guest's arrival.

The sulphur and nitrogen oxide emissions from the United States affect Canadian lakes, while British smokestack emissions cause problems in the forests of Northern Europe. The wastewater released from the manufacturing centres on the shores of the Great Lakes affect both Canadian and American waters. The radiation spewed into the atmosphere from the nuclear accident at Chernobyl could be found all over Europe, and even in North America.

These problems make us aware of the fact that we are all citizens of earth, and as such, we have a responsibility for its care, clean-up, and preservation.

CHECKPOINT

1. What is acid precipitation?

2. What effect has the soot and particulate matter in the atmosphere had on climate?

3. Name three household items which contain or were manufactured with CFCs.

14.10 Local Environmental Issues and the NIMBY Syndrome

The disposal of waste, whether sewage, sludge, domestic garbage, or hazardous waste, is a problem that most people will agree must be solved. It is understood that the site selection for a landfill has certain technical criteria which must be met. A landfill requires a commitment of land from a municipality. Approximately 1 ha of land is required for each 25 000 people for one year, if the landfill is filled to a depth of 3 m. The soil type, the water table level, drainage, and proximity to the community are all important considerations. If incineration is decided upon as a solution, it is expected that state-of-the-art emission controls will be in place when the incinerator is built.

After all of these technical problems have been solved, there is another problem to deal with – the **NIMBY syndrome** (*Not In My Back Yard*). Although most people will agree that some type of disposal system is necessary, they also want them located in some place other than near their homes. (See Figure 14.18.) Some believe that this is a no-win situation, that is, it is impossible to keep everyone happy when it comes to waste disposal. What do you think?

Figure 14.18
Not in my backyard

Other issues such as the zoning of land for housing developments, neighbourhood parks, commercial enterprises, factories, or nature preserves can also cause concerns about the effect on the environment and the community. What local environmental issues are currently being debated in your community? Are both sides of the issue given equal treatment in the media? Is this another no-win situation?

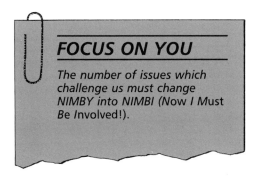

FOCUS ON YOU

The number of issues which challenge us must change NIMBY into NIMBI (Now I Must Be Involved!).

Activity 14C

Library Research Project

Select one of the following topics, and investigate and *prepare a written report*. Include in your report a description of the problem, its causes, its present status, possible solutions, and projections of its effect on the environment until the year 2050.

- the greenhouse effect
- the destruction of the northern and tropical forests
- the pollution of the oceans
- the effect of acid rain on the world's forests and lakes
- a particular pollution incident, such as an oil or chemical spill

Activity 14D

Focus on a Local Environmental Issue

Each local area has its own issues which directly affect the residents of the area. Work in a small group of four or five other students, and make a list of those issues which affect the people in your area. Each group should select one of these issues. Answer the following questions in your presentation.

1. What are the two sides of the problem?
2. (a) Who is involved?
   (b) Who supports each side?
3. Does anyone stand to gain or lose financially in this?
4. How will the environment be affected?
5. What are the possible alternatives?

- Some Suggested Issues

1. (a) Why are beaches polluted and closed in the summer?
   (b) Why can't something be done about this situation?

2. Why is hunting season limited to one or two weeks, and why are hunting licences required?

3. Should hunting be allowed in National and Provincial Parks?

4. Should home owners be forced to leave areas which are declared National or Provincial Parks?

5. Should the government impose quotas on commercial fisheries?

6. Why are licences necessary for those who fish for sport?

7. Should farmland be re-zoned for residential, commercial, or industrial use?

8. Should a factory be closed down if it cannot meet pollution control standards?

BIOLOGY AT WORK

DEMOGRAPHER
ECOLOGIST
ACTUARY
PLANT SCIENTIST
GENETICIST
FOOD CHEMIST
HYDROLOGICAL ENGINEER

Activity 14E

Conserving a Natural Area

Somewhere, not too far away from where you live, is an undeveloped area. It currently is not being used for a specific purpose, but it is considered to be special to those who live nearby. It is, in a sense, already a park or nature preserve.

Prepare a written/oral report which could be presented to your municipal government to support a bid to designate or zone the land as a park. Include in your report a discussion of the current and future real estate needs of local industry, and your reasons for supporting the zoning of the area as park land.

HEALTH CONCERNS

kwashiorkor
malnutrition
pesticide/herbicide poisoning
skin cancer
cataracts

Chapter 14 / Our Environment

CHAPTER FOCUS

Now that you have completed this chapter, you should be able to do the following:

1. Sketch a graph of the world's population growth, and compare it to exponential growth curves and stable population curves.
2. Interpret an age structure diagram.
3. Name four factors used to predict population growth.
4. Calculate the annual growth rate of a country given the necessary data.
5. Define and give an example of each of the following terms: ecosystem, population, community, producer, consumer, decomposer, habitat, niche, carrying capacity, environmental resistance.
6. Explain the difference between the biotic factors and the abiotic factors in an ecosystem and give two examples of each.
7. Define homeostasis and give an example of it in an ecosystem.
8. Explain the difference between a renewable resource and a non-renewable resource and give one example of each.
9. Describe the food consumption levels of the typical North American, and compare them to the consumption levels of people in less developed countries.
10. Give five reasons why there are food shortages.
11. List at least five ways in which more food could be made available throughout the world.
12. Explain how the destruction of a tropical rainforest can have long-term effects on the environment.
13. Describe at least three effects of dumping wastes in the oceans.
14. Explain how air pollution has changed the atmosphere to cause the greenhouse effect and destroy the ozone layer.
15. Describe some of the problems which are associated with the selection of new waste disposal sites.

SOME QUESTIONS TO ANSWER

1. (a) What is the earth's population estimated to be in the year 2000?
   (b) What do you think it might be in the year 2010?
2. Explain the meaning of a fertility rate of 2.9.
3. What does an age structure diagram indicate if its sides are almost vertical?
4. Why might food be unequally distributed to the members of a poor family?
5. What important event happened in the Green Revolution of the 1960s?
6. Emergency shipments of food are often sent to countries experiencing famines. Why might this food remain unused?
7. Why are some North American farmers paid not to grow crops?
8. What could you and your family do to reduce the amount of water you use?
9. Explain how some countries have planned to provide adequate water supplies in the future.
10. Explain why a living tree may be more valuable than the products which may be made from a tree which is cut down.
11. Describe three effects of plastic pollution in the oceans on animal life.
12. How much water would be wasted in your community if everyone left a tap dripping at a rate of 180 L per day for one week? Use the lowest estimate.
13. (a) Describe the greenhouse effect.
    (b) What causes this effect?
14. Explain how chlorofluorocarbons affect the ozone layer and why this may be dangerous to health.

Chapter 14 / Our Environment

SOME WORDS TO KNOW

Match each of the descriptions given in the left-hand column with a word shown in the right-hand column. DO NOT WRITE IN THIS BOOK.

1. Not in my backyard
2. Global warming trend
3. Where living things are found
4. Population specialist
5. Plant-eating animal
6. Growth by doubling
7. Place where an organism lives
8. All the living things in an ecosystem
9. Maintaining a steady state
10. Organisms which produce their own food
11. Organisms which break down dead or decaying matter
12. Average number of children born
13. Study of the interaction of organisms and the environment
14. Meat-eating animals
15. Group of organisms of the same species living together

A acid rain
B herbivore
C biotic potential
D demographer
E NIMBY
F homeostasis
G Green Revolution
H carnivores
I exponential
J biotic factor
K population
L environmental resistance
M decomposers
N producers
O greenhouse effect
P ecology
Q fertility rate
R chlorofluorocarbons
S biosphere
T habitat

Select any three of the unmatched words and, for each, give a proper definition or explanation in your own words.

SOME THINGS TO FIND OUT

1. Investigate the birth rate and death rate for your community and calculate its annual growth rate.
2. What are the immigration patterns and projections for Canada? How many people leave Canada each year?
3. Explain how a particular pesticide or a particular herbicide works.
4. Find out how much of the money raised by special fund-raising events, such as concerts, to buy food for famine victims is actually spent on food.
5. How much money is spent on pet food each year?
6. What is the status of the water diversion projects which have been proposed?
7. How large are the "windows" in the ozone layer, and where are the areas most affected by them?

APPENDIX

Some Commonly Abused Drugs and Their Effects

Drug	Short-Term Effects	Long-Term Effects
Alcohol Sedative Depressant	• initial relaxation, loss of inhibitions • slowing down of reflexes and reactions, impaired co-ordination • attitude changes, increased risk-taking to point of danger • acute overdose may lead to death	• regular, heavy use increases the possibility of: gastritis, pancreatitis, cirrhosis of the liver, oral cancer, certain cancers of the gastrointestinal tract, heart disease, brain damage • upon withdrawal following regular use, convulsions and delirium tremens may occur
Amphetamines Stimulants Benzedrine, Dexedrine, Neodrine	• reduced appetite, dilation of pupils • increased energy, alertness, faster breathing • increased heart rate and blood pressure which lead to increased risk of burst blood vessels or heart failure • risk of infection from unsterilized needles if injected WITH LARGER DOSES: • talkativeness, restlessness, excitation • sense of power and superiority • delusions and hallucinations • some frequent users become irritable, aggressive, paranoid, or panicky	• malnutrition • psychological dependence • after stopping, there usually follows a long sleep and then depression
Cocaine, Crack Stimulant	• same as amphetamines WITH LARGER DOSES: • stronger, more frequent "highs" • bizarre, erratic, sometimes violent behaviour • paranoid psychosis • sometimes a sensation of something crawling under the skin	• strong psychological dependence • destruction of tissues in nose if sniffed • other effects as amphetamines

Drug	Short-Term Effects	Long-Term Effects
Cannabis Modifier of mood & perception marijuana, hashish, "hash oil"	• a "high" feeling • increased pulse rate • reddening of the eyes • at later stage, user becomes quiet, reflective, and sleepy • impairs short-term memory, logical thinking, and ability to drive a car or perform other complex tasks • with larger doses, perceptions of sound, colour, and other sensations may be sharpened or distorted and thinking becomes slow and confused • in very large doses, the effects are similar to those of LSD and other hallucinogens – confusion, restlessness, excitement, and hallucinations	• a moderate tolerance • possible psychological dependence • loss of drive and interest • marijuana smoke contains 50% more tar than smoke from a high-tar cigarette: with regular use, risk of lung cancer, chronic bronchitis, and other lung diseases increases
LSD (Lysergic Acid Diethylamide)	• initial effects like those of amphetamines • later, distortions of perception – altered colours, shapes, sizes, distances producing exhilaration or anxiety and panic, depending on the user • feelings of panic or of unusual power may lead to behaviour that is dangerous to the user • occasionally, convulsions occur • strong tolerance develops very rapidly and disappears very rapidly	• the long-term medical effects of LSD are not known
Minor Tranquillizers Valium (Diazepam), Ativan, Halcion	• calms tension and agitation • muscle relaxation • lessened emotional responses to external stimuli, e.g., pain • reduced alertness • with larger doses, possible impairment of muscle co-ordination, dizziness, low blood pressure, and/or fainting	• physical dependence • withdrawal reaction (temporary sleep disturbances – abrupt withdrawal leads to anxiety, possible delirium, convulsions, and death)
Sedative Hypnotics Barbiturates (e.g., Amytal, Seconal, Nembutal) Non-Barbiturates (e.g., Placidyl, Dalmane, Doriden)	• small doses relieve anxiety, tension, producing calmness and relaxation • larger dose produces a "high" and slurred speech, staggering, etc. (similar to alcohol) • produces sleep in a quiet setting; otherwise, sleep may not occur • dangerous to drive a car or perform complex tasks • much larger doses produce unconsciousness • acute overdose can result in death	• tolerance and dependence if large doses are used • do not produce completely normal sleep – user may feel tired and irritable after sleeping • upon withdrawal, temporary sleep disturbances occur • abrupt withdrawal leads to anxiety and possible convulsions, delirium, and death

Drug	Short-Term Effects	Long-Term Effects
PCP (Phencyclidine) "Angel Dust", "peace pill", "crystal"	• euphoria – a "high" • faster, shallow breathing • increase in blood pressure and pulse rate • flushing and sweating • lack of muscle co-ordination • numbness of extremities WITH LARGER DOSES: • a fall in blood pressure, pulse rate, and respiration • nausea, vomiting, blurred vision, rolling movements and watering of the eyes, loss of balance, dizziness, convulsions, coma, and sometimes death • delusions, mental confusion, and amnesia are common	• possibility of flashbacks • possibility of prolonged anxiety or severe depression
Opiates Opium, Morphine, Codeine, Heroin	• relief from pain • produces a state of contentment, detachment, and freedom from distressing emotion • large doses create euphoria – a "high" • sometimes nausea and vomiting • acute overdose can result in death • risk of infection from unsterilized needles	• physical and psychological dependence • abrupt withdrawal results in moderate to severe withdrawal syndrome (cramps, diarrhea, running nose, etc.)

GLOSSARY

A

abatement equipment
Any device designed to reduce or eliminate the wastes that are released into the environment, especially from industrial processes

abiotic factors
The non-living things in a community

abscess (tooth)
Inflammation and infection of the pulp of a tooth, collection of pus at the base of a tooth

absorption
Taking up of substances as, for example, by capillaries, skin, or other tissues

accommodation
A change in the size of the pupil in response to a change in focal length

acid precipitation
Precipitation, usually rain or snow, which is acidic due to the presence of air pollutants

acne
Skin disorder of the sebaceous glands, common in adolescents

actin
Thin strand of protein in a muscle

activated sludge system
A system of wastewater treatment in which primary and secondary treatments are combined

adaptation
A shift in the reception of light from rods to cones, or from cones to rods, as light intensity changes

additives
Substances intentionally added to processed foods to serve several functions

adipose
Connective tissue composed of fat

aeration tank
A tank constructed with bubblers or sprayers to provide oxygen for the micro-organisms that break down the organic matter in wastewater

aerobic
Requiring the presence of oxygen for life functions

age distribution
The percentage of people in a population at each age level

age structure diagram
A profile of the percentage of people in a given population born within five-year intervals

air pollution
Airborne wastes

algal bloom
A rapid growth of algae

alveoli
Tiny air sacs in the lungs where exchange of gases between the lungs and capillaries takes place

amino acid
Organic molecules containing oxygen, hydrogen, carbon, and nitrogen; structural components of proteins

amniocentesis
Removal of some of the fluid from the amnion during pregnancy for tests and analysis

amnion
Membrane surrounding the fetus containing life-protecting fluid

amylase
Enzyme that breaks down starch

androgen
A male sex hormone

antagonistic
That which counteracts, or has the opposite action of, something else

antibody
Substance produced in the blood that provides immunity to certain types of micro-organisms

anti-diuretic hormone (ADH)
Hormone that acts in the kidney to control the amount of water retained or released

antigen
Substance that stimulates the production of antibodies

anus
Opening of the rectum and the digestive tract

aorta
The main artery of the body leaving the left ventricle of the heart

appendicular skeleton
Bones of the arms and legs, and the pectoral and pelvic girdles

appendix
Small structure attached to the caecum. It is a blind sac of little importance in human digestion.

areola
Ring of pigmented tissue around the nipple

artery
Any vessel carrying blood away from the heart to the tissues

arthritis
Inflammation of the joints

articular cartilage
Smooth, tough layer of cartilage covering the ends of most bones

articulation
A joint; where two bones come together

astigmatism
A condition caused by the uneven curvature of the lens or cornea

atrioventricular node (AV)
Clump of nerve cells beneath the right atrium that sends impulses from the pacemaker to the ventricles

atrium
One of a pair of upper chambers in the heart that receive blood from the veins

atrophy
A decrease in size of a muscle or organ

auditory canal
The channel leading into the head from the auricle (pinna)

auricle
The external ear; also called the pinna

Autonomic Nervous System (ANS)
Parts of the nervous system that control automatic responses

axial skeleton
The bones of the skull, vertebral column, and rib cage

axon
Part of the neuron which carries impulses away from the cell body

B

behaviour
Any observable muscular activity

bicuspid valve
Heart valve between the left atrium and left ventricle

bile
Secretion of the liver that helps the emulsification and digestion of fats

binocular vision
Vision with two eyes; permits depth perception

Biochemical Oxygen Demand (BOD)
A measure of how much oxygen is required to break down the organic material present in a sample of water

biodegradable
Able to be broken down or decomposed and made harmless by natural processes

biotic factors
The living things in a community

biotic potential
The maximum reproductive rate of a species

birthmark
Pigmented patches of skin present from birth

birth rate
The number of births per 1000 people per year in a given area

blackhead
Oxidized sebum that turns black in the glands of the skin

blastocyst
Hollow, fluid-filled sphere formed by the cells of the fertilized egg in early pregnancy

blind spot
The area where the optic nerve leaves the retina; no receptors present

blood type
Characteristic of red blood cells determined by the presence of one or more antigens

boil
Round, tender, inflamed area of skin with a core of pus and bacteria

bolus
A mass of soft food ready for swallowing

Bowman's capsule
Capsule around the glomerulus in the kidney, which aids in filtration of wastes from the blood

bronchi
The first two divisions of the trachea that enter the lungs and terminate in the bronchioles

bronchioles
Smaller divisions of the bronchi that lead to the alveolar ducts and air sacs

C

caecum
The blind pouch that is found at the start of the large intestine

Caesarean section
Delivery of a baby by cutting through the abdominal and uterine walls and removing the baby through the incision

calculus
Hard crust (tartar) that builds up on the surface of a tooth

callus
The tissue that develops from the procallus to seal the ends of a fractured bone

canine teeth
Pointed teeth next to the incisors

capillary
Small, thin-walled blood vessel that allows substances to pass in and out of the circulatory system

carbohydrates
Compounds composed of carbon, hydrogen, and oxygen; for example, sugars and starches

carotid arteries
Arteries leading to the head and neck

carpals
The small bones of the wrist

carrying capacity
The maximum number of individuals of a species that a particular ecosystem can support

cartilage
Strong, flexible type of connective tissue

celiac artery
Artery carrying blood to the stomach, spleen, and liver

cellular respiration
Process that takes place in cells that require oxygen and release energy.

cementum
Thin layer of tissue over the root of a tooth that helps to hold the tooth in the socket

Central Nervous System (CNS)
The brain and spinal cord

cerebellum
Part of the hindbrain; responsible for muscle co-ordination

cerebral cortex
Folded surface of the cerebrum

cerebral hemisphere
One half of the brain; the left and right halves are divided by a central fissure.

cerebrospinal fluid
Shock-absorbing fluid found around the brain and spinal cord

cerebrum
Largest part of the brain; controls voluntary movement, thought

cervix
The neck of the uterus

chemical digestion
Breakdown of foods by enzymes and digestive juices

chlorofluorocarbon (CFC)
A gas composed of carbon, chlorine, and fluorine that is used in many industrial processes. It rises to the upper levels of the atmosphere where it breaks apart the ozone molecules that form a protective shield around the earth.

cholesterol
A fatty compound found in animal fat, bile, blood tissues, etc.

chorion
Outer membrane around the embryo from which villi connect with the uterine lining, which in turn produces the placenta

choroid layer
Middle layer of the eyeball; contains blood vessels and pigment

chromosome
Small rod-shaped bodies in the nucleus that contain hereditary genetic material; 46 present in humans (23 pairs)

chyme
The mixture of partially digested foods and digestive juices found in the small intestine

circadian rhythm
The 24 h rhythm or pattern during which we are active, eat, and sleep

clarifier
An aerated settling tank used in primary wastewater treatment

clavicle
Collar-bone

clitoris
Small sensory structure at the entrance to the vagina

cochlea
A coiled tube of the inner ear; contains the receptors for hearing

colon
The portion of the large intestine from the caecum to the rectum

colour-blindness
The inability to distinguish some or all colours

comminutator
A device that shreds and grinds solids in primary wastewater treatment

common bile duct
Carries bile from the cystic and hepatic ducts to the small intestine

community
All of the different populations living together in the same area

compact bone
Hard, strong, very dense bone tissue

compost
A mixture of decaying organic material, such as leaves, waste from food preparation, and garden trash, which can be used for fertilizer

compound fracture
A broken bone that pierces the skin

conditioned reflex
A reflex that has been modified by training or learning

conduction
The transmission of a nerve impulse; transfer of heat or sound waves through a conducting medium

conduction deafness
Hearing loss caused by interference in the transfer of sound waves

conductor
A nerve that carries messages to and from the CNS

cones
Receptors in the retina responsible for detecting colour

conjunctiva
Mucous membrane covering the front of the eyeball and the inside of the eyelids

consumer
An organism that gets its food by eating other organisms

contraction
The shortening or tightening of a muscle

control system
A system that processes information and acts to maintain the body

convection
Transfer of heat in liquids or gases by means of currents

cornea
The transparent layer of tissue at the front of the eyeball

coronary
Term referring to the heart

corpus albicans
White scar tissue in the ovary; formed by atrophy of the corpus luteum

corpus cavernosum
Spongy body in the penis that fills with blood during erection

corpus luteum
Body of cells that fills the ovarian follicle following ovulation; produces progesterone

cortex
The outer portion or layer of an organ or structure

Cowper's gland
Two small glands in the male that produce some of the seminal fluid

crown
Part of the tooth visible above the gum

cystic duct
Connects the gall bladder to the hepatic duct

cytoplasm
The cellular substance between the cell membrane and the nucleus of a cell

D

death rate
The number of deaths per 1000 people per year in a given area

deciduous teeth
First set of primary teeth

demographer
A specialist in the study of human populations and statistics about birth rates, death rates, fertility rates, etc., in a given area

dendrite
Part of the neuron that carries impulses toward the cell body

dentine
A hard dense material that forms the main part of the tooth

deoxyribonucleic acid (DNA)
The genetic material carried by all organisms, composed of two complementary strands of nucleic acid wound in a helix

dermis
The true skin; skin tissue lying below the epidermis

dialysis
Process by which solutes move across a semi-permeable membrane. Kidney dialysis allows removal of harmful components from the blood.

diaphragm
The muscle and connective tissue partition between the thoracic cavity and the abdominal cavity; muscle of respiration

diastole
The period of relaxation for the heart muscles when the ventricles of the heart fill with blood

diet
A food eating pattern, a prescribed course of food

diploid
Cells having pairs of chromosomes present

disaccharide
A sugar that consists of two monosaccharides (single sugars) bonded together

dislocation
Condition in which the bones in a joint are pulled out of alignment and must be repositioned

domestic sewage
Waste from toilets, sinks, laundry tubs, etc.

domestic solid waste
Waste generated by households and small businesses

dominant
A trait that is always expressed when at least one allele is present

dorsal root
Sensory nerve that enters the spinal cord at the back or dorsal side

Down Syndrome
Congenital, chromosomal abnormality that causes mental retardation, slanted eyes, flattened forehead, poor muscle tone, etc.

drive
A condition of the body causing it to become active

drug
A non-food substance taken into the body that affects body cells

drug dependence
Drug addiction

duodenum
First part of the small intestine

E

ecology
The study of the interacting systems of organisms and the environment

ecosystem
The interacting system of living things and the non-living environment

ectoderm
Outer layer of cells in the embryo; develops into nervous system, sense organs, etc.

effector
A muscle or organ that responds to a particular stimulus

electrostatic precipitator
An air pollution abatement (reduction) device installed in a smokestack to collect small particles

emulsify
To prepare two liquids, which do not mix, so that one, in the form of small globules, is dispersed throughout the other

enamel
Hard outer layer that covers the teeth

end plate
Tissue at the end of a motor nerve that releases acetylcholine

endocrine system
Glands that produce and release secretions (hormones) directly into the bloodstream

endoderm
Inner layer of cells in the embryo that gives rise to digestive and respiratory organs

endometrium
The lining of the inner surface of the uterus

enriching substances
Substances rich in nutrients; usually refers to pollutants that enter waterways

enterokinase
Enzyme in the gastric juice that helps in the breakdown of proteins.

environmental resistance
Factors such as the availability of food or prevalence of disease that limit a population's biotic potential and slow its growth

enzyme
A protein able to change the rate of chemical reaction without being consumed in the reaction

epidermis
Outer layer of skin cells; contains no blood vessels

epididymis
A long, tiny tubule attached to the testes which collects and stores sperm.

epiglottis
Leaf-shaped cartilage that covers larynx during swallowing; helps deflect food into esophagus

epiphyseal cartilage
The growth plate of a bone; new bone tissue is formed here.

erepsin
An enzyme found in the intestinal juices that helps digest proteins

erythrocytes
Red blood cells; carry oxygen in the blood

esophagus
The tube leading from the pharynx to the stomach

essential amino acids
Amino acids that the body cannot make and must import from other proteins

estrogen
Female hormone that promotes sexual development

eustachian tube
The tube that connects the middle ear with the back of the throat

excrete
To separate and expel, to release waste

excretion
The separation and passing of waste from the body

expiratory reserve volume
Air forcibly expelled after normal expiration

extension
The act of straightening a limb

F

Fallopian tube
See oviduct

far-sightedness
Ability to see distant objects, inability to focus on nearby objects

fecal coliforms test
An indicator of the presence of pollution by human waste

feces
Undigested food waste discharged from the intestines

femur
Thigh bone; longest and strongest bone in the body

fertility rate
An estimation of the average number of children each woman will have in her childbearing years

fertilization
The union of an ovum (egg) with a sperm

fetus
The developmental stage from about the ninth week after conception to birth

fibre
Insoluble organic material, mostly carbohydrate; not digestible by humans

fibula
The smaller of the two bones of the lower leg

fimbriae
Fringed, funnel-shaped opening of the Fallopian tubes

fixed action pattern
Innate or unlearned behaviour

follicle
A small, hollow structure containing cells or a secretion; a layer of hormone-producing cells surrounding the ovum

follicle stimulating hormone (FSH)
Hormone produced by the pituitary gland; stimulates the development of a follicle in the ovary

fontanel
Small depression on the skull where three bony plates meet

Forel-Ule colour scale
An indicator of the dissolved and suspended materials in a sample of water

foreskin
Fold of skin that covers the head of the penis

fovea centralis
The area of the retina that is especially sensitive to detail

fracture
A broken or cracked bone

G

gall bladder
Small sac in the liver that stores bile

gamete
A male or female sex cell; sperm or ovum (egg)

gastric juice
Stomach secretions for the digestion of foods

gene
The unit of heredity responsible for transmission of a characteristic to the offspring; part of the DNA molecule

genitalia
External reproductive structures.

genotype
The hereditary make-up of an individual as determined by the genes

gestation
Period from fertilization to birth; pregnancy

glial cells
Cells that provide nourishment and support to the neurons of the brain

glomerulus
A small group of capillaries in Bowman's capsule of the kidney

glycogen
A complex carbohydrate (animal starch) stored in liver and muscle

greenhouse effect
The gradual warming of the earth's atmosphere which may accompany the rise in the carbon dioxide levels in the atmosphere. The carbon dioxide may act like a greenhouse, allowing the sun's rays in but not allowing the heat to escape.

green-stick fracture
A fracture in which the bone does not separate completely, but instead resembles a broken green stick, with some fibres intact

grey matter
Areas of unmyelinated fibres in the brain and spinal cord

H

habitat
The place where an organism lives in a community

halitosis
Chronic bad breath

hallucinogenic drug
A drug that changes sensory perceptions

haploid
Having one-half the normal number of chromosomes; sperm and eggs are haploid cells.

hard palate
The front part of the roof of the mouth

Haversian canal
Large tube-like vessel that carries blood supplies to bone cells

hazardous waste
Substances that are toxic or potentially harmful to humans and the environment

hemoglobin
A protein molecule found in red blood cells that is able to bond with oxygen

hemophilia
Hereditary blood disorder caused by lack of a blood-clotting factor; excessive bleeding from wounds; sex-linked recessive trait

hemorrhoids
Swollen veins in the rectal area

hepatic
Term referring to the liver

hepatic artery
Artery bringing blood to the liver

hepatic duct
Leads from the liver to the common bile duct

hepatic portal system
Veins that collect blood from the digestive tract and spleen and take it to the liver

hepatitis
An inflammation of the liver caused by an infection

heterozygous
Having one or more pairs of dissimilar genes or chromosomes for an inherited characteristic

homeostasis
The state of constancy of body composition and function

homologous
Similar in origin and structure

homozygous
Having two identical alleles for a trait

hormone
A chemical produced by an endocrine gland; affects target organs

Human Genome Project
New project to map all the human chromosomes

humerus
Bone of the upper arm

hydrochloric acid
Strong acid found in the stomach; aids in the digestion of proteins

hymen
A membranous fold which partly or completely closes the vaginal opening

hypertension
Prolonged high blood pressure

hypertrophy
An increase in size of a muscle or organ

hypothalamus
Part of the forebrain; controls the ANS and internal organs

hypothermia
Abnormally low body temperature caused by exposure; usually ranges from 32-35°C for the body core temperature

I

ileum
Last section of the small intestine

implantation
Embedding of the blastocyst (fertilized and developing ovum) in the uterine lining

incineration
The burning of waste under controlled conditions and high temperatures

incisors
Front, chisel-shaped teeth used for cutting and biting

industrial solid waste
Waste such as construction debris, machine scrap, mine tailings, and organic wastes produced by food processing plants

inferior vena cava
Large vein bringing blood to the heart from the trunk and lower limbs

insertion
The point of attachment of a muscle to the bone which moves when the muscle contracts

inspiration
Taking air into lungs

inspiratory reserve volume (IRV)
Amount of extra air you can forcibly draw in after a normal inhalation

insulin
Hormone produced in the pancreas that regulates sugar metabolism

intercostal muscles
Muscles between the ribs which aid in respiration

interstitial fluid
Fluid similar to plasma found between the cells. Carries nutrients from capillaries to cells

intervertebral discs
Cushions of compressible fibrous cartilage between vertebrae

iris
The coloured part of the front of the eye; contains muscles to control the size of the pupil

J

jaundice
Yellow pigmentation in the skin when bile pigment (bilirubin) in the blood is high

jejunum
Middle section of the small intestine

jugular veins
Blood vessels returning blood to the heart from the head and neck

K

karyotype
A mapping or arrangement of the 46 human chromosomes based on the size of the individual chromosome

keratin
A protein found in epidermal structures such as hair and nails

kidney
Pair of organs in the rear of the abdomen that filter out wastes from the blood

kiln
A furnace or oven for burning or drying

L

labia majora
Lips of the external female genitalia

labia minora
Lips of the inner female genitalia

lacrimal glands
Tear-producing glands of the eye

lactase
An enzyme that breaks down the sugar lactose

lactation
Production of milk

lactiferous duct
Small ducts within the breast that carry milk to the nipple

larynx
The "voice-box", located at the upper part of the trachea

leachate
Chemicals carried by water from a dump or landfill site and entering ground or surface water

learning disorder
A disorder in one or more of the processes involved in the understanding or use of spoken or written language

lens
A transparent, disc-shaped structure in the eye which focusses the image on the retina

leukemia
Cancer of the blood-forming organs

leukocytes
White blood cells; may be granular or non-granular in blood

light reflex
A change in the size of the pupil dependent on light intensity

limiting factors
Factors that slow the growth of a population

lipase
An enzyme that breaks down lipids (fats)

litter
Anything discarded and scattered about, such as rubbish

lymphatic system
System of one-way vessels that return fluids from the tissues to the veins; active in white cell formation, etc.

M

maltase
An enzyme that acts on the sugar maltose

marrow
Fatty tissue and blood vessels found in the centre of most long bones; site of red blood cell production.

medulla
Central portion of an organ, e.g., the medulla oblongata of the hindbrain

medulla oblongata
Part of the hindbrain; responsible for vital body functions

meiosis
A form of cell division that reduces chromosome number by one-half; occurs in the ovaries and testes during the production of eggs and sperm

melanin
Black or brown pigment found in skin, hair, and retina

meninges
Three protective membranes surrounding the brain

menstrual
Refers to the sloughing of the lining (endometrium) of the uterus; relating to the monthly flow of blood from the female genital tract

mesentery
A double-layered fold of the peritoneum attaching various organs to the wall of the body cavity

mesoderm
The middle germ layer of the embryo; gives rise to connective tissue, muscle, blood, and bone

messenger RNA
A single-stranded RNA molecule that carries the DNA coded sequence during protein synthesis

metacarpals
Bones of the hand

metatarsals
Large bones of the foot

micturition
Passing of urine

minerals
Inorganic substances found in food and water that supply vital components to the diet

mitosis
A type of cell division that results in production of daughter cells identical to the parent cell

mnemonic device
Any system of coding information to make it easier to remember

molars
Large teeth at the back of the jaw used for chewing and grinding food

moles
Small patches or circles of pigment in the skin

monocular vision
Vision with only one eye

mononucleosis
Acute infection believed to be carried by a type of herpes virus

monosaccharide
The simplest sugar

motor neuron
A neuron which conducts an impulse from the CNS to an effector, for example, a muscle

mucus
Protective fluid secreted by glands

myelin
Fatty protein sheath which covers the axons of some neurons; increases speed of impulse transmission along nerve

myofibril
Strands of actin and myosin proteins; combine to make muscle fibre

myometrium
Middle muscular wall of the uterus

myosin
Thick strand of protein in a muscle

N

nares
Nostrils

nasopharynx
The portion of the nasal cavity above the soft palate

near-sightedness
Ability to focus on nearby objects, inability to see at a distance

negative feedback control system
A method by which a change in one part of a system triggers a change in another part, which then adjusts to counteract the effect of the initial change and maintain a balanced state

nephron
The functional unit of the kidney that forms urine, regulates blood composition, and filters wastes

nerve deafness
Hearing loss resulting from damage to receptors or nerve cells

neurilemma
A thin, living membrane covering the myelinated fibres of the PNS

neuron
Nerve cell

niche
The role that an organism plays in a community

nictitating membrane
A thin membrane that serves as an inner eyelid in some animals

NIMBY syndrome
Not In My Back Yard; an individual's belief that although such things as incinerators and landfills are necessary, they should be located somewhere other than where he/she lives

nitrogen base
Four nitrogen bases form the steps of the DNA molecule and their sequence forms the genetic code.

non-biodegradable
Substances such as some plastics, glass, and metals, which resist being broken down by heat, light, and decomposer organisms, and remain for a very long time in the environment

nondisjunction
When chromosomes fail to separate during meiosis

non-renewable resource
A resource that once used cannot be replaced

nucleus
Small spherical structure within cell; contains DNA; centre of control and heredity

O

olfactory cells
Odour receptors found in the upper nasal cavity

optical illusion
A visually false sense impression of what actually exists

optic nerve
The nerve that carries impulses from the eye to the brain

oral cavity
The mouth

orbital cavities
Hollow depressions in the skull that contain the eyes

organ
Two or more tissues organized for a particular job

organelle
Small structure with specific function found within the cell

organ of Corti
Structure of the cochlea which contains the hearing receptors

orgasm
Climax of sexual excitement

origin
The beginning of a nerve; the fixed attachment of a muscle

ossification
The process of bone formation

otoliths
Tiny particles of calcium carbonate found in the inner ear; important in balance

oval window
A membrane-covered opening of the cochlea on which the stirrup rests

ovary
A female's egg-producing organ

oviduct
A tube that conducts an egg from the ovary to the uterus; also called the Fallopian tube

ovulation
Release of a mature egg from the ovary

ovum
Mature female reproductive cell

oxygen deficit
A condition in which the oxygen level of the tissues is depleted

oxyhemoglobin
Hemoglobin carrying oxygen

P

pacemaker (SA node)
Special nerve cells in the right atrium that initiate impulses that stimulate the heart beat

pancreas
Gland that secretes digestive enzymes and the hormone insulin

pancreatic duct
A duct carrying the pancreatic juices to the duodenum

pancreatic juice
The digestive fluid produced by the pancreas and released into the duodenum

parasympathetic nervous system
A division of the nervous system that controls normal body functions

parotid duct
The duct that carries saliva from the parotid salivary gland into the mouth

patella
Kneecap

pathogen
Any organism capable of causing disease

pectoral girdle
Attaches the arm to the trunk; made up of the clavicle and scapula

pedigree
A record of an individual's parentage through several generations; used to trace the inheritance of traits

pelvis
A basin-like skeletal structure that supports the spinal column and provides attachment for the legs

penis
The male reproductive organ through which semen and urine pass out of the body

pepsinogen
Secretion by stomach glands that aid in the digestion of proteins

peptic ulcer
Stomach ulcer

pericardium
The membrane that surrounds the heart

peridontal diseases
Diseases that affect the gums and soft tissues around the teeth

periosteum
A thin, double membrane covering a bone; controls bone development and contains nerves and blood vessels

Peripheral Nervous System (PNS)
The nerves that extend beyond the brain and spinal cord

peristalsis
The alternate contraction and relaxation of the walls of a tubular structure moving contents onwards

peritoneum
A layer of tissue that lines the walls of the abdominal and pelvic cavities, enclosing the organs

peritonitis
Inflammation of the peritoneum

personality disorder
A disorder that causes abnormal behaviours; e.g. delusions, hallucinations

phagocytosis
Engulfing of particles by cells

phalanges
The small bones of the toes and fingers

pharynx
The general area at the back of the mouth where the mouth joins the nasal cavity

phenotype
The visual appearance of an individual; the physical expression of genotype

physical digestion
Digestion that occurs in the mouth by the grinding and cutting action of teeth

pinna
See auricle

placenta
The structure attached to the inner uterine lining through which the fetus gets its nourishment and excretes its wastes

plaque
Decay-causing bacteria found on teeth.

plasma
The liquid portion of the blood

platelets
Tiny bodies found in blood which produce thromboplastin that is used in the clotting of blood

pleura
Refers to the membrane(s) lining the cavities of the thorax, or covering the lung

pollution index
A system used to alert people about potentially harmful levels of air pollutants in the atmosphere

polysaccharide
A carbohydrate consisting of many monosaccharides bonded together

pons
Part of the medulla bridging the medulla and the midbrain

population
A group of organisms of the same species living together in the same area

portal vein
Vein carrying blood from digestive organs to liver

primary treatment
The first of two main processes in wastewater treatment; designed to remove settleable solids

procallus
The connective tissue that forms around the ends of a broken bone

producer
An organism that produces its own food; for example, a green plant

progesterone
Hormone produced in the ovary that prepares the uterus for the implantation of the fertilized egg

prognosis
Forecast of the expected progress of a disease

prostate gland
Male gland; secretes seminal fluids

psychosomatic
Involving the relationship between mind and body

puberty
The time of life when both sexes become functionally capable of reproduction

pulmonary
Term referring to the lungs

pulmonary artery
Artery carrying blood from heart to lungs

pulmonary veins
Veins carrying blood from lungs to heart

pupil
Opening in the iris

R

R.I.C.E.
Treatment for a minor injury to the skeleton or muscles; Rest, Ice, Compression, Elevation

radiation
Sending out of rays from a source; heat, X-rays, radio-active, etc.

radius
The shorter of the two bones of the forearm

reaction time
Time elapsed while a nerve impulse travels from a receptor to the central nervous system and returns to a muscle

receptor
A sense organ responding to a particular type of stimulus

recessive
A characteristic that does not usually express itself in offspring because of suppression by a dominant gene

rectum
Last section of the alimentary canal; stores feces

red blood cells
Erythrocytes; carry oxygen in the blood

reflex act
An automatic or involuntary action

reflex arc
A series of neurons serving a reflex; produces involuntary actions

renal arteries
Vessels carrying blood to the kidneys

renal pelvis
The part of the ureter that enters the kidney

renal veins
Vessels carrying blood away from the kidneys

renewable resource
A resource that is renewed or replaced in a short period of time

rennin
An enzyme in the gastric juice that curdles and breaks down milk proteins

residual air capacity
Amount of air left in the lungs and airways after a forced exhalation

retina
Inner layer of the eyeball; contains rods and cones

Rh factor
One type of antigen found in blood; factor present in the blood of the child but absent in the mother; can cause hemolytic disease of the newborn

rhodopsin
Light-sensitive pigment found in the rods of the retina

ribonucleic acid (RNA)
Composed of nitrogen bases; carries the genetic code from DNA into the cell cytoplasm

rods
Light-sensitive receptors in the retina; sensitive to dim light

root (tooth)
The part of a tooth anchored in the jawbone

round window
A membrane-covered opening of the cochlea which permits fluid movement

S

saccharides
Any sugar

saccule
A fluid-filled sac of the inner ear responsible for balance

salivary gland
A saliva-producing gland in the mouth. Each gland is named according to its location; parotid (near the ear), submaxillary (under the upper jawbone), sublingual (under the tongue)

sanitary landfill
A waste management method that covers daily accumulated waste with soil in a series of cells and layers

sarcolemma
A sheath that wraps a group of muscle fibres

saturated
Holding all it can. A saturated fat has all the hydrogen it can hold on its chemical bonds.

scapula
Shoulder blade

sclera
White outer layer of the eyeball

scrotum
An external pouch of skin that contains the testes

scrubber
An air pollution abatement (reduction) device installed in a smokestack to collect harmful gases

sebaceous
Refers to sebum, a fatty secretion of the sebaceous (oil) glands

sebum
Substance secreted by sebaceous glands

Secchi disc
A white disc, with pie-shaped red or black segments, which is lowered into water to determine visually the relative amount of suspended material in the water

secondary treatment
The second of two main processes in wastewater treatment; designed to break down organic matter by aerobic bacterial action

semen
Fluid discharged from the penis containing sperm and the secretions of various glands

semicircular canals
Three tubes of the inner ear; respond to changes in movement

seminal vesicles
Pair of glands that produce seminal fluid to form a liquid vehicle for sperm

seminiferous tubules
The tubules in which the male sperm are produced in the testes

sensory neuron
A neuron that conducts impulses towards the central nervous system

septic system
A wastewater disposal system composed of an underground tank for decomposing solid wastes, and an absorption field, designed to allow the soil micro-organisms to break down the organic matter remaining in the liquid waste

sex-linked
Genetic traits carried on the sex chromosomes; since the female X chromosome is larger, most are found here

simple fracture
A broken bone that does not pierce the skin

sinoatrial node (SA)
See pacemaker

sludge
Solid materials collected during wastewater treatment. Raw sludge is removed during primary treatment, and digested sludge is collected during the secondary stage.

smog
A haze produced in the atmosphere by pollution. There are two main types: photochemical smog and industrial smog.

soft palate
The rear part of the roof of the mouth

solid waste reduction unit (SWARU)
An incinerator designed to reduce the volume of solid waste by about 95%, while providing a source of heat, or generating electricity

spermatic cord
Cord containing the spermatic artery, vein, nerve, and sperm-carrying duct, which leads from the testes through the opening between the abdomen and the scrotum

spermatogonia
Immature sperm

sphincter
A band of circularly arranged muscle that narrows an opening when it contracts

sphygmomanometer
Instrument used to measure blood pressure

spongy bone
Bone tissue full of tiny spaces, resembling a sponge

sprain
A temporary separation of the bones of a joint resulting in swelling and pain

stethoscope
Instrument for listening to sounds within the body

stimulus
Something that causes an action or reaction

stratum corneum
Outer layer of dead skin

stratum germinativum
Inner lining of skin which replaces worn out cells

sucrase
An enzyme that breaks down the sugar sucrose

sucrose
Table sugar; it is a disaccharide

sunblockers
Substances that prevent or control the amount of ultraviolet light penetrating the skin

sunscreens
Substances that allow some tanning to take place but absorb some ultraviolet light

superior vena cava
Large vein bringing blood to the heart from the head and arms

suture
An immovable joint uniting the bones of the skull or pelvis

sympathetic nervous system
A division of the nervous system that prepares the body in times of stress

synapse
The gap between neurons

synovial membrane
The smooth lining of a movable joint; secretes a lubricant

system
An organized group of related tissues with a common general function

systole
Contraction of the muscle of a heart chamber

T

target organ
The organ of the body that responds to a certain hormone

tarsals
Bones of the ankle

taste bud
An area on the tongue containing a taste receptor

taste papilla
A projection of tissue above the surface of the tongue which contains a taste bud

teat
A mammary papilla, or nipple; in the female, provides milk for the young

tendon
A tough, fibrous, inelastic tissue which attaches a muscle to bone

testes
A male's sperm-producing organ

testosterone
A male hormone produced in the testes

thalamus
Part of the forebrain; a sensory relay centre

threshold
The level or concentration at which a toxic substance presents a danger to people, plants, animals, or the environment

thyroid
Gland at the base of the neck that produces hormones affecting growth and metabolism

thyroid gland
Oval, reddish gland at the base of the larynx which produces thyroxin

tibia
The larger bone of the lower leg; the shin

tidal volume
The amount of air moving in and out of the lungs during normal breathing

trachea
Large tube leading from the larynx to the bronchial tubes; windpipe

transplant
An organ or tissue removed from a healthy invidual and surgically transplanted into a person to replace a diseased or non-functioning organ

tricuspid valve
Valve in the heart

triglyceride
A fat molecule composed of glycerol attached to three fatty acids

trisomy
Result of nondisjunction; when the cell carries an extra chromosome in addition to the normal homologous pair; for example, Down Syndrome

turbid
Not clear or transparent; clouded

turbinate bones
Shelf-like bones that extend into the nasal cavity to increase the surface area over which air passes

tympanic membrane
Eardrum

U

ulna
The main supporting bone of the forearm

umbilical cord
The structure connecting the fetus to the placenta

urea
A metabolic waste formed in the liver from protein breakdown

ureter
Duct taking urine from the kidney to the bladder

urethra
Duct carrying urine from the bladder to the exterior of the body

urinalysis
Chemical analysis of a sample of urine

urinary bladder
Extendible sac where urine is collected and temporarily stored

urine
The fluid formed by the kidney and eliminated from the body via the urethra

urogenital opening
A common opening for the urethra and the genital organs

urogenital sinus
The opening shared by both the vagina and the urethra in a young female animal

uterus
The womb; the organ in which the fetus develops before birth

utricle
A fluid-filled sac of the inner ear responsible for balance

uvula
Fleshy tissue suspended from the soft palate above the back of the tongue

V

vagina
Female genital organ that extends from the vulva to the cervix

vaginal opening
The opening in the female's body leading to the vagina

vas deferens
Tube that carries sperm from the testes to the seminal vesicles

vein
Blood vessel carrying blood towards the heart

ventral root
Motor nerve that leaves the spinal cord at the front or ventral side

ventricle
One of the two lower chambers of the heart

vertebrae
Small, irregularly shaped bones of the spine

vertebral column
The spine or backbone

villi
Small finger-like projections along the membranes of the small intestine

virus
Submicroscopic infectious organism that attaches to living cells to survive and reproduce

visual acuity
The sharpness with which detail can be seen

vital capacity
The amount of air contained in the lungs; the sum of the expiratory, reserve, tidal, and inspiratory reserve volumes

vitamin
An essential organic substance required for metabolic processes; it works with enzymes to control body functions.

vocal cords
Membranes inside the larynx that produce sound when air passes over them

W

waste
Any material that we do not want or which is discarded

wastewater
The mixture of domestic, commercial, and industrial wastes in the sewer system that arrives at a treatment plant

white blood cells
See leukocytes

white matter
Areas of myelinated fibres in the brain and spinal cord

Y

yolk sac
Small sac attached to the fetus that produces blood cells in the early stages of fetal development

Z

zygote
A cell produced by the union of sperm and egg: a fertilized ovum

INDEX

A

abiotic factors, 433
abscesses, in teeth, 257
acid precipitation, 414-419
  pH and, 415
acne, 18, 26
actin, 69-70
addiction, 104-105
additives, 235-236
adenoids, 188
aerobic exercise, 181
age structure diagram, 429-430
AIDS, 354-355
air pollution, 202, 409-414
  combustion, 409-410
  mechanical action, 410, 412
  reducing, 414
  smog, 412
  vaporization, 410
alcohol, in human body, 272-273
algal bloom, 385
alleles, 313
alveoli, 191
amino acids, essential, 219-220
amniocentesis, 320
amnion, 339
amylase, 258
anaerobic organisms, 401
analgesics, 105
androgens, 26
anemia, 159, 161
anorexia nervosa, 243
antagonistic pairs of muscles, 73-74
antibodies, 158
appendicitis, 275

appendicular skeleton, 45
appendix, 274-275
areola, 345-346
arteries, 164, 166, 168
  and blood pressure, 176
  hepatic, 267
  pulmonary, 169
  renal, 284
arterioles, 164
arthritis, 58
  rheumatoid, 58
articular cartilage, 41, 57
articulation, see joints
ascending colon, 275
association neuron, 97
associative memory, 93
astigmatism, 134
atherosclerosis, 217
Athlete's foot, 26
atmosphere, pollution of, 445-448
ATP molecules, 75
atrioventricular node, 174
atrioventricular valves, heart, 171
atrophy, 78
Autonomic Nervous System (ANS), 84, 100-101
  parasympathetic system, 101
  sympathetic system, 100
axial skeleton, 45
axons, 85

B

balance, organs of, 143-144
  otoliths, 143
  Romberg test, 145

  saccule, 143
  semicircular canals, 144
  utricle, 143
ball-and-socket joints, 58-59
barium X-rays, 264
behaviour, 84, 107-110
  innate, 108-109
bile, 267
binocular vision, 130-131
Biochemical Oxygen Demand (BOD), 385-386
biotic factors, 433
biotic potential, 433
birth, 343-345
birthmarks, 20
blackhead, 18
blastocyst, 338
blemishes, handling, 25-26
blind spot, retina, 126, 128-129
blister, 12
blood, 156
  elements of, 156-162
  plasma, 156
  platelets, 156, 160
  Rh factor, 164
  transfusions, 162-163
  transport of gases in, 200
  types, 162, 313-315
  viscosity of, 176-177
blood cells, 156
  amoeboid action, 158
  chemical properties, 158
  disorders of, 161-162
  phagocytosis, 158
  red, 157
  white, 157-158

473

blood cholesterol level, 217
blood clotting, 160
blood pressure
  factors affecting, 176-177
  measuring, 177, 179
blood sugar level, 211
body heat, conserving in
  emergency, 24
body mass, 243
  controlling, 241-242
body size, assessment of, 238-242
  controlling body mass, 241-242
  skinfold test, 238-239
body systems, 9-10
body temperature, regulating,
  22-24
boils, 26
bones
  as levers, 71
  broken, 61-63
  compact vs. spongy, 40
  cranial, 48
  facial, 48
  fine structure of, 44
  first aid for, 64
  formation, 43-44
  healing broken, 62-63
  lower limbs, 53
  marrow, 41
  nature and structure of, 40-41
  of arm and hand, 54-55
  skull, 46-48
  transplants, 47
  types of, 40
Borlaug, Dr. Norman, 440
Bowman's capsule, 285
brain, 87-93
  cerebellum, 88
  dissection, sheep, 94-95
  forebrain, 90-91
  hindbrain, 88
  medulla oblongata, 88
  midbrain, 89
  parts of, 88-91
  pons, 89
  protecting, 95-96
brain mapping, 89
breakfast, 231
breathing, see also respiratory
  system
  bronchial, 193
  exchange of gases, 197, 200-201
  expiration, 193
  inspiration, 193
  mechanism of, 192-196
  nervous control of, 201
  vesicular, 193
bronchi, 189, 191
bronchioles, 191
burns
  and artificial skin, 23
  treating, 25

C

caecum, 274
Caesarean section, 345
calculus (tartar), 254
callus, 63
Canada Food Guide, 232-233. See
  also daily food guide
*Canadian Environmental Protection
  Act*, 408
canines, 252
capillaries, 164-165
capsular ligament, 57
carbohydrates, 210-211
  sources of, 211-214
carbon monoxide, 409
cardiac sphincter, 262
carpals, 55
cartilage, 41
  torn, and arthroscopy, 60
cavities, 254
cell organelles, 5-7
cells
  exchange of gases in, 200-201
  that work together, 8-9
  types of, 5
  typical, 4
cementum, 254
Central Nervous System (CNS), 84
central neuron, 97
cerebral contusion, 96
cerebral cortex, 90
  motor activities of, 91-92
  sensory activities of, 92-93
cerebral hemispheres, 90
cerebrospinal fluid, 95
cerebrum, 90
cervix, 334
chemical composition of human
  body, 4
chemoreceptors, 146
chlorofluorocarbons (CFCs),
  446-447
cholesterol, 217
chorion, 338
chromosomes, 298
  heterozygous, 306
  homologous pair, 300
  homozygous, 306
  in combination, 306-307
  information carried by, 315-316
chyme, 263
circadian rhythm, 108
circulatory system, 10, 164-167, 172
  systemic system, 169-170
clavicle, 54
clitoris, 335
cocaine, 104, 107
colour-blindness, 134, 135
composting, 406
concussion, 95
conduction, 24
conductor, 120
connective tissue, 8
conservation, 441-442
contact lenses, 123
contraception, methods of,
  348-350

contraction of muscles, 68-70
control system, 120
convection, 24
copulation, 337
corpus albicans, 332
corpus cavernosum, 329
corpus luteum, 332
cortex, 284
cosmetics, 27
Cowper's gland, 328
crack, 107
crown, teeth, 254

D

daily food guide, 229-231
  breads/cereals group, 231
  fruits/vegetables group, 231
  meat group, 230
  milk group, 229-230
daily nutrient intake, recommended, 247
dandruff, 27
deafness, see also hearing loss
  conduction, 141
  nerve, 141
dendrites, 85
dentine, 254
depressants, 105
dermis, 11, 14
descending colon, 275
diabetes, 212-213
dialysis, 288-289
diaphragm, 192-193
diastole, heart, 174
diet
  absorption of molecules, 252
  absorption of nutrients, 271
  and dental health, 256
  and pregnancy, 347
  bolus, 258
  chemical vs. physical, 252
  defined, 210
  elimination, 274-275
  in stomach, 261-265
  peristaltic action, 261
  salivary glands, 258-259
  small intestine, 266
  swallowing, 260
  teeth, see teeth
  tongue, 258
digestive system, 10
  living with, 276-277
  major secretions of, 268
diploid cells, 304
disaccharides, 210
dislocation, bones, 61
dissection
  care of equipment and specimen, 360
  defined, 360
  directional terms, 360
  fetal pig, see fetal pig, dissecting
  materials required, 360
  safety considerations, 362
DNA, 298
  Watson-Crick model of, 299
Down Syndrome, 318, 319, 321
drive, 107
drowning, 198-199
drug abuse, 104-105
drug dependence, 104-105
drugs, and pregnancy, 346, 347
ducts, male reproductive system, 328
duodenum, 266

E

ear, 137-140
  auditory canal, 137
  auditory nerve, 139
  auricle, 137
  cochlea, 139-140
  eustachian tube, 139
  external, 137
  inner, 139-140
  middle, 138
  organ of Corti, 139
  oval/round windows, 139
  pinna, 137
  saccule, 139
  semicircular canals, 139, 144
  tympanic membrane, 138
  utricle, 139
eardrum, 138
ecology, 430
ecosystem, 430
  carrying capacity, 433
ectoderm, 341
effector, 97, 120
egg production, 304-305
ejaculation, 329
ejaculatory duct, 328
electroencephalogram (EEG), 92
elements found in human body, 4
enamel, teeth, 254
endocrine system, 10, 102-103
endoderm, 341
endometrium, 334
enterokinase, 266
environmental issues and NIMBY syndrome, 448-450
environmental standards, 447-448
enzymes, 258
epidermis, 11, 13-14
epididymis, 328
epiglottis, 189, 260
epilepsy, 92-93
epiphyseal cartilage, 41
epithelial tissue, 8
erepsin, 266
erythrocytes, 157
esophagus, 261
estrogen, 332
evening meals, 231
eye
  conjuctiva, 122
  dissection, 126-127
  external structure of, 120-123
  internal structure of, 124-129
  iris, 124

lacrimal glands, 122-123
lens, 125
muscles, 121
orbital cavities, 121
protecting, 121-123
pupil, 124
reflexes of, 129-130
eyeball
  blind spot, 126, 128-129
  choroid layer, 125
  cones, 125, 126
  cornea, 125
  fovea centralis, 126
  layers of, 125-126
  optic nerve, 126
  retina, 125-126
  rhodopsin, 126
  rods, 125, 126
  sclera, 125
eyebrows, 121
eyelashes, 122
eyelids, 121
excretion
  defined, 282
  through lungs, 282
  through skin and anus, 283
excretory system, 10, 282-293
  disorders affecting, 292-293
  nephron, 284, 285-287
  organs of, 283-284
  urine and ADH, 289-291
exercise, 242
exercise programs, 180-181
expiratory reserve volume, 194
exponential growth, population, 433

F

fallen arches, 53
Fallopian tubes, 332-334
far-sightedness, 131, 132
fats, 216-218
  in body, 218
  saturated/unsaturated, 216-217
fecal coliforms test, 390-391
feces, 275
femur, 53
fertilization, 337-338
fetal pig, dissecting
  circulatory system, 374-377
  digestive system, 365-371
  excretory system, 371-373
  external features, examining, 362-364
  parotid salivary gland, 365
  reproductive system, 377-380
  respiratory system, 373-374
  sublingual salivary gland, 365
  submaxillary gland, 365
fetus, 342
fibre, 228
fibula, 53
fingerprints, 14-16
fitness, and exercise, 77-78
fixed action patterns, 108, 109
follicles, ovarian, 330, 331
fontanel, 47
food consumption, 435-437
food energy, 215
food labels, 234-235
food supply
  crop losses, reducing, 438
  dietary changes, 440
  farmland, increasing, 438-439
  increasing, 438-440
  population control, 440
foot, bones of, 53-54
foreskin, 329
fracture, 61
  compound, 61-62
  green-stick, 61-62
  simple, 61-62
freckles, 20
frontal lobe, brain, 90
FSH (follicle stimulating hormone), 328
fully movable joints, 57-58, 61

G

gall bladder, 267
gametes, 303
gastric juice, 263
genes
  dominant vs. recessive, 307
  incomplete dominance, 312
  operation in cell, 301-304
  sex-linked, 315-316
  transmission of, 308-313
gene splicing, 317
genetic code, 298-300
  cracking, 300
genetic counsellors, 320, 321
genetics, 298-320. See also inheritance
  blueprint errors, 317-320
  defects, 319-320
  information carried by chromosomes, 315-316
  information for development, 298-305
  inheritance, 305-308
  transmission of genes, 308-313
genotype, 307
gestation, 342
gliding joints, 58-59
glomerulus, 285
glucose, and blood sugar, 211
greenhouse effect, 445-446
grey matter, 85

H

habitat, 431
hair, 17-18
  caring for, 28-29
halitosis, 257
haploid cells, 304
Haversian canals, 43
Haversian system, 43
hearing
  ear structure, 137-140

how sound travels, 138-139
loud sounds, effect on, 141-142
sound, 139-140
hearing loss, 141-142. See also deafness
heart, 168-176
  aorta, 169
  bicuspid valve, 169
  inferior vena cava, 169
  left atrium, 169
  left ventricle, 169
  muscle, 169
  nature of, 168
  parts of, 169-171
  right atrium, 169
  superior vena cava, 169
  tricuspid valve, 169
  ventricles, 169
heart attacks, 178
heart rate, regulating, 173-176
  factors affecting, 174-175
  nervous control of, 174
heart valves, 171
Heimlich manoeuvre, 190
hemoglobin, 157
hemophilia, 161
hemorrhoids, 275
hepatitis, 269
  infectious/serum, 269
hereditary disease, 319-320
hinge joints, 58-59
Holmgren test, 135
homeostasis, 22, 432
hormones, 102
  and male reproductive system, 328
  feedback, 335
  female, produced by ovaries, 332
  pituitary, 103
Human Genome Project, 300
human waste management, 384-421
  acid precipitation, 414-419
  airborne waste, 409-414
  alternatives to disposal, 405
  industrial hazardous waste, 407-408
  solid waste, 399-404
  waste defined, 384
  wastewater treatment, 394-398
  water pollution, 384-390
  water purification, 392-394
  water quality, testing, 390-391
humerus, 55
hydrochloric acid, 263
hymen, 334
hypertension, 180
hypertrophy, 78
hypothalamus, 22, 90
hypothermia, 23-24

I

ileocaecal valve, 274
ileum, 266
immovable joints, 57
implantation, 338
incineration, 403
incisors, 252
industrial hazardous waste, 407-408
information, processing, 120
inheritance, 305-308. See also genetics
  blood type, 313-314
  of two characteristics, 310
  pedigree diagram, 312-313
insertion of a muscle, 71
inspiratory reserve volume, 194
insulin, 211, 268
integumentary system, 9
intercostal muscles, 193
interpreter, 120
interstitial fluid, 167
intervertebral discs, 50-51
ISCH (interstitial cell stimulating hormone), 328
Ishora test, 135

isometric exercise, 75
isotonic exercise, 75

J

jaundice, 269
jejunum, 266
joints, 57-59
  types of, 57-59

K

karotypes, 318
keratin, 13
kidney, 283, 284
  and hypertension, 292-293
  artificial, 288-289
  kidney stones, 293
  kidney transplant, 287

L

labia majora/minora, 335
lacrimal ducts/glands, 122-123
lactase, 266
lactation, 345-346
lactic acid, 75
large intestine, 275
larynx, 189
leachate, 401
lead, 410
learning, 111-115
learning disorders, 112-113
lens, artificial, 131-132
lenses, 132
leukemia, 161
leukocytes, 157-158
ligaments, 57
lipase, 263, 266
liver, 267
lung capacity, 194-196
lungs, 192

exchange of carbon dioxide and oxygen in, 197
healthy, and air pollution, 202-203
residual air capacity, 194
vital capacity, 194
lymphatic system, 10, 167

M

maltase, 266
Malthus, Thomas, 433
mammary glands, 345-346
marijuana, 105
medulla, 284
meiosis, 303
meiotic nondisjunction, 319
melanin, 14
memory
 and habit, 112
 and learning, 111-112
meninges, brain, 95
meningitis, 96
menstrual cycle, 335-337
 luteal phase, 335
mesentery, 266
mesoderm, 341
messenger ribonucleic acid (mRNA), 301
metacarpals, 55
metatarsals, 53
micturition, 292
minerals, 226-227
mitosis, 302
mnemonic device, 111
molars, 252-253
molecules, 10
moles, 20
monocular vision, 130
mononucleosis, 162
monosaccharides, 210
mons pubis, 334
motor (efferent) neuron, 97
mouth, 260

mucin, 263
muscle fatigue, 75-76
muscles, 65-70
 "all or none" event, 68
 antagonistic pairs of, 73-74
 attachment to bones, 71
 cardiac, 65, 67
 contraction and extension, 68-70
 fibres, 68
 skeletal, structure of, 68-69
 skeletal/voluntary, 65, 66
 smooth, 65, 67
 stimulation, 70
 types of, 65-67
muscle tissue, 9
muscular system, 9
mutagenic agents, 317
mutations, 317
myelin, 85
myofibrils, 68
myometrium, 334
myosin, 69

N

nails, 17
nasal cavity, 186-187
near-sightedness, 131, 132
nephrons, 284, 285-287
nerve impulse, transmission of, 86-87
nerve tissue, 8
nervous system, 10
 and behaviour, 84-87, 107-110
 Autonomic (ANS), 84, 100-101
 brain, 87-93
 Central (CNS), 84
 endocrine system, 102-103
 foreign chemicals and, 104-105
 learning, 111-115
 parts of, 84
 Peripheral (PNS), 84
 reflex arc, 97
 spinal cord, 96-99

neurilemma, 85
neuron, 84-85
 motor, structure of, 85
neuroplasm, 84
niche, 431
nitrogen bases, 298
nitrogen dioxide, 410
non-renewable resources, 435

O

occipital lobe, brain, 90
ocean pollution, 442-443
 plastic, 443
olfactory cells, 146
oogonia, 304
optical illusions, 136
optic nerve, 126
organelles, 10
organs, 9-10
 and systems, 9-10
 origin of a muscle, 71
ossification, 43
osteocytes, 43
osteoporosis, 44
ovaries, 330
ovulation, 331
ovum, 330
oxygen deficit, 76
oxygen, need for, 186
oxyhemoglobin, 157
ozone layer, 446-447

P

pacemaker, heart, 174
pancreas, 267-268
pancreatic duct, 268
pancreatic juice, 267-268
parietal lobe, brain, 90
patella, 53
pathogens, 157
Pavlov, Ivan, 98

pectoral girdle, 54
pedigree diagram, 312-313
peer pressure, 107
pelvic girdle, 53
pelvis, 53
Penfield, Wilder, 89
penis, 329
pepsin, 263
pepsinogen, 263
peptic ulcer, 263
peridontal disease, 257
periosteum, 41
Peripheral Nervous System (PNS), 84
peripheral vision, 127-128
peritoneum, 292, 334
peritonitis, 275
personality disorders, 109, 110
phalanges, 53, 55
  birds, 56
pharynx, 188
phenotype, 307
physical dependency, 104
pituitary gland, 90, 102-103
pivot joints, 58-59
placenta, 339-341
plasma, 156
pleura, 192
pleurisy, 192
polar bodies, 304
pollution index, 409
polysaccharides, 210
population
  age distribution, 429-430
  and the environment, 430-432
  birth/death rates, 427-429
  changes, 427-429
  consumers/producers, 431
  environmental resistance to, 433
  factors affecting, 433
  fertility rate, 429
  human, 426, 435
  replacement rate, 429
  size, factors affecting, 433-435

pregnancy, 338-342
  embryo/fetus, development of, 341-342
  environmental factors affecting, 346-347
  planned, 347
premolars, 252-253
primary (deciduous), teeth, 253
procallus, 62
progesterone, 332
prostate gland, 328
protein, 219-221
  as energy source, 221
  body requirements, 220-221
  building, 301
  essential amino acids, 219-220
  sources of, 220
psychological dependency, 104
psychosomatic illness, 243
puberty, 328, 331
pulmonary circulation, 169
pyloric sphincter, 263

Q

quick-tanning products, 31-32

R

R.I.C.E. treatment, 64
radiation, 24, 317, 346
radius, 55
reaction time, 86-87, 120
receptor, 97, 120
rectum, 275
recycling programs, 405, 407
reflex act, 97
reflex arc, 97
reflexes, 97-98
  conditioned, 98-99
renewable resource, 435
renin, 263
reproduction, 326-355

reproductive system, 10
  birth and lactation, 343-346
  copulation and fertilization, 337-338
  female, 330-335
  male, 326-329
  menstrual cycle, 335
  pregnancy, 338-342
  role of environment, 346-347
  taking charge of sexuality, 347, 352, 354-355
rescue breathing, 198-199
respiratory system, 10. See also breathing
  air-conducting structures, 186-192
rheumatoid arthritis, 58
Rh factor, 164
ribs, 52
  floating, 52
root, tooth, 254

S

salivary glands, 258-259
sarcolemma, 68
scapula, 54
schizophrenia, 109
scrotum, 326
sebaceous glands, 18
sebum, 18
semen, 328
seminal vesicles, 328
seminiferous tubules, 327
sense receptors, 21-22
senses
  hearing, 137-140
  smell, 146-147
  taste, 145-147
  touch, 148
  vision, 129-136
sensory (afferent) neuron, 97
sewage, domestic, 394
sex chromosomes, 315

sex determination, 315
sex-linkage, 315-316
sex organs
  female, 330-335
  male, 326-329
sexually transmitted disease, 351-354
sinoatrial node, heart, 174
skeletal system, 9, 40
  adaptations, 56
  bones, 40-41
  cartilage, 41
  joints, 57-59
  lower limbs, 53-54
  muscle, 65-70
  muscle-bone connection, 70-74
  pelvis, 53
  sternum and ribs, 52
  upper limbs, 54-55
skeleton, 40
  major bones of, 45
  skull bones, 46-48
  support, 45-55
  vertebral column, 48-51
skin
  artificial, and burns, 23
  caring for, 25-26
  cold vs. heat receptors, 22
  dermis, 14
  epidermis, 13-14
  features of, 20-21
  functions of, 11
  glands, 18-19
  goose flesh, 22
  health and hygiene, 25-29
  nature of, 11
  overexposure, 30-32
  patterns of, 14
  problems, 26-27
  regulating body temperature, 22–24
  sense receptors, 21-22
  structure of, 11-12
skin cancer, 27

skull, bones of, 46-48
slightly movable joints, 57
sludge disposal, 442
small intestine, 266
smell, sense of, 146-147
smog, photochemical vs. industrial, 412
smoking, 203-205
  and lung disease, 204-205
  during pregnancy, 205
Snellen eye chart, 134
solid waste, 399-404
  biodegradable, 401
  domestic/industrial, 399
  dumps, 401-403
  hazardous, 399
  incineration, 403
  landfill, 401-403
  non-biodegradable, 402
solid waste reduction unit (SWARU), 403
sperm production, 304, 327
spermatogonia, 304
sphincter, 262
sphygmomanometer, 177
spinal cord, 96-99
  functions of, 97
  injury to, 99
spine, see vertebral column
sprain, 61
stereoscopic, 130
sternum, 52
stimulus, 84, 107, 120
stomach, 262-263
stratum corneum, 13
stratum germinativum, 13-14
striations, 68
sucrase, 266
sulphur dioxide, 409
sunblockers, 32
sunscreens, 32
sun tan/burn, 30-32
sutures, 46
sweat glands, 18-20

synapse, 87
synovial fluid, 57-58
synovial membrane, 57-58
systole, heart, 174

T

tarsal bones, 53
taste buds, 145
taste, sense of, 145-147
tears, artificial, 123
teeth
  maintaining health of, 254-256
  plaque, 254
  problems with, 257
  structure of, 254
  types, and nature of, 252-53
temporal lobe, brain, 90
tendonitis, 72
tendons, 72
testes, 326, 327
testosterone, 327
thalamus, 90
thalidomide, 346
thyroid cartilage, 189
tibia, 53
tidal volume, 194
tissues, 8-10
  exchange of gases in, 200-201
tobacco smoke, effects of, 204
tongue, 258
tonsils, 188
touch, sense of, 148
trachea, 189
transfer ribonucleic acid (tRNA), 301
transfusion, blood, 162-163
transplant, kidney, 287
transverse colon, 275
triglyceride, 216, 217
trisomy, 319
tropical rainforests, destruction of, 443-445
turbinate bones, 186

U

ulcer, 264
ulna, 55
ultraviolet radiation, 30-31
umbilical cord, 341
ureters, 283, 291-292
urethra, 283, 292, 328
urinalysis, 290-291
urinary bladder, 283, 292
urinary system, 10
urinary tract infections, 292-293
urination, 292
urine
  and antidiuretic hormone (ADH), 289-291
  collection and release of, 291-292
  composition of, 290
uterus, 334
  structures within, 339-341
uvula, 260

V

vagina, 334
valves, veins, 166
varicose veins, 166
vas deferens, 328
veins, 166, 168
  portal, 267
  pulmonary, 169
  renal, 284
venules, 164
vertebra, 48
vertebral column, 48-51
villi, 271
viruses, and pregnancy, 346
vision, see also eye; eyeball
  accommodation, 129
  adaptation, 129
  binocular, 130-131
  characteristics of sight, 129-131
  light reflex, 129
  measuring ability to see, 131-136
visual acuity, 134
visual illusion, 136
visually-impaired students, aids for, 133
vitamins, 224
  fat-soluble vs. water-soluble, 223
  sources of, 222-223
  storage of, 223

W

waste, see also human waste management; solid waste
  airborne, 409-414
  defined, 384
  four R's of reducing, 405
  industrial hazardous, 407-408
wastewater
  activated sludge system, 395, 396
  composition of, 394-395
  primary treatment, 395, 397
  secondary treatment, 397
  septic systems, 395
  sludge, 398
  tertiary treatment, 398
  treatment of, 394-398
water, 225-226
  saving, 441
water pollution, 385-390
  enriching substances, 385
  litter, 385
  toxic substances, 387-388
water purification, 392-394
water quality standards, 398
water quality, testing
  aquatic organisms, 390
  clarity and colour, 391
  fecal coliforms test, 390-391
white matter, 85
wrinkles, 20-21

Y

yolk sac, 341

Z

zygote, 305, 338

PHOTO CREDITS

All photos not specifically credited to another source were taken by Gordon S. Berry or are courtesy of John Wiley & Sons Canada Ltd.

CHAPTER 1
Chapter opener: Birgitte Nielsen.
Figures 1.2c, g: Ontario Science Centre.
Figure 1.6: Bela Nagy.
Figures 1.13, 1.15, 1.16, 1.18: Crystal Image Photography.
Figure 1.20: Barry McCammon.

CHAPTER 2
Chapter opener: Mario Scattolini.
Figure 2.5: Osteoporosis Society of Canada.
Figure 2.8: St. Joseph's Hospital, Peterborough.
Figures 2.28, 2.29: Crystal Image Photography.
Figure 2.34: Biology Department, Trent University.
Figure 2.40: Crystal Image Photography.
Figure page 60: The Wellesley Hospital.

CHAPTER 3
Chapter opener: Crystal Image Photography.
Figure 3.10: Harold S. Gopaul.
Figure 3.12: Crystal Image Photography.
Figure 3.16: Gerhard Gscheidle/The Image Bank.
Figures 3.21, 3.22, 3.23, 3.25, 3.26: Crystal Image Photography (3.25: man with goose by C.B. Hannell).
Figure page 89: Dr. L. Heier.

CHAPTER 4
Chapter opener: Ontario Ministry of Transport.
Figures page 133: Shamus Nesling, Peel Board of Education.

CHAPTER 5
Chapter opener: MDS Health Group Limited.
Figure 5.13: Ontario Science Centre.
Figure 5.21: Vernon J. Freer.
Figure 5.22: F. Cole, McMaster University.
Figure page 181: Crystal Image Photography.
Figures page 199: Don Galbraith.

CHAPTER 6
Chapter opener: Toronto General Hospital.
Figure 6.5: Dr. James Hogg, Faculty of Medicine, University of British Columbia.
Figure 6.13: Canadian Cancer Society.
Figure page 203: Crystal Image Photography.

CHAPTER 7
Chapter opener: New Brunswick, Industry Science Technology Canada.
Figures 7.1, page 213, 7.2, 7.5, 7.6, Table 7.7 (milk & breads), page 243: Crystal Image Photography.

CHAPTER 8
Figure page 273: Peterborough Police Department.
Figure page 276: Crystal Image Photography.

CHAPTER 9
Chapter opener: Parks Canada.
Figures 9.6, page 288: The Kidney Foundation of Canada.
Figure 9.7: Crystal Image Photography.

CHAPTER 10
Chapter opener: Crystal Image Photography.
Figures page 318, 10.11: Department of Medical Genetics, Faculty of Medicine, University of British Columbia.

CHAPTER 11
Chapter opener: Crystal Image Photography.

CHAPTER 12
Chapter opener: Toronto General Hospital.
Figures 12.2, 12.8, 12.10, 12.14: Crystal Image Photography.

CHAPTER 13
Chapter opener (scenic, faucet): Crystal Image Photography.
Figure 13.2: Elaine Freedman.
Figure 13.5: Waters Canada Ltd.
Figures 13.9, 13.10, 13.12, 13.16, 13.17, 13.20: Crystal Image Photography.
Figure 13.21: Canapress.
Figure 13.23: Ontario Ministry of the Environment.
Figure 13.24: Ontario Hydro.
Figure page 417: Hamilton Spectator.
Figures page 420: Crystal Image Photography.

CHAPTER 14
Chapter opener: NASA.
Figure 14.1 Mark Antman.
Figure 14.10: TV Ontario.
Figure 14.11: Crystal Image Photography.
Figure 14.13: Canapress.
Figure 14.14: Adrian Dorst/Sierra Club of Western Canada.
Figure 14.15: World Wildlife Fund.
Figure 14.18: Ontario Waste Management Corporation.